Advances in wool technology

The Textile Institute and Woodhead Publishing

The Textile Institute is a unique organisation in textiles, clothing and footwear. Incorporated in England by a Royal Charter granted in 1925, the Institute has individual and corporate members in over 90 countries. The aim of the Institute is to facilitate learning, recognise achievement, reward excellence and disseminate information within the global textiles, clothing and footwear industries.

Historically, The Textile Institute has published books of interest to its members and the textile industry. To maintain this policy, the Institute has entered into partnership with Woodhead Publishing Limited to ensure that Institute members and the textile industry continue to have access to high calibre titles on textile science and technology.

Most Woodhead titles on textiles are now published in collaboration with The Textile Institute. Through this arrangement, the Institute provides an Editorial Board which advises Woodhead on appropriate titles for future publication and suggests possible editors and authors for these books. Each book published under this arrangement carries the Institute's logo.

Woodhead books published in collaboration with The Textile Institute are offered to Textile Institute members at a substantial discount. These books, together with those published by The Textile Institute that are still in print, are offered on the Woodhead website at: www.woodheadpublishing.com. Textile Institute books still in print are also available directly from the Institute's website at: www.textileinstitutebooks.com.

A list of Woodhead books on textile science and technology, most of which have been published in collaboration with The Textile Institute, can be found at the end of the contents pages.

Woodhead Publishing in Textiles: Number 72

Advances in wool technology

Edited by
N. A. G. Johnson and I. M. Russell

The Textile Institute

CRC Press
Boca Raton Boston New York Washington, DC

WOODHEAD PUBLISHING LIMITED
Cambridge, England

Published by Woodhead Publishing Limited in association with The Textile Institute
Woodhead Publishing Limited, Abington Hall, Granta Park, Great Abington
Cambridge CB21 6AH, England
www.woodheadpublishing.com

Published in North America by CRC Press LLC, 6000 Broken Sound Parkway, NW, Suite 300, Boca Raton, FL 33487, USA

First published 2009, Woodhead Publishing Limited and CRC Press LLC

British Library Cataloguing in Publication Data
A catalogue record for this book is available from the British Library.

Library of Congress Cataloging in Publication Data
A catalog record for this book is available from the Library of Congress.

Woodhead Publishing ISBN 978-1-84569-332-9 (book)
Woodhead Publishing ISBN 978-1-84569-546-0 (e-book)
CRC Press ISBN 978-1-4200-7968-5
CRC Press order number: WP7968

The publishers' policy is to use permanent paper from mills that operate a sustainable forestry policy, and which has been manufactured from pulp which is processed using acid-free and elemental chlorine-free practices. Furthermore, the publishers ensure that the text paper and cover board used have met acceptable environmental accreditation standards.

Typeset by Replika Press Pvt Ltd, India
Printed by TJ International Limited, Padstow, Cornwall, England

Contents

Contributor contact details *xi*

Woodhead Publishing in Textiles *xv*

Preface *xxi*

Part I Advances in wool fibre and fabric production

1 Improvement of wool production through genetic manipulation 3
G. E. Rogers, The University of Adelaide, Australia; and C. S. Bawden, South Australian Research and Development Institute, Australia

1.1 Introduction 3
1.2 The techniques of genetic manipulation for animals 4
1.3 What aspects of wool production might be amenable to genetic manipulation? 5
1.4 Increasing the yield of wool per animal 6
1.5 How might wool quality be modified by affecting fibre molecular structure? 10
1.6 Some other aspects of wool production that could be achieved by transgenesis 17
1.7 Conclusion and future trends 19
1.8 References 19

2 The objective measurement of wool fibre quality 22
P. Sommerville, Formerly Corporate Development Manager, Australian Wool Testing Authority Ltd, Australia

2.1 Introduction 22
2.2 Definitions 23
2.3 Principles of wool metrology 23
2.4 Applications of wool metrology 23

2.5 Statistical methods in wool metrology 26
2.6 Development of wool measurement systems 30
2.7 Standardisation of measurement systems 31
2.8 Application of commercial regulations 35
2.9 Advances in the measurement of wool 35
2.10 Latest developments in wool metrology 55
2.11 Future trends 57
2.12 Sources of further information and advice 58
2.13 References 59

3 Wool as a natural renewable fibre 61
I. M. Russell, CSIRO Materials Science and Engineering, Australia

3.1 Introduction 61
3.2 The environmentally aware retailer and consumer 62
3.3 The growing importance of European legislation 63
3.4 Environmental claims 65
3.5 EU eco-label processing in practice 77
3.6 Future trends 82
3.7 Sources of further information and advice 83
3.8 References 83

4 Advances in wool spinning technology 86
M. W. Prins, CSIRO Materials Science and Engineering, Australia

4.1 Introduction 86
4.2 Brief review of advances in topmaking 87
4.3 Yarnspec™: predicting spinning performance 90
4.4 Worsted spinning 91
4.5 Future trends 100
4.6 Sources of further information and advice 104
4.7 Acknowledgements 104
4.8 References 104

5 Advances in wool weaving and knitting 106
S. Gupta, National Institute of Fashion Technology, India

5.1 Introduction 106
5.2 Advances in yarn preparation 107
5.3 Warp preparation 107
5.4 Advances, benefits and applications of new weaving technologies 113
5.5 Advances, benefits and applications of new knitting technologies 119
5.6 Future trends 127
5.7 Sources of further information and advice 128
5.8 Further reading 128

6 Advances in the manufacture of nonwoven wool 130
N. Finn and R. Wood, CSIRO Materials Science and Engineering, Australia

6.1 Introduction 130
6.2 Nonwoven processes and recent innovations 131
6.3 Bonding systems 137
6.4 Finishing 145
6.5 Future trends 145
6.6 Sources of further information and advice 145
6.7 Acknowledgement 146

7 Wool finishing and the development of novel finishes 147
J. Shen, De Montfort University, UK

7.1 Introduction 147
7.2 Dimensional stability and shrink-resist finishing for machine-washable wool 148
7.3 Stain-repellent and stain-release finishing 158
7.4 Insect-resist and insect-repellent treatments 161
7.5 Flame-retardant finishing 165
7.6 Enzyme applications in wool finishing 167
7.7 Future trends 176
7.8 Acknowledgements 177
7.9 References 177

8 The coloration of wool 183
D. M. Lewis, The University of Leeds, UK

8.1 Introduction 183
8.2 The theoretical basis of wool dyeing 184
8.3 Environmental impact of wool coloration processes 191
8.4 Reactive dyes for wool 198
8.5 Wool protection during dyeing by using reactive dyes and/or anti-setting agents 200
8.6 Ink-jet printing of wool fabrics 207
8.7 References 209

Part II New wool products and applications

9 Improving the whiteness and photostability of wool 217
K. R. Millington, CSIRO Materials Science and Engineering, Australia

9.1 Introduction 217

9.2	Wool colour	218
9.3	The effect of wool processing on colour	222
9.4	The chemistry of wool colour and photostability	226
9.5	Methods for improving the whiteness and photostability of wool	239
9.6	Future trends	241
9.7	Sources of further information and advice	243
9.8	References	244
10	**Enhancing wool products using nanotechnology** R. DENNING, CSIRO Materials Science and Engineering, Australia	248
10.1	Introduction	248
10.2	Nanotechnology application to wool	256
10.3	Current commercial applications of nanotechnology to wool	260
10.4	Future trends	261
10.5	Sources of further information and advice	261
10.6	References	262
11	**Wool performance apparel for sport** B. HOLCOMBE, University of Wollongong, Australia	265
11.1	Introduction	265
11.2	The human system	266
11.3	Textile properties relevant to sports apparel	269
11.4	Textiles for sports apparel	277
11.5	The Sportwool™ concept	279
11.6	Future trends	280
11.7	The role for wool	282
11.8	References	283
12	**High-performance wool blends** M. MIAO, CSIRO Materials Science and Engineering, Australia	284
12.1	Introduction	284
12.2	Wool blend products, their benefits and applications	285
12.3	Processing systems for wool blend yarns	296
12.4	Spinning wool blend on the long staple worsted system	297
12.5	Processing wool and wool blend on the cotton spinning system	298
12.6	Wool–filament composite yarns	301
12.7	References	304

13 Intelligent wool apparel 308
S. R. Collie, AgResearch Limited, New Zealand

13.1 Introduction 308
13.2 Definitions and categories of smart textiles 308
13.3 Wool's intrinsic 'smartness' 310
13.4 Current applications of wool in intelligent textiles 311
13.5 Opportunities for the use of wool in intelligent textiles 319
13.6 References 319

14 Application of wool keratins ranging from industrial materials to medical devices 323
R. Kelly, RJK Developments Limited, New Zealand

14.1 Introduction 323
14.2 Overview of the structure and characteristics of keratin protein 324
14.3 Extraction of keratin from wool for new product development 325
14.4 Industrial trends relating to sustainable polymers 326
14.5 Advanced applications of wool keratins 326
14.6 Future trends 329
14.7 Sources of further information and advice 329
14.8 References 329

Index 332

Contributor contact details

(* = main contact)

Editors' contact details

Dr Nigel A. G. Johnson and Dr Ian M. Russell*
C/-CSIRO Materials Science and Engineering
Henry Street
Belmont
Victoria 3216
Australia

E-mail: Nigel.Johnson@csiro.au; Ian.Russell@csiro.au

Chapter 1

Professor George. E. Rogers*
School of Molecular and Biomedical Science
The University of Adelaide
SA 5005
Australia

E-mail: george.rogers@adelaide.edu.au

Dr C. Simon Bawden
South Australian Research and Development Institute
Roseworthy,
SA 5371
Australia

E-mail: Bawden.Simon@saugov.sa.gov.au

Chapter 2

Peter Sommerville
11 Balfour St
Doncaster
Victoria 3108
Australia

E-mail: psommerv@bigpond.net.au

Chapter 3

Dr Ian M. Russell
CSIRO Materials Science and Engineering
Henry Street
Belmont
Victoria 3216
Australia

E-mail: Ian.Russell@csiro.au

Chapter 4

Mr Martin W. Prins
CSIRO Materials Science and Engineering
Henry Street
Belmont
Victoria 3216
Australia

E-mail: martin.prins@csiro.au

Chapter 5

Professor Sanjay Gupta
Department of Fashion and Textiles
National Institute of Fashion Technology
NIFT Campus
Hauz Khas
New Delhi – 110016
India

E-mail: sgupta34@hotmail.com

Chapter 6

Dr Niall Finn* and Mr Raymon Wood
CSIRO Materials Science and Engineering
Henry Street
Belmont
Victoria 3216
Australia

E-mail: niall.finn@csiro.au

Chapter 7

Dr Jinsong Shen
Textile Engineering and Materials Research Group
De Montfort University
The Gateway
Leicester
LE1 9BH
UK

E-mail: jshen@dmu.ac.uk

Chapter 8

Professor David M. Lewis
Department of Colour Science
The University of Leeds
Leeds
LS2 9JT
UK

E-mail: d.m.lewis@leeds.ac.uk

Chapter 9

Dr Keith R. Millington
CSIRO Materials Science and Engineering
Henry Street
Belmont
Victoria 3216
Australia

E-mail: keith.millington@csiro.au

Chapter 10

Dr Ron Denning
CSIRO Materials Science and Engineering
Henry Street
Belmont
Victoria 3216
Australia

E-mail: ron.denning@csiro.au

Chapter 11

Dr Barry Holcombe
Intelligent Polymer Research Institute
University of Wollongong
Northfields Avenue
North Wollongong
NSW 2500
Australia

E-mail:
barry.holcombe@bigpond.com

Chapter 12

Dr Menghe Miao
CSIRO Materials Science and Engineering
Henry Street
Belmont
Victoria 3216
Australia

E-mail: menghe.miao@csiro.au

Chapter 13

Dr Stewart R. Collie
Textile Science and Technology
AgResearch Limited
Lincoln Research Centre
Cnr Springs Road and Gerald Street
Private Bag 4749
Christchurch 8140
New Zealand

E-mail:
stewart.collie@agresearch.co.nz

Chapter 14

Dr Rob Kelly
RJK Developments Limited
115 Hackthorne Road
Christchurch 8022
New Zealand

E-mail: Rob.kelly@keratec.co.nz

Woodhead Publishing in Textiles

1 **Watson's textile design and colour Seventh edition**
Edited by Z. Grosicki

2 **Watson's advanced textile design**
Edited by Z. Grosicki

3 **Weaving Second edition**
P. R. Lord and M. H. Mohamed

4 **Handbook of textile fibres Vol 1: Natural fibres**
J. Gordon Cook

5 **Handbook of textile fibres Vol 2: Man-made fibres**
J. Gordon Cook

6 **Recycling textile and plastic waste**
Edited by A. R. Horrocks

7 **New fibers Second edition**
T. Hongu and G. O. Phillips

8 **Atlas of fibre fracture and damage to textiles Second edition**
J. W. S. Hearle, B. Lomas and W. D. Cooke

9 **Ecotextile '98**
Edited by A. R. Horrocks

10 **Physical testing of textiles**
B. P. Saville

11 **Geometric symmetry in patterns and tilings**
C. E. Horne

12 **Handbook of technical textiles**
Edited by A. R. Horrocks and S. C. Anand

13 **Textiles in automotive engineering**
W. Fung and J. M. Hardcastle

14 **Handbook of textile design**
J. Wilson

15 **High-performance fibres**
Edited by J. W. S. Hearle

16 **Knitting technology Third edition**
D. J. Spencer

17 **Medical textiles**
Edited by S. C. Anand

18 **Regenerated cellulose fibres**
Edited by C. Woodings

19 **Silk, mohair, cashmere and other luxury fibres**
Edited by R. R. Franck

20 **Smart fibres, fabrics and clothing**
Edited by X. M. Tao

21 **Yarn texturing technology**
J. W. S. Hearle, L. Hollick and D. K. Wilson

22 **Encyclopedia of textile finishing**
H-K. Rouette

23 **Coated and laminated textiles**
W. Fung

24 **Fancy yarns**
R. H. Gong and R. M. Wright

25 **Wool: Science and technology**
Edited by W. S. Simpson and G. Crawshaw

26 **Dictionary of textile finishing**
H-K. Rouette

27 **Environmental impact of textiles**
K. Slater

28 **Handbook of yarn production**
P. R. Lord

29 **Textile processing with enzymes**
Edited by A. Cavaco-Paulo and G. Gübitz

30 **The China and Hong Kong denim industry**
Y. Li, L. Yao and K. W. Yeung

31 **The World Trade Organization and international denim trading**
Y. Li, Y. Shen, L. Yao and E. Newton

32 **Chemical finishing of textiles**
W. D. Schindler and P. J. Hauser

33 **Clothing appearance and fit**
J. Fan, W. Yu and L. Hunter

34 **Handbook of fibre rope technology**
H. A. McKenna, J. W. S. Hearle and N. O'Hear

35 **Structure and mechanics of woven fabrics**
J. Hu

36 **Synthetic fibres: nylon, polyester, acrylic, polyolefin**
Edited by J. E. McIntyre

37 **Woollen and worsted woven fabric design**
E. G. Gilligan

38 **Analytical electrochemistry in textiles**
P. Westbroek, G. Priniotakis and P. Kiekens

39 **Bast and other plant fibres**
R. R. Franck

40 **Chemical testing of textiles**
Edited by Q. Fan

41 **Design and manufacture of textile composites**
Edited by A. C. Long

42 **Effect of mechanical and physical properties on fabric hand**
Edited by Hassan M. Behery

43 **New millennium fibers**
T. Hongu, M. Takigami and G. O. Phillips

44 **Textiles for protection**
Edited by R. A. Scott

45 **Textiles in sport**
Edited by R. Shishoo

46 **Wearable electronics and photonics**
Edited by X. M. Tao

47 **Biodegradable and sustainable fibres**
Edited by R. S. Blackburn

48 **Medical textiles and biomaterials for healthcare**
Edited by S. C. Anand, M. Miraftab, S. Rajendran and J. F. Kennedy

49 **Total colour management in textiles**
Edited by J. Xin

50 **Recycling in textiles**
Edited by Y. Wang

51 **Clothing biosensory engineering**
Y. Li and A. S. W. Wong

52 **Biomechanical engineering of textiles and clothing**
Edited by Y. Li and D. X-Q. Dai

53 **Digital printing of textiles**
Edited by H. Ujiie

54 **Intelligent textiles and clothing**
Edited by H. Mattila

55 **Innovation and technology of women's intimate apparel**
W. Yu, J. Fan, S. C. Harlock and S. P. Ng

56 **Thermal and moisture transport in fibrous materials**
Edited by N. Pan and P. Gibson

57 **Geosynthetics in civil engineering**
Edited by R. W. Sarsby

58 **Handbook of nonwovens**
Edited by S. Russell

59 **Cotton: Science and technology**
Edited by S. Gordon and Y-L. Hsieh

60 **Ecotextiles**
Edited by M. Miraftab and A. Horrocks

61 **Composite forming technologies**
Edited by A. C. Long

62 **Plasma technology for textiles**
Edited by R. Shishoo

63 **Smart textiles for medicine and healthcare**
Edited by L. Van Langenhove

64 **Sizing in clothing**
Edited by S. Ashdown

65 **Shape memory polymers and textiles**
J. Hu

66 **Environmental aspects of textile dyeing**
Edited by R. Christie

67 **Nanofibers and nanotechnology in textiles**
Edited by P. Brown and K. Stevens

68 **Physical properties of textile fibres Fourth edition**
W. E. Morton and J. W. S. Hearle

69 **Advances in apparel production**
Edited by C. Fairhurst

70 **Advances in fire retardant materials**
Edited by A. R. Horrocks and D. Price (forthcoming)

71 **Polyesters and polyamides**
Edited by B. L. Deopura, R. Alagirusamy, M. Joshi and B. S. Gupta

72 **Advances in wool technology**
Edited by N. A. G. Johnson and I. M. Russell

73 **Military textiles**
Edited by E. Wilusz

74 **3D fibrous assemblies: Properties, applications and modelling of three-dimensional textile structures**
J. Hu

75 **Medical textiles 2007**
Edited by J. Kennedy, A. Anand, M. Miraftab and S. Rajendran (forthcoming)

76 **Fabric testing**
Edited by J. Hu

77 **Biologically inspired textiles**
Edited by A. Abbott and M. Ellison

78 **Friction in textiles**
Edited by B. S. Gupta

79 **Textile advances in the automotive industry**
Edited by R. Shishoo

80 **Structure and mechanics of textile fibre assemblies**
Edited by P. Schwartz

81 **Engineering textiles: Integrating the design and manufacture of textile products**
Y.E. El Mogahzy

Preface

In 2007, the Australian wool industry celebrated the 200th anniversary of the beginning of the Australian wool trade, supplying merino-based raw wool and wool products to the global textile industry. A photographic competition held as part of the anniversary celebrations was won by a stark and iconic image of an ancient wool press in a 100-year-old darkened wool shed. The image described 'A relic that epitomises the resilience of the Australian Wool Industry. Scarred, but still solid and stoic, it is a timeless reminder of the tough people, and tough times behind a proud and resilient industry' (Merino 200, 2007).

That image looks back nostalgically to a time when wool was a major textile fibre, but those times have long gone. This book looks forward optimistically, reviewing how recent advances in science and technology will help the wool industry meet the challenges and opportunities it faces in its 202nd year.

The reality is that wool is now a minority fibre, with only a 1.9% share of a global fibre market. It does, however, retain a greater market share in some significant areas such as apparel and interiors and a few industrial niches. Wool's market share began to decline from the 1950s, to some extent in absolute tonnages, but more importantly in percentage terms as the total textile market increased in size through global population growth and as textile consumption per head increased, especially in developed countries.

Synthetic fibres were created and enhanced through scientific and technological breakthroughs which show no sign of abating. They were cheap to produce and began to dominate the market and to outstrip even cotton production. In the 1960s the holy grail of the synthetic fibre industry was to develop a fibre that could mimic the comfort, appearance and moisture-retention properties of the natural fibres. That research effort was doomed to

---

Merino 200 (2007), 'Iconic image announced winner of 'Australian Wool to the World' Photographic Competition', http://www.merino200.com/lo/index.html, accessed 14 May 2008.

fail, largely because wool is a highly complex biological material that has challenged scientists over many decades, and has proved impossible to replicate.

However, the synthetic fibres continued to innovate and to develop fibres with their own unique and distinctive properties, both physical and chemical, that now claim their own performance advantages. A cursory examination of the active sporting and outdoor clothing markets will quickly identify the multitude of properties that have been engineered into the synthetic fibres as a result of this intensive research and development.

Wool has not stood still throughout that same period. Research into the complex structure of wool was intense and the knowledge gained has been at the forefront of our understanding of complex biomaterials. Its complexity also offers opportunity as we understand it better, and there are new scientific tools and techniques to explore and modify it to create new features and levels of performance. Wool remains a unique and almost magical fibre, with its ability to absorb a high level of moisture but with a simultaneous ability to provide protection against showers, with resilient mechanical properties and with natural odour resistance. With the fundamental knowledge of the structure and chemistry of wool that we now possess, wool fabrics can be engineered to be either warm or cool and to function superbly in extreme sporting situations or in the most elegant of formal occasions.

This book highlights the breadth of sciences that have been applied in recent years to create advances in wool across the full spectrum, from genetics of wool production, to processing, to finishing, and to its use as a source of proteins for a range of novel applications.

Unfortunately, as wool has declined in importance, the global research effort backing its further development has also declined. This will limit the scope for the future advances in wool that will be needed if this wonderful fibre is to continue to compete with the inevitable advances in alternative materials, and not be further relegated to the status of a rare novelty fibre. Marketing will remain important, but it will only take the fibre so far. Consumers are constantly seeking improved performance and exciting new innovations that only research and development can provide.

On the other hand, a great opportunity on wool's horizon is the growing consumer interest in natural and sustainable materials. This aspect will require further research effort if wool is to fully exploit this opportunity, and not fall foul of the counter-claims that wool production is unsustainable, that some animal husbandry aspects are unethical and that its processing uses unfriendly chemicals. Significantly the wool industry has begun to recognise the need to be proactive in this area and the theme of the most recent IWTO Congress, notably in Beijing, was 'Wool – The Environmentally Friendly Fibre'. IWTO President Günther Beier 're-emphasised the importance of capitalising on the ecological aspects of the wool fibre and its inherent properties as a textile fibre, which can bring wool back onto the retail floor prominently'.

Many commentators and indeed many in the wool industry feel that wool is currently at a crossroad; however, the Editors strongly believe that the challenges ahead can be addressed by continued focused research, and some progress is reported in the following pages.

Part I of this book deals with advances in wool fibre and fabric production. Rogers and Bawden identify opportunities that might be provided to grow wool fibres with tailored properties through advances in genetic modification by targeting the skin follicle, the 'engineering room' where the fibre is produced. The opportunities are great but the wool fibre has evolved to its current optimised state over millions of years. As a result, progress is slow and difficult.

One of the great challenges for wool is that it is a natural fibre with great inherent variability. Sommerville describes developments in presale objective wool testing and in modelling that allow consignments of wool to be prepared which will have predictable properties and processing characteristics at top and yarn stages. These variability issues are not faced by the synthetic fibres, and the growth of objective wool testing has greatly improved reproducibility and has significantly reduced early stage costs.

Russell describes the development of credible and workable definitions of eco-wool and organic wool by the wool industry and notes the increasing importance of European consumers and European legislation in establishing good environmental practice through textile supply chains. An important development is the increasing availability in the marketplace of eco-wool tops and yarns available for spinners, weavers and knitters to purchase; these products significantly shorten the supply chain for the late stage product manufacturers and the availability of eco-wool garments is expected to increase rapidly.

Chapter 4 by Prins and Chapter 5 by Gupta describe the latest developments in wool spinning, weaving and knitting. These are areas of intense innovation by researchers, machinery manufacturers and processors. Wool benefits by adapting generic advances made principally for use with other fibres. This is where costs can be reduced, but more importantly, this is where the new and attractive wool fabrics are generated that will excite the consumers of the future.

Finn and Wood examine the possibilities for wool to bypass the traditional methods of fabric formation via yarns and to be processed directly into fabrics and other products through the various non-woven routes that have had such an impact on synthetic fibre consumption. The authors identify the problems and the opportunities encountered for wool in these products. Many developments in this area are being driven by the cost savings available and this is important for wool; however, the products are unique and may fill potential market niches, not just in technical applications but in certain apparel opportunities.

To complete Part I, new developments in the areas of finishing and coloration are reviewed by Shen and Lewis. While these areas also stimulate innovation in aesthetics and performance to entice the consumer, many of the recent developments in these areas are driven by environmental problems of the past and the present. Many of these problem areas, such as chlorine for shrink resistance, heavy metals in dyes and insect-resist agents are unique to wool. It is essential that wool can continue to be converted into garments that meet all of the performance and environmental needs of the modern consumer, and still meet the essential aesthetic needs of drape, appearance retention, handle and deep subtle colours that are unique to wool.

Part II of this book examines possibilities for new wool products and applications. Wool has long been thought of as an autumn/winter fibre, prevented from entering lightweight spring/summer markets because it has been considered in its traditional role as a warm, heavy fabric, unavailable in light pastel shades. Developments in spinning, in the weaving of singles yarn and the 'CoolWool' innovations have countered the first misconception; however, the fact remains that wool is less white than cotton and the synthetics. Millington describes research aimed at producing stable white wool fabrics. Progress has been made in understanding the fundamental reasons for wool's yellowing behaviour, but importantly, this understanding has allowed the technical development of stable white wool blend fabrics.

The 21st century has seen very rapid growth in the field of nanotechnology and most researchers consider that the area is still in its infancy. Denning reviews the field and examines current applications and possible future applications of the technology to wool. There are several nanotechnology treatments applied to cotton that have not yet been used on wool and there are many opportunities on the near horizon.

An important step in changing the perception of wool from its old-fashioned and traditional role was the development of 'Sportwool', a wool blend fabric that took full advantage of wool's unique moisture buffering capacity to produce a very high-performance fabric for elite sports. Holcombe describes the requirements of moisture management fabrics and examines wool's position in this high-profile and rapidly developing area.

Wool has many unique properties as a textile fibre; however, there are occasions when it can be blended with other fibres with specific and complementary properties to further improve the performance of the blend. Miao reviews the objectives of blending of wool with other fibres and the techniques for obtaining optimum performance for the different blend yarns and fabrics.

The field of intelligent garments is expanding rapidly, but again, the area is still in its infancy. These technologies can add the ability to 'sense-and-respond' to the most basic textile materials, posing a threat to wool's superior position that has been gained through its natural ability to respond to changing environment. Collie reviews the opportunities for wool, by making it electrically conductive, but also by exploiting its rich protein chemical functionality.

Kelly provides a fitting and stimulating conclusion to the book by viewing wool not necessarily as a fibre, but as a chemical repository containing families of unique and complex keratin compounds and potential structural materials. The focus is on novel and high-value materials where the microstructure of the wool is disassembled and rebuilt.

N. A. G. Johnson
I. M. Russell

Part I

Advances in wool fibre and fabric production

1

Improvement of wool production through genetic manipulation

G. E. ROGERS, The University of Adelaide, Australia; and
C. S. BAWDEN, South Australian Research and Development Institute, Australia

Abstract: This review describes the manipulation of the sheep genotype by genetic engineering, which is defined as the introduction or elimination of specific genes in the existing cells of an organism through modern biological techniques for the purpose of changing one or more of its phenotypic characteristics. This can be achieved by direct injection of a DNA sequence for a gene or gene-related function or by transfer of an intact nucleus from a somatic cell containing a set of desirable genes into an enucleated egg cell. The potential of transgenesis over conventional breeding is that single gene changes can be made for a particular phenotype without accompanying multiple gene effects that can be disadvantageous. Several examples of engineered sheep are described. The potential for making other transgenic changes are discussed but transgenic sheep research has virtually ceased because of high cost and the perceived low public acceptance of transgenic animals at the present time.

Key words: wool properties, genetic engineering, gene transfer, transgenesis, wool keratin.

1.1 Introduction

Over many decades remarkable improvements in wool production in terms of quality (diameter, length, staple strength, crimp and colour) and yield (rate of wool growth) have been achieved through selective breeding and nutrition. Sheep breeding methodologies can be regarded as genetic engineering and will undoubtedly be of benefit in the future through the advent of genome characterisation and gene discovery that will provide molecular markers for more accurate selection of sheep and wool characters. The term genetic manipulation used in the present discussion is usually referred to as genetic engineering and is defined as the introduction or elimination of specific genes in the existing cells of an organism through modern molecular biology techniques for the purpose of changing one or more of the organism's phenotypic characteristics. This can be achieved by direct injection of a DNA sequence characterised for a gene or gene-related function into a fertilised egg (transgenesis) or by transfer of an intact nucleus from a somatic cell containing a set of desirable genes into an enucleated egg cell, a process that yields an animal clone, or by variations of these. The potential advantage of transgenesis is that single gene changes can be made rather than selectively breeding sheep for a particular phenotype, in which case there are accompanying multiple gene effects that can be disadvantageous.

This review describes the manipulation of the sheep genotype by these procedures, utilising tested and prospective genes for improving aspects of wool growth and wool quality, and some of the problems to be confronted.

1.2 The techniques of genetic manipulation for animals

Genetic manipulation techniques enable the insertion of single genes as DNA into the genomes of farm animals and to obtain thereby new phenotypes with changes in production characters. A recent excellent review (Wheeler, 2007) covers the general field of production of transgenic livestock achieved to the present time. In the case of sheep, single gene variations can be targeted for wool growth and properties. There are two major strategies: one is to increase a particular character, a 'gain of function', and the other is to decrease or totally obliterate a production character or 'loss of function'.

There are several methods for achieving transgenesis, the transference of selected genes into the genome. The major method used is microinjection of DNA containing the gene of interest into the pronucleus of a fertilised ovum. The injected one-cell embryo is then transferred to a recipient female (Wheeler, 2007). A mouse was the first transgenic animal to be produced by microinjection of the gene for human growth hormone, in the 1980s (Brinster *et al.*, 1981). The result was an increased growth rate and a larger mouse. Since that time there have been extensive studies of transgenesis for animal production including many novel ideas as yet not investigated. Other methods that avoid microinjection for inserting genes into fertilised ova have been established including sperm-mediated DNA in which the DNA is adsorbed onto the sperm surface and transferred into the egg by fertilisation *in vitro* followed by implantation into a recipient female (Lavitrano *et al.*, 2006). Similar procedures include liposome-mediated transfer into the ovum or electroporation of the transgene DNA into sperm or egg (Ogura, 2002). Retroviruses can be used as efficient carriers for gene transfer into ova. A development with potential for transgenic animal production is retroviral gene transduction into spermatogonia (the precursor cells of spermatozoa) stem cells and then transplantation to the animal's testes (Nagano *et al.*, 2001). Transgenic progeny have been successfully produced in mice by this method but application by whatever retroviral route in sheep is problematic in the present climate of opinion on acceptability of transgenic techniques in sheep production.

The ultimate in gene transfer is probably nuclear transfer in which the nucleus of a somatic or a stem cell is transferred into an enucleated ovum by microinjection or by cell–cell fusion. The result is a clone of the animal that donated the somatic or the stem cell nucleus. The technique of therapeutic cloning not only could lead to insertion of a nucleus from cells of a sheep carrying a gene or genes for a desirable phenotype but could be a nucleus from a cultured somatic or stem cell after insertion of a specific gene in a recombination process called 'knock in'. Alternatively, a gene can be removed by recombination, so-called gene 'knock out'. Such procedures in sheep would be expensive and difficult and have not been investigated, although a successful specific cloning would not then necessitate frequent repetition.

All transgenic procedures require the selected gene to be linked to DNA sequences that can control expression, in particular a promoter, an essential DNA sequence that can direct expression of the gene in an appropriate tissue or tissues. The examples for influencing wool growth to be discussed involve genetic manipulations that could enable genes to function systemically (in all or most tissues) including the wool follicle or to have expression of new or modified genes targeted directly to one or more of the seven cellular layers of the wool follicle (Fig. 1.1).

In order to affect wool growth and wool properties by inserting novel genes and directing expression to the wool follicle, a promoter is needed that is follicle-specific, and one that can even target expression to particular cell lineage of the follicle. Six different cell layers that are formed in the follicle from different cell lineages, namely the outer root sheath, three layers of inner root sheath (IRS), the fibre cuticle and the fibre cortex, differentiate from the follicle bulb (Fig. 1.2). Each of the cell lineages expresses specific genes that are potential sources of gene promoters for targeting the different cell layers of the follicle.

1.3 What aspects of wool production might be amenable to genetic manipulation?

Theoretically, genetic manipulation could change many aspects of wool growth and properties given the necessary knowledge of what genes are involved. The following list comprises phenotypic characteristics that are important for wool usage. Most of them could be judged from current knowledge to be within range of manipulation of wool follicle function by transgenesis or by specific gene-related molecular factors:

- increasing yield;
- improving fibre strength;
- decreasing the diameter;
- decrease fibre stiffness;
- altering crimp;
- increasing lustre;

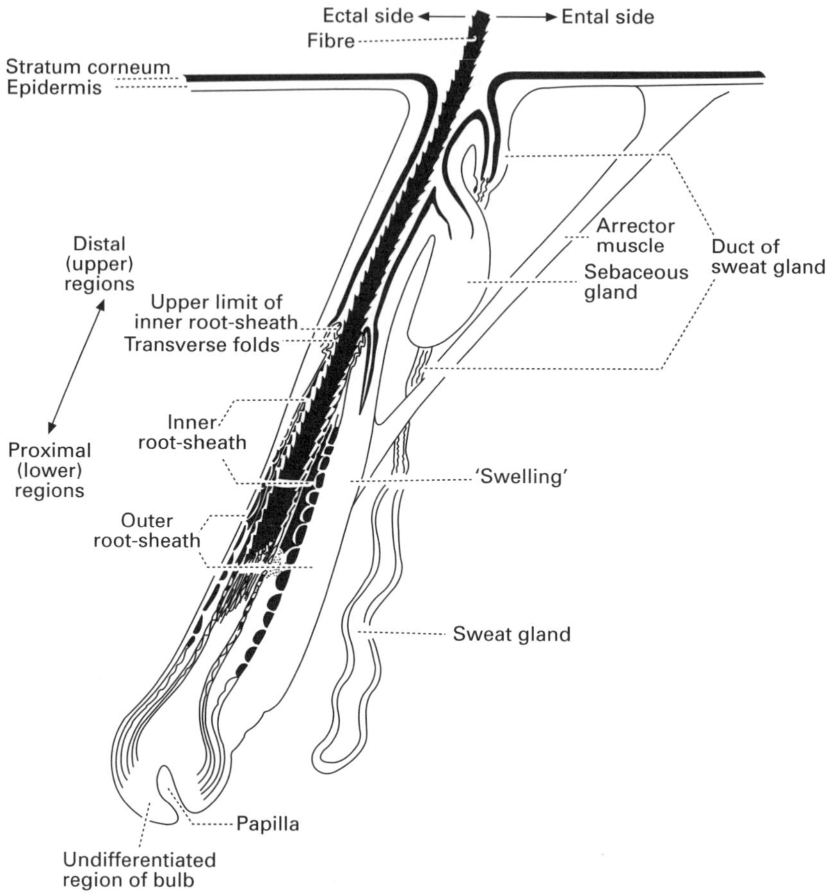

1.1 The wool follicle showing the main layers and associated sebaceous and sweat glands (Auber, 1951). Reproduced with permission.

- modifying surface properties for special uses;
- eradication of dark fibres from the fleece;
- prevention of blowfly strike;
- improving the technique of biological wool harvesting.

1.4 Increasing the yield of wool per animal

The quantity of wool fibre that grows on a sheep is determined by a complex of cellular events but primarily depends on the initiation of primary (P) follicles and secondary (S) follicles and most importantly on the additional population of follicles that derives from the secondaries, so-called secondary-derived (S^D) follicles. Moore *et al.* (1998) proposed that embryonic skin

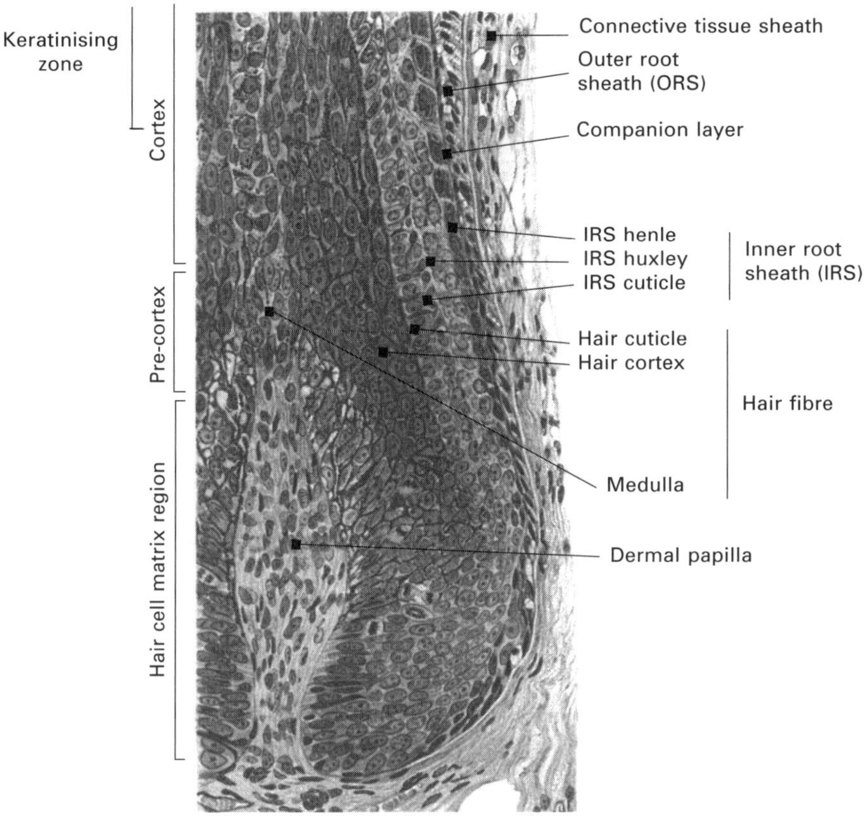

1.2 The human hair follicle showing the main layers with cellular detail of the seven cell lineages that differentiate from the bulb (Rogers *et al.*, 2006). Reproduced with permission.

generates initiation sites in skin with a *defined number* of specialised dermal (pre-papilla) cells. These cells induce follicle formation and finally form the dermal papilla in each follicle. According to that proposal and supported by sheep selection data, if a fixed population of the pre-papilla cells induces more primary follicles, the total number of secondary follicles induced would be less and then the secondary/primary (S/P) ratio would be lower. Conversely, if a lower number of primaries are induced, more secondaries would form and the S/P ratio would be higher. Such a process could account for the apparent relationship between follicle population, wool yield and fibre diameter that is the basis of a selection methodology ('soft-rolling skin', SRS™) that is used by some woolgrowers.

It is broadly accepted that reaction–diffusion mechanisms in which diffusing morphogens (Nagorcka and Mooney, 1989) act across the epidermal/dermal basement membrane in the follicle bulb are responsible for the establishment of the pattern of follicle distribution. Studies mainly in mice have identified many signalling factors that can influence the initiation and pattern follicles in the skin, and undoubtedly such factors operate similarly for wool follicles. The Wnt signalling pathway for example is essential to the dermal–epithelial initiation of hair follicles (Andl *et al.*, 2002; Reddy *et al.*, 2004). Further experimental evidence for the importance of WnT and its inhibitor for the establishment of follicle spacing in the skin has recently been provided (Sick *et al.*, 2006). Over-expression in transgenic mice of β-catenin, a component in the Wnt pathway, and of another factor, Sonic Hedgehog, can initiate formation of follicles and also induce fusion of follicles (Lo Celso *et al.*, 2004). Initiation of the follicle pattern also depends on the localisation of lymphoid enhancer factor (Lef-1) at focal points in the epidermis (Zhou *et al.*, 1995). A reaction–diffusion mechanism involving the factor ectodysplasin-A1 (Eda) and its receptor Edr (Barsh, 1999) with another factor BMP for establishing patterning of hair follicles has been discussed by Barsh (1999). Furthermore, when the factor Eda was over-expressed in transgenic mice many hair follicles became fused at the level of the infundibulum (the neck of the follicle) instead of being separated by interfollicular epidermis (Zhang *et al.*, 2003), a feature resembling the formation of S^D follicles in sheep. It can be concluded from these examples that if the interplay of such signalling molecules in developmental pathways in the growth of wool were better understood, it is likely that their function could be manipulated by transgenesis to change wool follicle population and wool production.

Another feature of wool growth is that measurements of cells differentiating from the follicle bulb have indicated that up to 80% of cells that leave the bulb differentiate into the inner root sheath layers and not wool fibre (Hynd *et al.*, 1986). Theoretically, more fibre could be produced if that differentiation pathway could be biased to allow differentiation into cortical and cuticle cells. Considering what is known of the involvement of specific regulatory genes in establishment of inner root sheath cell fate in mouse follicles, notably the transcription factors GATA3 (Kaufman *et al.*, 2003) and TCF3 (Merrill *et al.*, 2001), it might be possible to divert the commitment of wool follicle bulb cells from an inner root sheath cell fate towards a fibre fate by transgenesis. This would improve the so-called production ratio, diverting protein synthesis to the fibre-producing cells and would lead to increased yields of wool per follicle. The desirability of such a manipulation would depend on how much it would influence the IRS structure that is part of the fibre growth process and whether the effect might result in an increase in fibre diameter.

The possibility of increasing the rate of wool growth by over-expressing growth factors such as growth hormone (GH) or insulin-like growth factor (IGF) has been investigated. Transgenic sheep have been produced expressing a transgene for ovine GH by microinjection of a gene construct possessing the coding region linked to a metal-activated (metallothionein) promoter for controlling expression. The promoter was activated by providing zinc to the sheep via drinking water. Some of the transgenic sheep grew faster and produced more wool than control sheep but the growth effects varied and it was clear that the expression of the gene depended on its location in the genome (Adams and Briegel, 2005; Adams *et al.*, 2002). A serious disadvantage of over-expression of GH is its potential to cause health problems in the animal later in life. In other experiments transgenic sheep have been produced by pronuclear microinjection with a mouse keratin gene promoter linked to an ovine insulin-like growth factor 1 (IGF1) cDNA. Although it was reported that fleece weight increased in transgenic animals compared with non-transgenic, the results were not reproducible (Damak *et al.*, 1996; Su *et al.*, 1998).

Certain biochemical pathways that do not normally function in sheep have been suggested as a means to increasing wool production and have been subjected to initial investigation. If these pathways could be introduced by transgenesis using appropriate genes they could provide metabolites that are essential for wool growth but can be in short supply. For example, adequate supplies of cysteine to the follicles are essential for wool growth. Inadequate supply results in depressed fibre growth and 'breaks' in the fibres appear with consequent fibre weakness. That supplying cysteine, or its precursor amino acid methionine, can increase fleece weight has been demonstrated empirically (Reis, 1979).

The proteins of plant material ingested by sheep are degraded to amino acids by the microflora of the rumen and a significant amount of the cysteine is further degraded with the release of sulphide into the ruminal fluid and excreted after conversion to sulphate. It was postulated that if the genes for two enzymes that utilise sulphide for cysteine synthesis, serine transacetylase and *O*-acetyl serine sulphydrylase, were targeted to, and expressed in, the rumen wall then it might be possible to 'capture' the sulphide as cysteine that would then be absorbed into the blood and increase the amount of available cysteine. The necessary enzymes constitute the cysteine synthesis pathway in bacteria and the two genes have been characterised. These genes were linked with promoters and their expression successfully tested in cultured cells and mice (Ward, 1994). Microinjection of similar genes into sheep under the control of a viral promoter (Rous sarcoma virus long terminal repeat) had limited success in producing transgenic sheep with the two enzymes expressed in the tissues (Bawden *et al.*, 1995). Experiments have not been conducted in targeting to the rumen although candidate promoters are known. It should be noted that if general or systemic expression of the genes were to occur in the transgenic sheep it would not be a problem since cysteine synthesis is dependent on sulphide, which is present only in the rumen.

Another novel pathway suggestion that has been investigated is based on the knowledge that cellulose digestion in the sheep's rumen produces high levels of acetate. If acetate could be converted to glucose by a metabolic process not normally present, it could provide a useful metabolic supplement for the energy requirements of wool growth. Preliminary investigations have demonstrated that gene constructs possessing the bacterial genes encoding the two enzymes needed for the pathway can be expressed in mammalian cells and also in transgenic mice, but their functioning in sheep has not been pursued (Ward, 2000).

1.5 How might wool quality be modified by affecting fibre molecular structure?

1.5.1 The protein structure of wool and its relation to physical properties

The physical properties of wool fibres such as elasticity and strength and the distinctive thermal attributes of wool textiles are a consequence of their cellular structure, their specific proteins and the organisation of the proteins within the cells of the cortex of the fibre. The detailed structures of the wool fibre cortex and cuticle and the proteins that compose them have been studied at the chemical and electron microscopic levels for decades (see Powell and Rogers, 1997). Primarily the cortical cells are filled with 8 nm diameter keratin filaments (referred to as keratin IF or KIF) that are members of the large intermediate filament family found in animal cells. In the wool cortex the KIFs are separated by a matrix that appears under the electron microscope as dense material between individual IFs. The matrix is a complex of non-fibrous protein families and referred to as keratin-associated proteins or KAPs. They fall into two main types, one with a composition characterised by a high content of amino acids glycine and tyrosine ('high glycine/tyrosine proteins') and the other of cystine ('high-sulphur proteins').

In fine wool fibres the cells of the orthocortex and paracortex that were originally distinguished by their differing degrees of access to dyestuffs, display a distinct organisation of the intermediate filaments, the KIFs, and matrix. In the paracortical cells the KIFs are strongly aligned with the fibre axis and in cross-section they show a quasi-regular packing, whereas in the orthocortical cells the KIFs are inclined to the fibre axis and packed into fibre aggregates called macrofibrils (Fig. 1.3). In cross-section view the macrofibrils have the appearance of 'whorls' because of the tilt and individual KIFs become indistinguishable out from the centre.

The protein structure of a KIF is complex in that it requires two distinct protein types, Type I and Type II, in order for it to form. The two types are only distantly related in amino acid sequence but nevertheless very similar

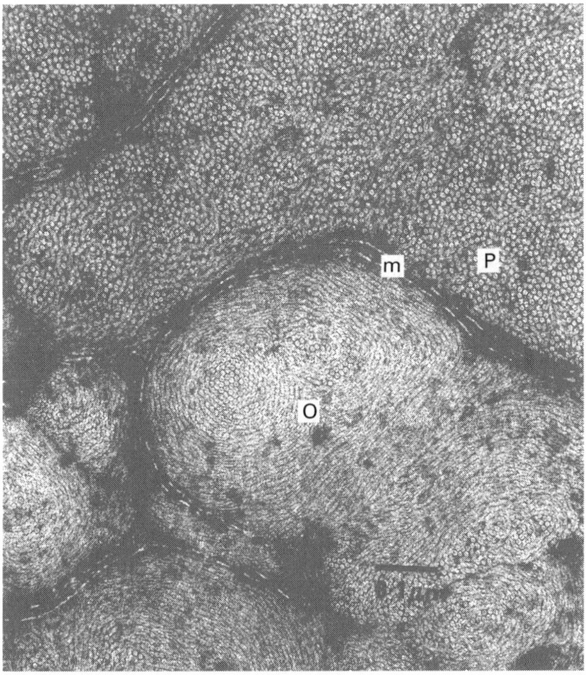

1.3 Transmission electron micrograph of a cross-section of a fine Merino wool fibre showing the ortho- (O) and paracortical (P) cortical cell types with differing organisation of the keratin intermediate filaments. The cells are separated by a boundary consisting of the cell membrane complex (m) present between all of the cortical cells.

in their secondary structure. The stability of the cortex is dependent upon chemical bonds that are present between the KIFs and also between the KIFs and matrix molecules. They include hydrogen bonds of the α-helix and disulphide bonds that are present within the KIFs and connect them to the matrix molecules. Very little is known about these intermolecular bonds in terms of what particular amino acids in the protein sequences are involved. Presumably, overall there is a high degree of specificity in the intermolecular bonding. A diagram illustrating these aspects is given in Fig. 1.4.

The typical behaviour of a wool fibre when stretched in the presence of water molecules (the stress/strain curve) results from the breakage of the hydrogen (non-covalent) bonds as the α-helical proteins of the keratin IFs unwind (the yield region of the stress/strain curve). The yield region ends when the disulphide (covalent) bonds resist further extension until the breakage of the fibre (see Feughelman, 1997, for a detailed discussion of these physical properties).

The contents of the cells of the cuticle consist of amorphous proteins in three layers that are distinguishable by electron microscopy (Fig. 1.5). The

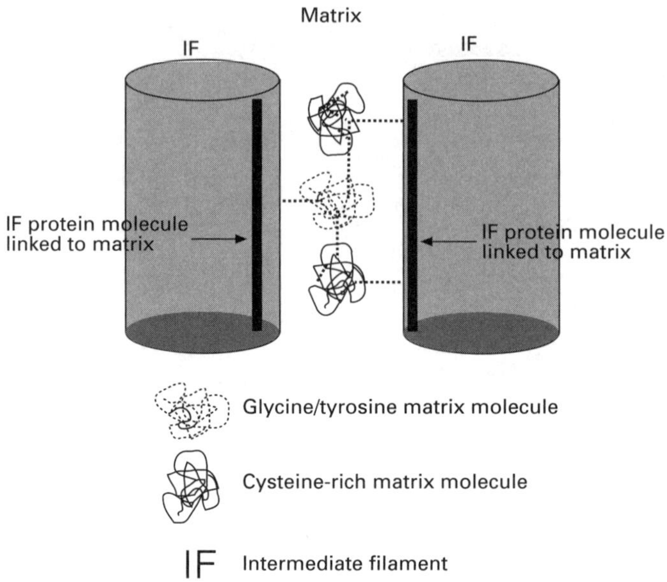

1.4 A diagram illustrating the linkages by disulphide bonds (dotted lines) present between globular-type matrix molecules and between the matrix molecules and the rod-like α-helical protein molecules in neighbouring intermediate filaments.

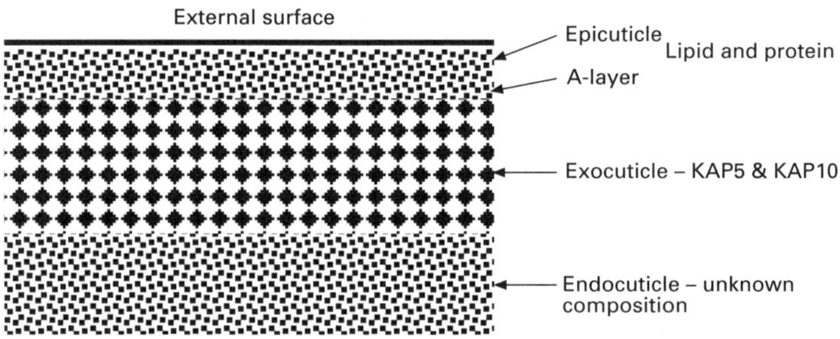

1.5 Diagrammatic representation of the layers constituting the protein layers in a cuticle cell.

layer beneath the epicuticle on the surface of each cuticle cell is the exocuticle that has an outer or A-layer. An innermost layer in each cell is the endocuticle. The proteins of the cuticle have been only partially characterised in the exocuticle that has two proteins identified as KAP-type proteins and related to the sulphur-rich KAPs of the matrix in the fibre cortex. Proteins of the endocuticle have not been identified but are usually regarded as cytoplasmic proteins that remain after cell differentiation. The exocuticle KAP proteins are crosslinked by disulphide bonds and produce a stiff protective covering for the fibre. The influence of the overlapping flattened cells of the cuticle (Fig. 1.6) on the elastic properties of wool fibres is minimal but they play a major role on fibre migration in concert with the elasticity of the fibre cortex in the shrinkage of wool textiles. The effect of the cuticle is significantly diminished by chemical treatments that flattens the cells and reduces the ratchet-like profile.

For example, the chlorination step of the commercial Hercosett process partially degrades the exocuticle and A-layer (Fig. 1.5) of the cuticle cells, resulting in a marked lowering of the differential friction. Binding a polymer to the outside surface of the fibre increases this effect.

1.5.2 The expression of keratin and novel proteins in the fibre by transgenesis

Any genetic manipulation of the fibre contemplated to alter the physical properties in some way, such as increasing fibre strength, has to be based on

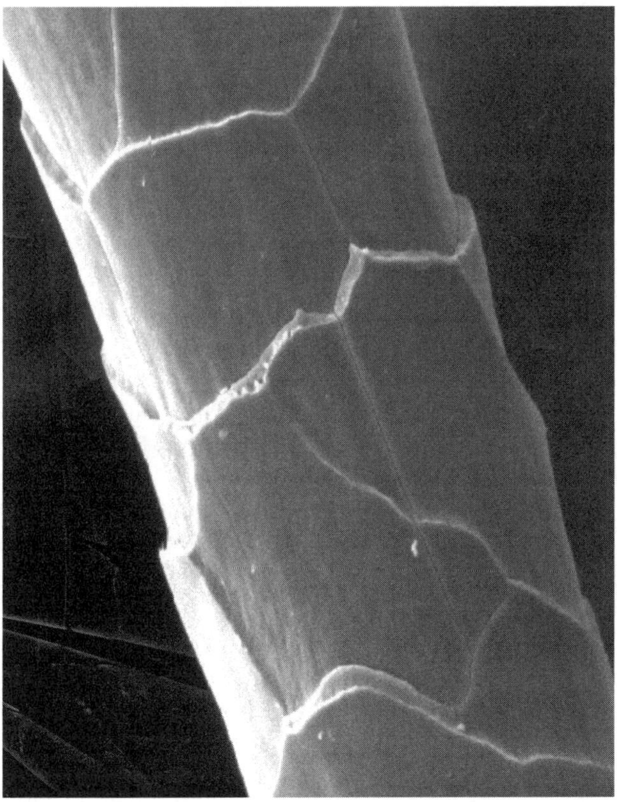

1.6 The cuticle of a wool fibre showing the overlapping flattened cells and their edges in profile.

the known structure of cortex and cuticle as outlined here. The protein composition of the cortex or the cuticle could be modified by either increasing or decreasing the abundance of one of the proteins. According to genomic studies of human hair there are 10 functional genes for Type I and 10 for Type II keratin IFs (Langbein and Schweizer, 2005) and for the matrix proteins there are 17 glycine/tyrosine KAP genes and about 67 genes for the sulphur-rich KAPs (Rogers *et al.*, 2006). All of the genes are present in the genome as large domains in several chromosomes and apparently evolved through gene duplication and mutation. Their numbers and domain organisation could be expected to be similar in sheep according to protein analyses and current knowledge from gene characterisation. The multiplicity of members in these protein families and their similarity of sequence suggest that the molecular structure of the filament and matrix can tolerate variations in the relative abundance of the different chains. Indeed, wools from different sheep breeds maintain the basic structure of filament and matrix despite variation in protein composition. Hence it can be concluded that transgenic manipulation is a valid approach to obtaining changes in fibre composition by over-expression or under-expression of selected keratin genes.

At present, only over-expression of a limited number of single genes in the wool cortex have been experimentally tested in transgenic sheep. The timing of gene expression in the cortex depends on the promoter that is linked to the gene to be expressed. The choice of promoters from KIF and KAP genes, in principle, could be utilised to allow a transgene to be expressed at stages of wool fibre formation that are either early, midway or late and would be expected to have differing effects on fibre properties. An important aspect to be noted is that experiments conducted so far have indicated that promoters of the KIF genes are more active than promoters of the KAP genes. It is known that the global control of gene expression occurs through certain loci in genomes. Hair and wool keratin genes exist in domains in the chromosomes, and characterisation of such regions could be important for the fine-tuning of transgenic gene expression and thus for the modulation of consequent changes to wool structure and growth.

The first objective in the transgenesis investigations supported by the Australian Cooperative Research Centre (CRC) for Premium Quality Wool and the agencies supporting the CRC was to devise strategies that might improve the strength and/or elasticity of the fibre. In the initial experiments, a sheep gene for a Type II KIF called K2.10 was chosen to provide the promoter for expressing transgenes targeted to the wool follicle cortex that activates expression relatively early in cortical cell differentiation. This gene promoter had already been shown to be functional in mouse hair follicles (Powell and Rogers, 1990) and in merino wool follicles (Bawden *et al.*, 1998). A promoter–gene construct was made, comprising the K2.10 promoter linked to other regulatory sequences with a cloning site to enable insertion of chosen DNA sequences adjacent to, and downstream from, the promoter. In order to reduce interference to gene expression by host chromosomal DNA, a functional insulator sequence was also included in the basic construct (Bawden *et al.*, 1998, 2000).

The experiments to enhance intrinsic strength of wool fibres involved introducing more cysteine-rich KAP proteins into the cortical cell structure on the basis that they could increase the level of crosslinking of disulphide bonds. An alternative choice was to introduce another crosslink, the isopeptide crosslink, by expressing the transglutaminase enzyme responsible for catalysing the formation of these links between lysine and glutamine side-chains in proteins. The other aspect considered for change was fibre elasticity and the route chosen for investigation was to attempt to convert the cortex to a predominantly orthocortical ('soft') type by increasing the level of one of the glycine/tyrosine-rich proteins or by increasing the proportion of KIFs in the cortex by co-expression of Type I and Type II wool KIF transgenes. These manipulations were shown by *in situ* hybridisation to achieve the expected expression of the proteins in the wool cortex. Furthermore, the presence of the expressed proteins could be visualised by electron microscopy except in the case of co-expression of the Type I and Type II genes. These transgenic experiments demonstrated that the phenotypes could be inherited by following generations. The load-bearing capacity and the extensibility of fibres from the transgenic animals were evaluated and in all cases (except for the Type I and Type II wool KIF co-expression) the overall result was that these physical properties were reduced in comparison with fibres of age-matched non-transgenic siblings. However, preliminary tests on several transgenic wools woven into fabric indicated that there was a significant increase in comfort (M. Huson, personal communication). It is possible that the prickle property was reduced by increasing the flexibility of the fibres.

The expression data obtained so far indicates that the level of transgene expression is such that even if the transgene product inserts into the native KIF-matrix structures excess proteins are also 'dumped' in the cells as amorphous masses. This situation was observed for instances when the matrix protein KAP6 or a Type I IF protein (without its Type II partner) were over-expressed (Fig. 1.7).

Another consequence of transgene expression that has been observed is that expression of endogenous genes is reduced, presumably because the protein synthetic machinery can be overwhelmed by the transgene expression. Future problems in transgenic research to be solved are to produce transgene constructs that:

- have promoters that are active at the correct time of cortical differentiation;
- include DNA elements (microRNAs) found in the genome that can control gene expression. Synthetic anti-sense gene sequences also can be designed to target specific genes and reduce the level of their expression and

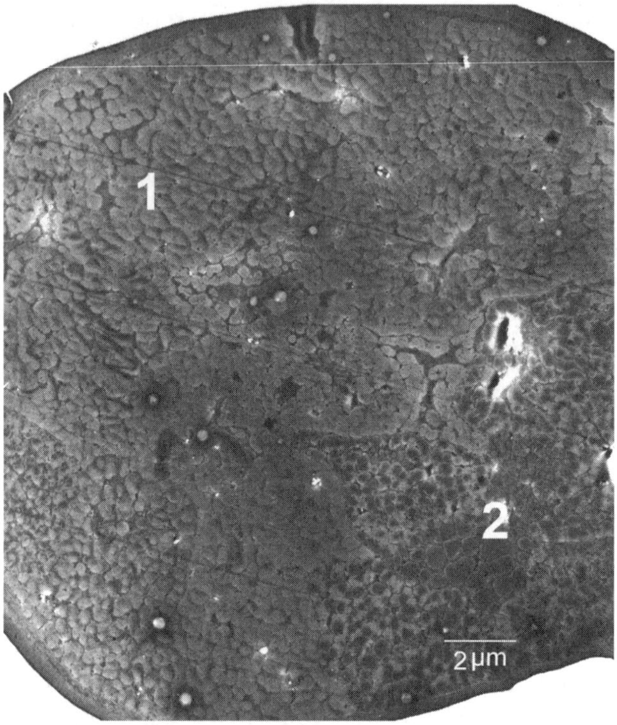

(a)

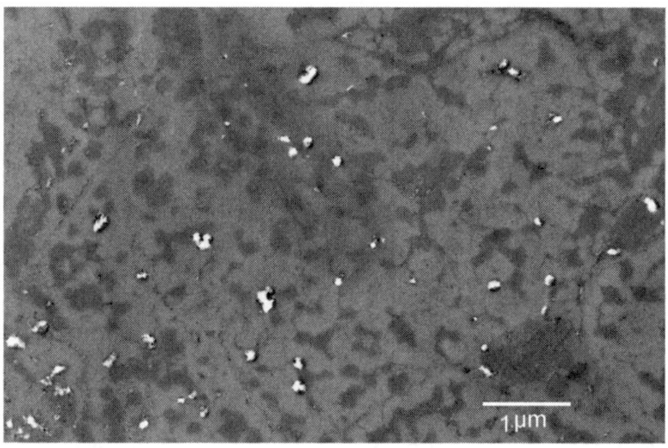

(b)

1.7 (a) Transmission electron micrograph of a wool fibre from a transgenic sheep expressing a transgene for KAP6 protein directed to the wool cortex. Normal orthocortex '1' and paracortex '2' that has an orthocortex appearance resulting from the deposition of KAP6 protein. (b) Transmission electron micrograph of a wool fibre from a transgenic sheep expressing a transgene for a Type I IF protein directed to the wool cortex. Deposits of IF protein darker than the normal keratin are present throughout the cortex.

constitute part of that approach. It is pertinent to mention that although fibre stiffness mentioned above is a function of the keratin proteins in the cortex, the cells of the cortex are enveloped by membranes (m in Fig. 1.3) that appear to consist of stiff crosslinked proteins, one of them might be involucrin and another, loricrin. It is possible that any 'stiffening' effect could be could be reduced or eliminated using anti-sense technology on these proteins and thereby increase fibre flexibility.

The modification of wool properties by expressing foreign protein in wool, especially those that have a fibrous molecular structure, has been proposed but awaits investigation. Collagen, elastin and silk are candidate fibrous proteins and if expressed at sufficiently high levels in the cortex they would be expected to alter physical properties depending on whether they were deposited in the normal fibrous form or as amorphous deposits. Of note is spider drag-silk, a particularly strong silk. The gene for this has been isolated and characterised, and the monomeric form of the protein has been expressed in the milk of goats (Williams, 2003). The expressed protein has been tested for fibre regeneration with the aim of producing materials with high tensile strength (Seidel *et al.*, 2000). The prospect of expressing the silk gene in the wool of transgenic sheep to strengthen it was considered by the CRC for Premium Quality Wool several years ago but it was not considered to be sufficiently commercially viable (C. S. Bawden and G. E. Rogers, unpublished data).

Two cell layers that play a role in the determination of the structure and properties of the fibre are the cuticle itself and the inner root sheath of the follicle that is involved in shaping the cuticle surface as the fibre differentiates in the follicle. The edges where cuticle cells overlap (Fig. 1.6) favour monodirectional fibre movement and thereby promote garment shrinkage. No transgenic experiments have been undertaken to reduce this cuticle behaviour but some strategies can be considered to that end, such as reducing the edge profile to resemble the flatter profile seen in cashmere fibres. One strategy would be to target transgenic changes to the specific proteins of the exocuticle layer (Fig. 1.5) by reducing their abundance or their level of crosslinking, or both, or by introducing novel proteins or synthetic polypeptides that do not crosslink. Another approach could be a disruption of the cells of the inner root sheath cuticle cell layer that might result in a modification of the imprinting effect of that layer on the cuticle cell edges.

1.6 Some other aspects of wool production that could be achieved by transgenesis

A central problem of wool production is the increasing cost of the harvesting of wool by mechanical shearing and alternative procedures have been sought for many years, especially biological ones. When the discovery of epidermal growth factor (EGF) and its remarkable effect on keratinisation were reported, wool biologists made the major finding of the dramatic effect of exogenous EGF isolated from mouse salivary glands on wool growth (Moore *et al.*, 1982). It was shown that the EGF inhibited DNA synthesis in the follicle cells that led to cell death (apoptosis) and fibre shedding (Panaretto *et al.*, 1984). These remarkable findings were ultimately incorporated into a commercial procedure in which EGF is produced by factory-scale recombinant DNA technology involving expression of the human EGF gene in bacterial culture and purification of the factor. Although administration of a single subcutaneously injected dose of EGF can successfully replace shearing and is used by some wool growers, the impact on the industry has seen only partial adoption. The technique (Bioclip™) requires special equipment for harvesting the wool and is not applicable to sheep weight over 60 kg or to pregnant ewes. There may be other growth factors known and unknown that might be found to act more effectively than EGF in temporarily disrupting wool growth, but discovering such factors would require extensive basic and applied research. There may be a factor or factors that could be used in conjunction with EGF to improve its efficacy and application. Researchers in transgenic sheep technology have considered the possibility of making sheep transgenic to EGF and controlling its expression in the skin when needed but it is not conceivable that such a procedure with so many complex issues associated with it could replace the direct simple injection of EGF. Nevertheless 'gene discovery' research for example, into the normal follicle mechanisms in Wiltshire sheep that bring about seasonal shedding (Nixon *et al.*, 2002) could bring forth new information that might be utilised in developing new molecular methods for the harvesting of wool.

External parasite infestation such as blowfly strike is another problem. Attention has been focused on the possibility of controlling it by secreting insecticidal proteins into the follicle areas around the breech of the sheep. Known insecticidal proteins that could be candidates for investigation through sheep transgenesis are wheat agglutinin (Giovanini *et al.*, 2007) and proteins from *Bacillus thuringiensis* (Carlini and Grossi-de-Sa, 2002). A strategy for targeting such molecules to make them secrete onto the skin of the sheep could be to link the insecticide gene to a gene promoter of a gene from the sebaceous or sweat glands. Strategies of this kind were initiated by CSIRO within the CRC for Premium Quality Wool but ceased on closure of the CRC.

Molecular methods to control the problem of dark fibres in wool would appear to have some potential since the genetics, biochemistry and cellular mechanisms of melanin formation are understood and coat colour has been induced by expression of the tyrosinase gene in transgenic mice (Lavado *et al.*, 2005; Tanaka and Takeuchi, 1992). Studies to reduce the expression of the tyrosinase gene in the melanin pathway by utilising anti-sense mRNAs were initiated by CSIRO during the life of the CRC for Premium Quality Wool. Since that time the discovery of microRNAs that are normally used by cells to regulate genes (Scherr and Eder, 2007) provides more tools for investigating the dark fibre problem.

1.7 Conclusion and future trends

Transgenic sheep research in recent years has raised many possibilities for modifying wool and its production. Intense work in this area has ceased for several reasons, including the current low level of acceptance of molecular genetic manipulation as a farm production procedure. However, a major reason is also the high cost of 'blue-sky' research and the economics of applying any proven genetic result. Research in Australia during the 1985–2000 period has certainly demonstrated that transgenes can be inserted into the sheep genome and are actively expressed. Although sheep transgenic research has not yielded any clearly defined application so far, the research has demonstrated that over-expression of several wool keratin genes produces a marked change in the balance of IF/KAP structural components of the wool fibre altering the elastic behaviour of fibres, and inserted metabolic-type genes can also partially function. These achievements provide support for the opinion that rapid advances in the understanding of gene regulation in follicle differentiation and further refinement of transgenesis techniques will further enable the development of transgenic sheep with new and novel products. To this end a study of the genes functioning in the sheep during development and in adults producing wool is being conducted in Australia in the Sheep Genomic Program (see http://www.sheepgenomics.com). These studies should provide new clues to selection of the best genetics for wool production and also some potential targets for genetic enhancement (GM), possibly by transgenesis and likely to be focused on regulators that dictate development of higher follicle densities, finer fibres and improved staple strength. The adoption of any outcomes from such studies will probably depend largely on the demands of the wool market and competition from other fibres, especially the range of synthetic fibres that are fossil fuel-based.

1.8 References

ADAMS, N. R., and BRIEGEL, J. R. (2005). Multiple effects of an additional growth hormone gene in adult sheep. *J Anim Sci* **83**, 1868–1874.

ADAMS, N. R., BRIEGEL, J. R., and WARD, K. A. (2002). The impact of a transgene for ovine growth hormone on the performance of two breeds of sheep. *J Anim Sci* **80**, 2325–2333.

ANDL, T., REDDY, S. T., GADDAPARA, T., and MILLAR, S. E. (2002). WNT signals are required for the initiation of hair follicle development. *Dev Cell* **2**, 643–653.

AUBER, L. (1951). The anatomy of follicles producing wool fibres with special reference to keratinization. *Trans Roy Soc Edinburgh* **62** (part I), 191–254.

BARSH, G. (1999). Of ancient tales and hairless tails. *Nature Genet* **22**, 315–316.

BAWDEN, C. S., SIVAPRASAD, A. V., VERMA, P. J., WALKER, S. K., and ROGERS, G. E. (1995). Expression of bacterial cysteine biosynthesis genes in transgenic mice and sheep: toward a new *in vivo* amino acid biosynthesis pathway and improved wool growth. *Transgenic Res* **4**, 87–104.

BAWDEN, C. S., POWELL, B. C., WALKER, S. K., and ROGERS, G. E. (1998). Expression of a wool intermediate filament keratin transgene in sheep alters fibre structure. *Transgenic Res* **7**, 1–15.

BAWDEN, C. S., MCLAUGHLAN, C. J., PENNO, N. M., WALKER, S. K., HYND, P. I., and ROGERS, G. E. (2000). Genetic manipulation to modify wool properties and fibre growth rate. *Proc NZ Soc Anim Production* **60**, 147–154.

BRINSTER, R. L., CHEN, H. Y., TRUMBAUER, M. E., YAGLE, M. K., SENEAR, A. W., WARREN, R., and PALMITER, R. D. (1981). Somatic expression of herpes thymidine kinase in mice following injection of a fusion gene into eggs. *Cell* **27**, 223–231.

CARLINI, C. R., and GROSSI-DE-SA, M. F. (2002). Plant toxic proteins with insecticidal properties. A review on their potentialities as bioinsecticides. *Toxicon* **40**, 1515–1539.

DAMAK, S., SU, H., JAY, N. P., BARRELL, G. K., and BULLOCK, D. W. (1996). Improved wool production in transgenic sheep expressing insulin-like growth factor 1. *Biotechnology (N Y)* **14**, 185–188.

FEUGHELMAN, M. (1997). *Mechanical Properties and Structure of Alpha-Keratin Fibres* (Sydney, UNSW Press).

GIOVANINI, M. P., D. SALTZMAN, K., PUTHOFF, D. P., GONZALO, M., HERBERT, W. O., and WILLIAMS, C. E. (2007). A novel wheat gene encoding a putative chitin-binding lectin is associated with resistance against Hessian fly. *Molecular Plant Pathol* **8**, 69–82.

HYND, P. I., SCHLINK, A. C., PHILLIPS, P. M., and SCOBIE, D. R. (1986). Mitotic activity in cells of the wool follicle bulb. *Aust J Biol Sci* **39**, 329–339.

KAUFMAN, C. K., ZHOU, P., PASOLLI, H. A., RENDL, M., BOLOTIN, D., LIM, K. C., DAI, X., ALEGRE, M. L., and FUCHS, E. (2003). GATA-3: an unexpected regulator of cell lineage determination in skin. *Genes Dev* **17**, 2108–2122.

LANGBEIN, L., and SCHWEIZER, J. (2005). Keratins of the human hair follicle. *Int Rev Cytol* **243**, 1–78.

LAVADO, A., MATHEU, A., SERRANO, M., and MONTOLIU, L. (2005). A strategy to study tyrosinase transgenes in mouse melanocytes. *BMC Cell Biol* **6**, 18.

LAVITRANO, M., BUSNELLI, M., CERRITO, M. G., GIOVANNONI, R., MANZINI, S., and VARGIOLU, A. (2006). Sperm-mediated gene transfer. *Reprod Fertil Dev* **18**, 19–23.

LO CELSO, C., PROWSE, D. M., and WATT, F. M. (2004). Transient activation of beta-catenin signalling in adult mouse epidermis is sufficient to induce new hair follicles but continuous activation is required to maintain hair follicle tumours. *Development* **131**, 1787–1799.

MERRILL, B. J., GAT, U., DASGUPTA, R., and FUCHS, E. (2001). TCF3 and LEF1 regulate lineage differentiation of multipotent stem cells in skin. *Genes Development* **15**, 1688–1705.

MOORE, G. P., PANARETTO, B. A., and ROBERTSON, D. (1982). Inhibition of wool growth in merino sheep following administration of mouse epidermal growth factor and a derivative. *Aust J Biol Sci* **35**, 163–172.

MOORE, G. P., JACKSON, N., ISAACS, K., and BROWN, G. (1998). Pattern and morphogenesis in skin. *J Theor Biol* **191**, 87–94.

NAGANO, M., BRINSTER, C. J., ORWIG, K. E., RYU, B. Y., AVARBOCK, M. R., and BRINSTER, R. L.

(2001). Transgenic mice produced by retroviral transduction of male germ-line stem cells. *Proc Natl Acad Sci USA* **98**, 13090–13095.

NAGORCKA, B. N., and MOONEY, J. R. (1989). The reaction–diffusion system as a spatial organizer during initiation and development of hair follicles and formation of the fibre. In *The Biology of Wool and Hair*, G. E. Rogers, P. J. Reis, K. A. Ward, and R. C. Marshall, eds. (London, New York, Chapman and Hall), pp. 365–379.

NIXON, A. J., FORD, C. A., WILDERMOTH, J. E., CRAVEN, A. J., ASHBY, M. G., and PEARSON, A. J. (2002). Regulation of prolactin receptor expression in ovine skin in relation to circulating prolactin and wool follicle growth status. *J Endocrinol* **172**, 605–614.

OGURA, T. (2002). *In vivo* electroporation: a new frontier for gene delivery and embryology. *Differentiation* **70**, 163–171.

PANARETTO, B. A., LEISH, Z., MOORE, G. P., and ROBERTSON, D. M. (1984). Inhibition of DNA synthesis in dermal tissue of merino sheep treated with depilatory doses of mouse epidermal growth factor. *J Endocrinol* **100**, 25–31.

POWELL, B. C., and ROGERS, G. E. (1990). Cyclic hair-loss and re-growth in transgenic mice overexpressing an intermediate filament gene. *EMBO* **9**, 1485–1493.

POWELL, B. C., and ROGERS, G. E. (1997). The role of keratin proteins and their genes in the growth, structure and properties of hair. In *Formation and Structure of Hair*, P. Jolles, H. Zahn, and H. Hocker, eds. (Basel, Birkhauser Verlag), pp. 59–148.

REDDY, S. T., ANDL, T., LU, M. M., MORRISEY, E. E., and MILLAR, S. E. (2004). Expression of Frizzled genes in developing and postnatal hair follicles. *J Invest Dermatol* **123**, 275–282.

REIS, P. J. (1979). Effects of amino acids on the growth and properties of wool. In *Physiological and Environmental Limitations to Wool Growth*, J. L. Black, and P. J. Reis, eds. (New South Wales, University of New England Publications Unit), pp. 223–242.

ROGERS, M. A., LANGBEIN, L., PRAETZEL-WUNDER, S., WINTER, H., and SCHWEIZER, J. (2006). Human hair keratin-associated proteins (KAPs). *Int Rev Cytol* **251**, 209–263.

SCHERR, M., and EDER, M. (2007). Gene silencing by small regulatory RNAs in mammalian cells. *Cell Cycle* **6**, 444–449.

SEIDEL, A., O. LIIVAK, CALVE, S., ADASKA, J., JI, G., YANG, Z., GRUBB, D., ZAX, D. B., and JELINSKI, L. W. (2000). Regenerated spider sillk: processing, properties and structure. *Macromolecules* **33**, 775–780.

SICK, S., REINKER, S., TIMMER, J., and SCHLAKE, T. (2006). WNT and DKK determine hair follicle spacing through a reaction–diffusion mechanism. *Science* **314**, 1447–1450.

SU, H. Y., JAY, N. P., GOURLEY, T. S., KAY, G. W., and DAMAK, S. (1998). Wool production in transgenic sheep: results from first-generation adults and second-generation lambs. *Anim Biotechnol* **9**, 135–147.

TANAKA, S., and TAKEUCHI, T. (1992). Expression of tyrosinase gene in transgenic albino mice: the heritable patterned coat colors. *Pigment Cell Res* **5**, 300–303.

WARD, K. A. (1994). Preventing hair loss in the mouse. *Nature* **371**, 563–564.

WARD, K. A. (2000). Transgene-mediated modifications to animal biochemistry. *Trends Biotechnol* **18**, 99–102.

WHEELER, M. B. (2007). Agricultural applications for transgenic livestock. *Trends Biotechnol* **25**, 204–210.

WILLIAMS, D. (2003). Sows' ears, silk purses and goats' milk: new production methods and medical applications for silk. *Med Device Technol* **14**, 9–11.

ZHANG, M., BRANCACCIO, A., WEINER, L., MISSERO, C., and BRISSETTE, J. L. (2003). Ectodysplasin regulates pattern formation in the mammalian hair coat. Genesis **37**, 30–37.

ZHOU, P., BYRNE, C., JACOBS, J., and FUCHS, E. (1995). Lymphoid enhancer factor 1 directs hair follicle patterning and epithelial cell fate. *Genes Development* **9**, 700–713.

2 The objective measurement of wool fibre quality

P. SOMMERVILLE, Formerly Corporate Development Manager, Australian Wool Testing Authority Ltd, Australia

Abstract: This chapter discusses the measurement and specification of raw wool and wool sliver, providing an overview of the current status of wool fibre metrology, highlighting the major advances that have occurred in the past 50 years. The chapter first provides definitions of relevant terms, outlines the principles of wool metrology, its applications within the wool industry and the statistical methods employed. Development, standardisation and regulation of the measurement systems are then described. Finally, advances, latest developments and future trends in wool metrology are discussed in the context of the important physical characteristics of raw wool and sliver.

Key words: wool fibre diameter, wool staple length and strength, wool colour, wool quality, quality control in wool processing.

2.1 Introduction

Put simply, fitness for use defines quality, of which there are two general aspects: quality of design and quality of conformance. Raw materials and products are generally available in various grades or levels of quality. These variations are often intentional, and consequently the appropriate technical term in such instances is quality of design. Quality of conformance is how well the product conforms to the specifications and tolerances required by the design.

Every product, including wool, possesses a number of elements that jointly describe its fitness for use. These elements are often called quality characteristics. Quality characteristics may be of several types. In the case of wool and wool products these include:

- *physical*: length, weight, fineness, yarn evenness;
- *sensory*: handle, feel, appearance, colour;
- *time orientation*: reliability, durability, serviceability.

Generally it is difficult (and expensive) to provide customers with raw materials and products that have flawless quality characteristics. A major reason for this difficulty is variability. Wool is an extremely variable material. It varies along the fibre, among fibres, among staples, among animals, among mobs, among bloodlines and among regions.

The quality characteristics of wool were traditionally estimated subjectively.

Today, at all levels of the industry, technology providing objective measurements has almost completely replaced the senses of vision and touch because now most of the quality characteristics can be assigned a numerical value. This chapter focuses upon the measurement and specification of raw wool and wool sliver, providing an overview of the current status of wool fibre metrology, highlighting the major advances that have occurred in the past 50 years.

2.2 Definitions

The number of technical terms used by wool metrologists is very large, and consequently a complete glossary will not be included here. A comprehensive glossary is published by the Australian Wool Testing Authority Ltd.[1] However, there are some key terms that occur frequently throughout this chapter, and their definition can be found in Table 2.1.

2.3 Principles of wool metrology

Put quite simply, metrology is the science of measurement. Metrologists are generally focused upon developing (evaluating) technologies and systems for objectively measuring the quality attributes of raw materials and of manufactured products. The objective of wool metrologists is to develop measurement systems and associated standardised testing regimes with repeatable and numerically defined levels of precision.

2.4 Applications of wool metrology

2.4.1 Commercial trading

The transformation of wool from a fibre growing on a sheep into a textile product occurs via a long pipeline where ownership of the fibre may pass through several pairs of hands. Metrology has a major role to play in this exchange of ownership since objective measurements now provide most of the information that determines the value of the fibre. Over the past 50 years the objectively measured parameters for wool have been incorporated into the contracts and specifications utilised by the buyers and sellers along this pipeline. The quality characteristics of raw wool are ranked in order of their economic importance in Table 2.2. Wool metrology has also facilitated the streamlining of selling systems, resulting in very significant increases in productivity and reductions in cost.

Table 2.1 Definitions of key terms

Accuracy	A measure of the closeness of a test result to the true value. The difference between accuracy and precision should be noted. (Also see bias, precision and confidence limits.)
Ash content	The residue of a scoured wool subsample after it has been subjected to charring followed by heating to 800 °C.
Bias	A constant or systematic difference between a true value and corresponding test results.
Card sliver	A continuous strand of opened and loosely assembled scoured wool fibres, together with variable amounts of vegetable matter. Its linear density is approximately constant and it is without twist
Coefficient of variation (CV)	A statistical measure of the variability exhibited within a set of values. It expresses the standard deviation as a percentage of the mean; the higher the CV, the greater the variability.
Commercial yield	Any of the yields, calculated from the wool base, vegetable matter base and hard heads and twigs base, as specified in Section 3.0 of the IWTO Core Test Regulations. Yields may be expressed either as net clean mass of a lot or delivery, or as the net clean mass expressed as a percentage of the net greasy mass.
Confidence limits	An expression of the precision of the mean of a set of values, usually associated with a stated probability, most often 95%. It is the interval around the mean within which, with the stated probability, the true value is expected to lie.
Core sample	A representative sample of raw wool obtained from each bale in the lot by coring techniques.
Core test	The series of measurements, typically of wool base, vegetable matter base and mean fibre diameter, carried out on a core sample.
Distribution	Distribution can be expressed in the form of a frequency table or as a frequency histogram with data grouped into classes of 1 µm size, and integer micrometre values as midpoints of the class intervals.
Fibre diameter	The thickness of individual fibres; it is customary to quote an average value (mean fibre diameter or MFD) in micrometres.
Fibre diameter	The distribution of fibre diameter in a wool sample.
Grab sample	A sample taken from within a bale of wool by a mechanically driven jaw sampling device.
Greasy wool	Wool from the sheep's back or sheepskins which has not been scoured, solvent degreased or carbonised or otherwise processed. It contains grease and suint extruded from the follicles in the skin and dirt and vegetable matter picked up from grazing.
Mean fibre diameter	The arithmetic mean of all fibre diameter readings in a sample.
Precision	An indicator of the repeatability of a measurement; it is often expressed in terms of confidence limits.

Table 2.1 Continued

Raw wool	Wool fibre together with variable amounts of vegetable matter and extraneous alkali-insoluble substances, mineral matter, wool waxes, suint and moisture. It includes (a) greasy wool; (b) wool which has been scoured, carbonised, washed or solvent degreased; (c) scoured skin wools; and (d) slipe wools.
Scoured wool	Wool that has been aqueously washed or solvent washed to remove most of the grease, wax and suit and dirt.
Standard deviation	A statistical measure of dispersion of individual results.
Staple	A well-defined bundle of fibres which has been removed from a mass of greasy wool as a unit.
Variance	The variance is the square of the standard deviation and is a measure of the distribution of values around the mean.
Vegetable matter	Burrs (including hard heads), twigs, seeds, leaves and grasses present in wool.
Vegetable matter base (VMB%)	The oven-dry mass of ash-free, ethanol-extractives-free burrs (including hard heads), twigs, seeds, leaves and grasses present, expressed as a percentage of the mass of the sample.
Wool base (WB%)	The oven-dry mass of wool fibre free from all impurities, i.e. ash free from, ethanol extractives and all vegetable matter and other alkali-insoluble impurities, expressed as a percentage of the mass of the sample.
Yield	The amount of clean fibre, at a standard regain, that is expected to be produced when a delivery of raw wool is processed. The yield may be expressed both as a clean mass in kilograms and as a percentage of the mass of raw wool prior to processing.

Table 2.2 Significance of single fibre and staple characteristics in processing

Characteristic	Importance
Yield	****
Vegetable matter	****
Mean fibre diameter	****
Staple length	****
Staple strength/position of break	****
Colour (yellowness)	****
Coloured fibres	****
Fibre diameter variability	**
Staple length variability	**
Cots	**
Crimp/resistance to compression	**
Staple tip	*

**** Major ** Secondary * Minor

2.4.2 Quality control

Objective measurements provide the means to monitor and adjust wool processing machinery to minimise variation within the product. They also provide tools for monitoring breeding programmes conducted by wool producers, to ensure that the objectives of these programmes are being met. Measurements are now integrated into the quality control systems at all the levels of the industry, extending from the farm to the completion of early stage processing and beyond.

2.4.3 Research

Modern science and engineering depend upon the ability of scientists and engineers to measure critical characteristics of the system they are examining. Wool measurements have been and continue to be used in research programmes directed towards improving wool production, wool processing and fabric characteristics.

2.5 Statistical methods in wool metrology

Statistical methods are a key component of wool metrology, and play a significant role in the trading of wool. The industry has developed comprehensive specifications defining the statistical methods to be employed in development of standards,[2] maintenance of quality control and resolution of disputes.[3–9]

Metrologists define systems that measure a particular characteristic by direct reference to primary reference standards such as length or weight, as primary systems. Systems that measure the same characteristic, but require calibration by reference to a primary system are secondary systems. These distinctions are important because different instruments or methods based on primary systems should be expected to give the same answers, whereas those based on secondary systems may not, particularly if they define the quality characteristic in a different way. Table 2.3 lists the elements that need to be quantified statistically.

Objective determination of defined characteristics of materials usually involves measurements based on a small proportion of the total material of interest. In materials that are homogeneous, obtaining a representative sub-sample of the whole is a relatively simple problem. Where there is heterogeneity, obtaining a sub-sample that is representative of the whole is a much more difficult task. Wool is a heterogeneous material. The sampling procedures for sale lots or consignments of wool must be carefully developed using statistical techniques to ensure that the sample represents the bulk with a predictable degree of error.

Table 2.3 Statistical criteria for evaluating measurement systems

Criterion	Numerical measure
Precision	Absolute standard deviation Relative standard deviation Coefficient of variation Variance
Bias	Absolute systematic error Relative systematic error
Sensitivity	Calibration sensitivity Analytical sensitivity
Detection limit	Blank plus 3 times the standard deviation of the blank
Range	Limit of quantitation (LOQ) to limit of linearity (LOL)
Selectivity	Coefficient of selectivity

Source: P.J. Sommerville, *Technical & Commercial Requirements of Wool Testing Systems*, AWTA Ltd Newsletter, September 2001.

2.5.1 Accuracy and precision

Precision describes the reproducibility of results – that is, the agreement between numerical values of two or more replicate measurements, or measurements that have been made in exactly the same way. Generally, the precision of a testing system can be obtained simply by repeating the measurement, using the same technique, a number of times. The determination of precision generally requires all the sources of variation to be identified and quantified.

Precision is often confused with accuracy. Accuracy simply describes the correctness of a result and must always be determined by reference to a primary system.

Ideally, the accuracy and the precision of any measurement system will be identical, but frequently they are not. It is quite possible to have a very precise secondary measurement system (the answers are highly reproducible), which differs consistently from the 'true' value. This does not limit its usefulness, provided it is used in all instances where comparisons must be made. Terms that are widely used to describe the statistical significance of a set of replicate data are defined, together with some related terms, in Table 2.4.

2.5.2 Bias

Metrologists are concerned with two types of error:

- random or indeterminate errors; and
- systematic or determinate errors.

Table 2.4 Defining the precision of analytical methods

Terms	Definition
Absolute standard deviation	$s = \sqrt{\dfrac{\sum_{i=1}^{N}(x_i - \bar{x})^2}{N-1}}$
Relative standard deviation	$\text{RSD} = \dfrac{s}{\bar{x}}$
Standard deviation of the mean	$s_m = \dfrac{s}{\sqrt{N}}$
Coefficient of variation	$\text{CV} = \dfrac{s}{\bar{x}} \times 100$
Variance of the mean	s_m^2
Confidence level (95%)	$\text{CL} = \pm 1.96\sqrt{s_m^2}$

x_i = numerical value of the ith measurement.

$\bar{x}$ = the mean of N measurements = $\dfrac{\sum_{i=1}^{N} x_i}{N}$

Source: P.J. Sommerville, *Technical & Commercial Requirements of Wool Testing Systems*, AWTA Ltd Newsletter, September 2001.

The error in the mean of a number of replicate measurements is equal to the sum of these two errors. Random or indeterminate errors impact upon precision. Bias may have little or no effect on precision, but it has a significant effect upon accuracy.

Bias is a result of systematic or determinate errors. Systematic errors always act in one direction, resulting in a consistently larger or a consistently smaller result than that provided by the reference measurement. In general, bias can be determined only by reference to measurements provided by primary measurement systems (i.e. systems based on direct reference to primary metric standards such as length and weight). Bias can exist between measurements provided by secondary measurement systems, but unless the bias can be confirmed by reference to a primary measurement system, the metrologist may never be sure whether one or both of the secondary measurement systems are responsible for the bias. Bias can result from several causes. Generally, these are classified into one of six groups:

1. *Sampling errors*: invariably these result from inadequate design of sampling systems.
2. *Differences in fundamental assumptions*: in the case of wool fibre fineness, different assumptions about the geometry of the fibre by different instrumental methods may lead to bias.
3. *Personal errors*: bias can also be the result of blind prejudice. Most of

us, however honest, have a natural tendency to estimate readings in a direction that improves the precision of a set of results, or causes the results to fall closer to a preconceived notion of the true value.

4. *Instrumental errors*: bias can be caused by instrument drift, or by assumptions made by the technology used in the instrument.
5. *Method errors*: generally this type of bias is the result of a failure to maintain rigid control over the environmental conditions that impact upon the measurement.
6. *Interferences*: bias can also be caused by interferences arising from other constituents in the sample.

Bias may be constant over the range of variation of the characteristic being measured, or it may vary over this range. One of the objectives of standardising wool testing systems is the elimination or at least the minimisation of bias. Where bias cannot be eliminated, provided it is not level dependent, the measurement technology may still be useful.

2.5.3 Sensitivity

The sensitivity of an instrument or a testing system refers to its ability to discriminate between small differences in the material being analysed. Commonly three factors limit sensitivity:

- the slope of an instrument's calibration curve;
- the precision of the instrument; and
- the error in the sampling system.

2.5.4 Detection limit

The detection limit is a minimum value of the characteristic being measured that can be detected at a known confidence level. This is not an important issue, for example, when measuring mean fibre diameter, because wool fibres never approach zero fineness, and most measurements are conducted within ranges that exceed the probable detection limit by factors greater than three.

2.5.5 Range

The useful range of an analytical method can be defined as the lowest point at which a measurement can be made (the detection limit or the LOQ), to the point at which the calibration departs from linearity (LOL). However, some measurement systems have non-linear calibration functions. The useful range in these instances is more difficult to define.

2.5.6 Selectivity

Selectivity refers to the degree to which the analytical method is free from interferences by other species in the sample matrix. This is generally not a major issue when testing wool. However, as indicated previously, it may be an issue for measurement of fibre diameter if extraneous synthetic fibres or very fine vegetable matter is present in the sample.

2.6 Development of wool measurement systems

The use of objective measurements in wool processing goes back more than two centuries. The first recorded attempt to objectively measure wool's most important characteristic, fibre fineness, was in 1777, when Daubenton measured the width of some wool fibres under a microscope by comparison with lines drawn on a piece of quartz.[10] However, it was the development of core testing in the United States after 1937 that made the representative sampling of bales of greasy or scoured wool possible, and led to the development of test methods, which were able to provide more accurate and precise estimates of the important value-determining characteristics of greasy wool.

A major breakthrough in wool metrology occurred in the period 1945 to 1955 with the development of the Airflow instrument, which provided a relatively inexpensive but indirect method for measuring the fineness of wool sliver. In the United Kingdom, WIRA and the British Wool Federation developed yield and fineness test methods, and both the United States and Europe established different processing and moisture tolerances to estimate commercial yields of fibre obtained from greasy wool. CSIRO in Australia developed manual pressure core equipment and with the AWTA Ltd in Australia and NZWTA Ltd in New Zealand designed mechanical core sampling equipment suitable for large volume sampling.

By the late 1960s methods for preparing and measuring the fineness of greasy wool by Airflow had also been developed, and this instrument, despite its known limitations, became the primary technology for measuring the fineness of wool. Contiguously, testing methods for determining yield, wool base and vegetable matter base of greasy wool were finalised.

Once wool metrologists had developed the sampling and testing technologies and standard test methods became available, the impetus for the growth in objective measurement of mill consignments and then pre-sale farm lots was driven by the users of wool, to assist them in ensuring the quality of conformance of the yarns, fabrics and garments they produced. Wool producers were initially reluctant to adopt objective measurement and sale by sample of auction lots, but adoption accelerated throughout the 1970s as price differentials between untested and tested lots became evident.

At the same time scientists at CSIRO and AWTA Ltd developed grab sampling systems and measurement instruments that enabled the objective measurement of staple length and staple strength. Large-scale industry trials, designated TEAM (trials to evaluate additional measurements), were conducted throughout the 1980s, developing a database and associated prediction formulae which enabled these staple measurements to be related to processing outcomes. By the beginning of the 1990s most of the key quality parameters of raw wool (see Table 2.2) could and were being measured and certified.

2.7 Standardisation of measurement systems

The most important issue with any measurement, or measurement technology, is to achieve a predictable consistency of the results. Consistency can only be obtained after standardisation of sampling and testing processes, and to do this, data must be collected on:

- sampling and sample preparation variation;
- calibration techniques for test instruments;
- between-instrument comparisons;
- repeatability of measurements on the same wools; and
- between-laboratory effects.

With this data, a test method can be established, the precision of the test calculated and, if required by industry, a standard test method issued.

The International Wool Textile Organisation or IWTO (also known as the Federation Lanière Internationale) was established in 1928 as an arbitration body for the international trade of wool and wool products, and was born out of an arbitration agreement signed between the representative bodies of the British and French wool-textile industries in 1927.

IWTO soon realised that in order for contracts to be fully specified, objective techniques for measuring wool characteristics and invoice weights were required, thereby minimising the likelihood of a dispute but also providing an objective basis for arbitration should a dispute arise. Consequently IWTO now plays a central role in fostering the development of internationally accepted test specifications for raw wool, wool sliver, yarns and fabrics.

Membership of IWTO is based on national committees from member countries, each of which appoints delegates to attend each meeting. The membership of each national committee is decided by the industry in the particular country it represents but it is normally based on representative organisations from different sectors of the industry within that country. However, in some cases private companies, as distinct from sectional industry organisations, are also members of national committees. IWTO also offers associate membership to organisations or companies.

IWTO meets annually. These meetings, attended by some 300 delegates, provide the forum for the various committees established by IWTO to conduct their business, and for discussion and exchange between members on issues of interest to the industry, thereby making it possible to adopt convergent positions to overcome specific problems. The meetings are structured around technical and commercial streams, allowing contributions from metrologists to be reviewed by the commercial players and vice versa.

Over the past 50 years IWTO has overseen the establishment of standard test methods (or Test Specifications) that are universally used around the world. Test houses utilising these test methods, provided they are appropriately accredited by national accreditation bodies, can issue IWTO Test Certificates. These documents then form an integral part of the contractual arrangements between buyers and sellers.

IWTO has developed procedures for the development, review, progression or regulation of IWTO Test Methods and Draft Test Methods.[2] Prior to 1971, all approved Specifications were known as IWTO Test Methods (TM). In November 1971, regulations were laid down to establish a new category of Test Method called a Test Method under Examination (TME), to facilitate control over the commercial application of developing Specifications.

In December 1991 detailed consideration was given to the procedures for the review and withdrawal of standards, together with the need for guidance on the format for presenting supporting technical information. A document was prepared and it was agreed to replace the TME title with that of a Draft Test Method (DTM). The main difference between an IWTO TM and a DTM is that the latter has not yet demonstrated sufficient reproducibility to meet the technical standards for acceptable inter-laboratory variation. Although DTMs define the standard methodology being developed, they have no official status for commercial usage, unless agreed between the contracting parties.

DTMs represent the first formal approval stage in the development of IWTO TMs. The responsible technical group normally continues the work to upgrade them to full IWTO TM status as quickly as possible. DTMs provide an opportunity for both technical and commercial evaluation of the developing methodology, during its logical progression to full standardisation. They are normally held at this status for a minimum of two years.

Normally DTMs precede the full TM status. However, in exceptional circumstances, such as when important weaknesses are identified, full TMs can be downgraded to DTM status.

IWTO-0-01,[2] while defining the process to be followed in developing IWTO TMs, includes four key appendices:

- Appendix A documents which DTMs and full TMs are the responsibility of each of the technical groups operating under the umbrella of the Technology and Standards Committee.
- Appendix B describes how to present supporting data for a new method, and includes a comprehensive description of the statistical analysis required, particularly when another method already exists. The analysis considers the equivalence of the new method to the existing method.

- Appendix C provides guidelines on drafting and presenting a TM.
- Appendix D describes a number of statistical techniques that can be applied to estimate precision limits for IWTO TMs. However, these procedures assume that the variances for the particular parameter determined by the method are independent of the level of the parameter. In situations where this is demonstrated not to be the case, Appendix B describes the appropriate analysis to be applied.

The important parameters that can now be measured for greasy wool are:

- wool base (and hence yield);
- vegetable matter base and hardheads and twigs;
- mean fibre diameter, standard deviation and coefficient of variation of diameter;
- staple length, strength and position of break;
- colour (brightness and yellowness); and
- bulk.

The key methods that are used to specify greasy wool are listed in Table 2.5.

The information provided by these test specifications is applied in two ways:

- to predict the performance of the wool during processing so that it meets the specifications provided by the processor; and
- to establish market value of the wool and thereby determine the price to be paid to the producer.

Table 2.5 IWTO test specifications for greasy wool

IWTO-0	Introduction to IWTO Specifications: Procedures for the Development, Review, Progression or Relegation of IWTO Test Methods and Draft Tests Methods
IWTO-12	Measurement of the Mean and Distribution of Fibre Diameter Using the SIROLANLASERSCAN Fibre Diameter Analyser
IWTO-19	Determination of Wool Base and Vegetable Matter Base of Core Samples of Raw Wool
IWTO-28	Determination by the Airflow Method of the Mean Fibre Diameter of Core Samples of Raw Wool
IWTO-30	Determination of Staple Length and Staple Strength
IWTO-31	Calculation of IWTO Combined Certificates for Deliveries of Raw Wool
IWTO-38	Method for Grab Sampling Greasy Wool from Bales
IWTO-47	Measurement of the Mean and Distribution of Fibre Diameter of Wool Using and Optical Fibre Diameter Analyser (OFDA)
IWTO-56	Method for the Measurement of Colour of Raw Wool

The key parameters that can be certified are shown in the context of their predictive applications in wool processing in Table 2.6.

A small proportion of the Australian clip is also certified for colour. This is in contrast with the New Zealand clip where almost 100% is measured for colour. Similarly 70% of Australian wool is measured for staple length and strength whereas only a minor proportion is measured in New Zealand. Some of these parameters assume different importance depending upon the breed of sheep from which the wool is harvested and the commercial requirements in the country of origin.

Other, non-certified information, such as mean fibre curvature (MFC), vegetable matter base (VMB) broken down into burrs, seed and shive (as well as hard heads and twigs), along fibre fineness profile, and dark and medullated fibre is also available.

Essentially the same parameters can be measured for scoured wool as can be measured for greasy wool. The major exception is staple length and

Table 2.6 Information provided by greasy wool measurements

Certified parameters	Predictive application	Test method(s)
Mean fibre diameter	Top diameter	IWTO-28 IWTO-12 IWTO-47
Wool base	Schlumberger Dry Top & Noil Yield	IWTO-19
Vegetable matter base	IWTO Clean Scoured Yield Japanese Clean Scoured Yield Australian Carbonising Yield	
Staple length	TEAM formulae for:	IWTO-30
Staple strength	• hauteur	
Position of break	• coefficient of variation of hauteur • romaine	
Colour	Whiteness of top Brightness of top	IWTO-56

Note 1: The TEAM formulae also include terms for mean fibre diameter and vegetable matter base. A new TEAM formula arising out of the TEAM-3 Trial adds terms including coefficient of variation of diameter and coefficient of variation of staple length.

Note 2: Measurement of colour in greasy wool has not achieved a significant level of adoption in Australia, in part due to the fact that Australian Merino wool is renowned for its whiteness and brightness. However, some lines of wool, such as crutchings, skirtings and bellies, are often significantly more coloured than fleece wools. There have also been seasons, when conditions have been wet and humid where significant discoloration of fleece wools has occurred. In New Zealand, where cross-bred wool predominates, sale lots are routinely measured for colour. Research has shown that there is a strong relationship between the clean colour of greasy wool and the clean colour of the resultant top.

strength as once greasy wool is scoured the conformation of the merino wool staple is destroyed. Instead fibre length of scoured wool can be measured as 'length after carding'.

Commonly, measurements are also made of the moisture content (expressed as regain), grease content and ash content. Occasionally the pH of the scoured wool is also measured.

The key parameters that can be measured on wool sliver are:

- hauteur;
- barbe;
- colour;
- contamination by dark fibres and/or neps;
- moisture content expressed as regain; and
- grease content.

All IWTO TMs and DTMs are published in the '*IWTO Red Book*'.[11]

2.8 Application of commercial regulations

IWTO has developed a series of other documents that are used in conjunction with these test methods to resolve disputes between trading partners:

- *IWTO Blue Book*;[9] and
- *IWTO Regulations*.

The International Wool Textile Arbitration Agreement, set out in the *Blue Book*, provides for the resolution of disputes arising between partners from different countries. Such a procedure ensures a fair and efficient settlement by the involvement of committed and experienced arbitrators from the industry. The *Blue Book* is regularly reviewed to reflect the current commercial practice and requirements.

Associated with each IWTO TM are Regulations which govern their application, interpretation and other procedures associated with their use. These regulations include formal procedures and technically determined criteria for assessing the validity of test results subject to dispute or suspected to be in error.[3–8]

2.9 Advances in the measurement of wool

2.9.1 Sampling

Grab sampling

Grab sampling equipment was first developed by CSIRO as part of the Australian Objective Measurement Program (AOMP), conducted during the early 1970s. This was an Australian Government-funded programme, which enabled the industry to move towards a new selling system based on the display of samples taken from bales rather than the display of the bales themselves.

Grab samples are taken to create a representative sample of a sale lot, and are used for the following purposes:

- further sub-sampling for the determination of average staple length and staple strength;
- the subjective appraisal of unmeasured characteristics; and
- as a display sample for perusal by prospective buyers.

Grab samples are obtained using a mechanical grab apparatus. They are taken from each bale in a sale lot in such a way that every portion of wool in the sale lot has an equal chance of being selected and each bale within a sale lot is equally represented in the sample.

Wool is generally packaged in bales made from synthetic material. Openings or slits are made from above through the side of the bale to allow access. The openings are made using a pack slitter, which may be either:

- a sharp knife or blade used to cut the pack material; or
- a heated metal bar used to melt the synthetic pack material.

A set of hydraulically operated jaws that operate with a pincer action are mounted on a powerful hydraulic arm that moves vertically. The grab machine operates by driving this arm through the slit made in the pack material by the pack slitter and into the densely packed bale of wool. The jaws close, grasping a sample of wool, then the arm withdraws from the bale and the jaws open to release the sample. Grab machines may be automated or manually operated. They may have from one to nine grab arms.

Core sampling

A core sample is a representative sample taken from a sale lot comprising one or more bales. Initially core samples were taken using a manual hand core device consisting of a long tube, with a slit along the length of the tube and a replaceable cutting tip. Bales must be weighed at the same time that core samples are taken, and these samples are sealed in airtight bags to prevent any change in the moisture content of the sample during the interval between sampling and testing.

The statistically based systems for obtaining a representative sample using this device were developed in the United States and refined during the 1960s within Australia and New Zealand. Mechanical machines were developed by CSIRO, AWTA Ltd and New Zealand during the late 1960s. They consist of a coring chamber that encloses a bale during core sampling. The bale is presented in the chamber base uppermost. During core sampling a platen lifts the bale, compressing it against the top of the chamber. Hollow core tubes with removable sharpened tips or cutters are driven downwards through the base of the bale by hydraulic rams, penetrating almost the entire length of the bale. Flexible tubes connected to the core tubes evacuate the core sample material, depositing it in a plastic bag.

The first machines used a single core tube, the position of which could be varied. Later, core machines with multiple core tubes were built, and the sampling system was controlled by computer. Modern sampling lines consist of a grab machine contiguous with a core sampling machine (see Fig. 2.1).

2.9.2 Moisture

Measurement of moisture content of scoured wool is conducted to provide an estimate of the quantity of wool present in a consignment. This was one of the first measurements used for commercial trading. The test method simply involves removing the moisture from a weighed sample obtained from the consignment by core sampling using a stream of air heated to 105 °C.

Electrical capacitance, near infrared reflectance spectroscopy (NIRS) and microwave radiation have all been explored as alternative secondary

2.1 Grab and core sampling in an Australian wool store.

measurement systems for determining moisture. While such systems have found applications for on-line measurements in wool scours for quality control, they are not as yet sufficiently precise or accurate for use in IWTO Certification and consequently laboratory-scale instruments have not been produced.

2.9.3 Yield

Early measurements of yield were based on 'washing yields'. A sample of greasy wool was scoured in a laboratory scour to remove impurities such as wax, suint, dirt and grease (Fig. 2.2). The scoured sample was then dried and

2.2 Early laboratory scour (circa 1958).

weighed, the dry weight being expressed as a simple percentage of the greasy weight.

This simple procedure, particularly for Australian wool, provided a biased overestimate of the actual yield, as it included any vegetable matter and remaining dirt and grease not removed in the scouring process.

By the end of the 1960s IWTO had established IWTO-19, which is a much more comprehensive Test Method, and which enables compensation of the yield estimates for these biases. IWTO-19 involves the following steps:

- *Blending*: the core sample is blended mechanically to ensure that the individual cores from the bales are mixed as uniformly as possible. In the early days of yield testing this was done by hand. Today, rapid mechanical systems are used.
- *Sub-sampling*: the blended core sample is sub-sampled to remove two or more 150 gram samples that are each tested separately. This requirement was introduced in IWTO-19 to improve precision (the average of n results improves precision by a factor equal to the square root of n) and also to provide some internal quality control by building comparison checking into the test method.
- *Scouring*: laboratory scouring machines have evolved significantly since the very early days of yield testing. The first scours were multi-bowl devices that simulated commercial scouring operations. The New Zealand Wool Testing Authority developed single bowl machines in the late 1960s and these were rapidly adopted by AWTA and the Wool Testing Bureau South Africa. These machines underwent considerable automation enhancement during the 1980s in line with improvements of electronic control systems that were evolving during this decade. As a consequence the time taken to scour a single sub-sample has decreased from between 15–20 minutes to 4.5 minutes. A modern laboratory scouring line is shown in Fig. 2.3.
- *Drying*: forced air drying systems were developed by CSIRO and others during the 1960s. In simple terms these consisted of a heating and temperature control system that forced hot air (105 °C) through a laboratory scoured wool sample packed into container with a porous top and base. This process is preceded by centrifuging the sample to remove as much of the residual moisture as possible. In the 1970s AWTA designed multi-head driers capable of simultaneously drying 12 sub-samples. A further enhancement introduced to IWTO-19 in the early 1980s was the automation of the detection of the drying end point, which had previously been detected by successive weighings.
- *Determination of residuals*: IWTO requires the amount of residual amount of dirt, grease and vegetable matter remaining after scouring to

2.3 Modern laboratory scouring line (AWTA Ltd Melbourne).

be measured and subtracted from the oven-dried scoured weight so that the wool base can be accurately calculated.

Residual grease is measured on a 10 gram specimen using solvent extraction with alcohol. In 1990 IWTO-19 was amended to allow a secondary method, NIRS to replace the solvent extraction technique. NIRS must be calibrated by the solvent extraction method, but once this is done the measurement is very rapid (requiring only a few seconds compared with more than 1 hour), non-destructive and requiring far less labour input. Although the technology was approved for use in 1990 it was a further 10 years before the costs of the required instrumentation reduced to a level that made replacement of the solvent extraction technique economic. In all AWTA Ltd laboratories NIRS is now used routinely for this measurement.

The residual dirt or mineral matter is measured by burning a 10 gram specimen at 800 °C, then weighing the residual ash. In 2001 it was demonstrated by SGS New Zealand that residual mineral matter could also be measured by NIRS[12] and IWTO-19 was amended accordingly in 2003. Managing interferences caused by the presence of dags in some farm lots in Australia delayed immediate implementation of NIRS for this purpose. However, AWTA Ltd has successfully resolved this problem and NIRS is currently being deployed to replace the ashing method.

A method for determining vegetable matter was developed in 1943, where the wool was dissolved by boiling a specimen in a solution of hydrogen peroxide, copper sulphate and sodium bicarbonate.[13] This process degraded the wool such that it could be removed from the liquid and placed in a 1% solution of sodium carbonate maintained at 95–100 °C. This completely dissolved the wool, leaving the intact vegetable matter behind. It could then be recovered using a 40 mesh sieve, washed, dried and weighed.

Subsequently, as IWTO-19 was developed, it was found that immersing a 40 gram scoured wool specimen in hot 10% solution of sodium hydroxide also dissolved the wool specimen. This technique was largely manual and highly labour intensive. By 1984 AWTA Ltd had developed equipment that automated this process.

In 2007/08 AWTA Ltd deployed machines developed by the company that automatically segment the scoured sub-sample into the separate specimens required for measurement of residuals, dramatically reducing the labour input required.

- *Calculation of wool base and vegetable matter base*: calculation of wool base and vegetable matter base is straightforward once the required data is available. The calculations are described in detail in IWTO-19.

The determination of yield substantially emulates the processes that occur in early stage processing. Attempts have been made to streamline the process using chemical and spectroscopic technologies by dissolving the wool and measuring the products in the solution using NIRS.[14] While the answers obtained were highly correlated with the results produced by IWTO-19, the precision of the method was inferior.

2.9.4 Fibre diameter

There is an extensive literature on developments in fibre diameter measurement available on AWTA Ltd's website.[15] After 1960 the projection microscope, although still required for calibration of more modern secondary measurement systems, was largely supplanted by the Airflow instrument. Interestingly this instrument was first developed by the cotton industry. Its first application to wool began in the 1940s with an IWTO Test Method for wool sliver (IWTO-6) finally becoming available in 1960.

The Airflow instrument (Fig. 2.4) measures porosity of a compressed plug of fibres of a fixed mass, conditioned to constant moisture content. The

2.4 The Airflow instrument.

instrument can be set up for constant flow, where variations in pressure correlate with diameter, or for constant pressure, where variation in flow correlate with diameter. IWTO elected to develop its methods using both approaches but ultimately the constant pressure instrument became the instrument of preference. The first application was the measurement of top.

Extending Airflow to measure raw wool required machinery to convert a matted sample of short fibres from a scoured core sample into a sliver. Again technology developed for use in the cotton industry was adapted. Trash separators or 'Shirley Analysers' are small-scale carding machines for separating cotton fibres from the vegetable matter trash that is also collected by the pickers. These machines were adapted to remove vegetable matter in samples of laboratory scoured wool, producing a loose 'sliver' of clean fibre.

A standard test method for the measurement of the diameter of raw wool (IWTO-28) was approved by IWTO in 1971. Thereafter there was little attempt at further improvements other than the Sonic Airflow (Fig. 2.5), developed by CSIRO in the early 1970s which used sound-generated oscillating pressure waves instead of a constant stream of air. During the 1970s there was also an effort by AWTA Ltd to automate some of the measurement functions by using electronics but this was abandoned when it became apparent that the technology of the time was not up to the task.

The first significant next step was the development of the Fibre Fineness Distribution Analyser (FFDA) by CSIRO (Fig. 2.6), commencing in 1976. The first commercial instruments became available in 1979. The basis of the instrument was the reduction in the intensity of light transmitted by a laser and subsequently detected by a photo-detector when the beam was transected by a fibre carried in a transport fluid through a cell through which the laser beam was directed. The magnitude of this drop is directly related to the diameter of the individual fibres.

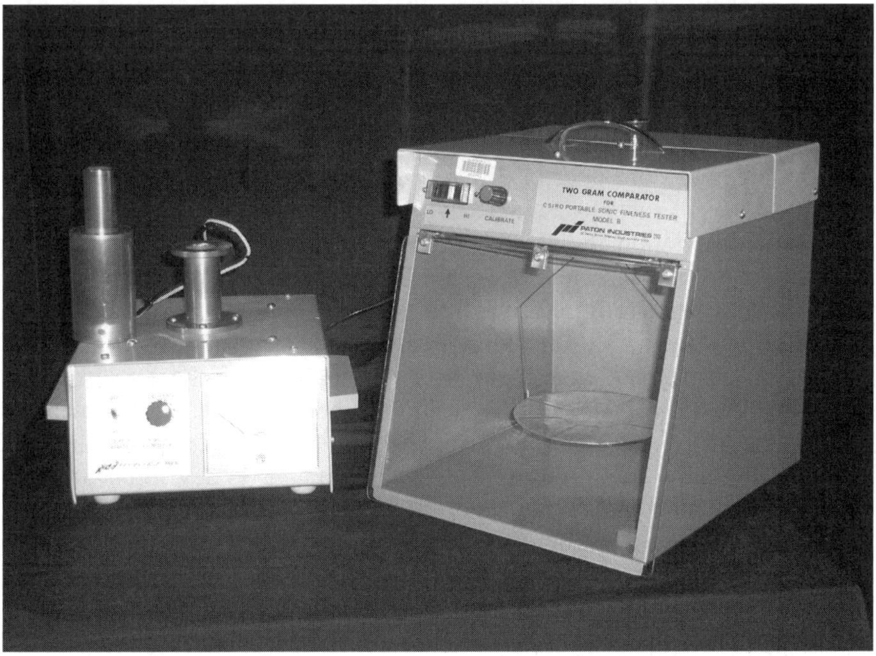

2.5 The Sonic Airflow.

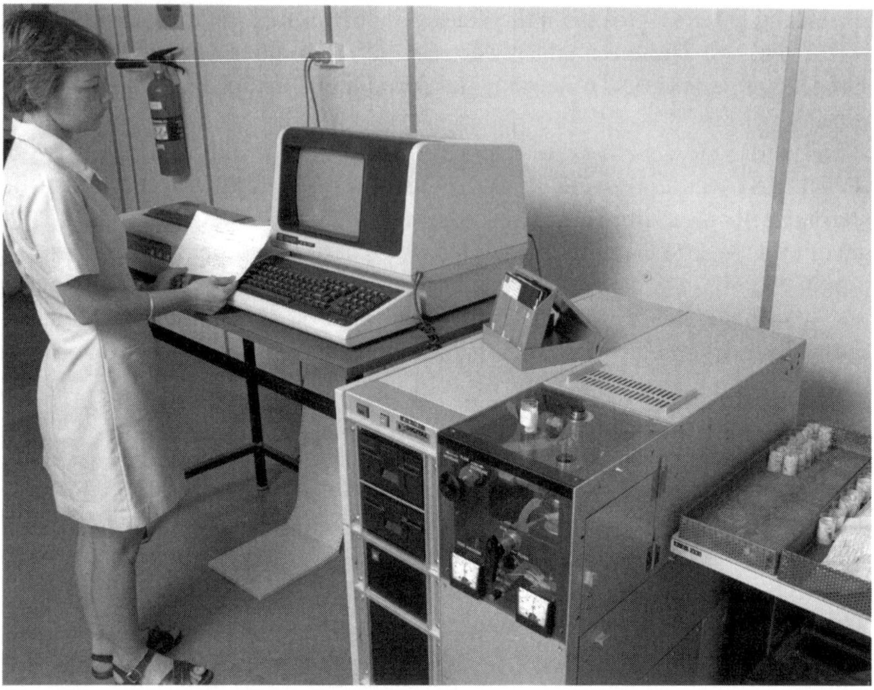

2.6 Fibre fineness distribution analyser.

The FFDA (or FDA) offered information about the distribution of diameter as well as the mean. Potentially it was not affected by biases that were known to affect the Airflow arising from differences in fibre diameter distributions and fibre density differences between samples. However, the instrument was plagued by a diameter-dependent bias in mean fibre diameter and standard deviation of diameter. These biases arose from the failure of the systems incorporated in the instrument to consistently discriminate between events when only one fibre transected the beam and when multiple fibres transected the beam. Modifications to the shape of the measurement cell used by the instrument were made in 1985 but although these reduced the diameter bias, they did not eliminate it and the bias in standard deviation remained.

In parallel with this AWTA Ltd in 1982 commenced development of an entirely new technology. This involved using a video camera and frame grabber to capture a 40× magnified digital image of fibre snippets distributed on a microscope slide. Computer algorithms analysed the image to locate and measure the widths of the fibre snippets. Development of this instrument (FIDAM – Fibre Imaged Display And Measurement) continued until 1990, when AWTA Ltd announced that it would not continue further development.

By then CSIRO had solved the issues that had plagued the FFDA instrument, through a substantial improvement in the fibre discrimination and selection system, and were about to release a new enhanced version trademarked 'Sirolan™ Laserscan' (Figs 2.7 and 2.8). For technical and commercial reasons AWTA Ltd decided to pursue the implementation of this technology instead of the FIDAM technology.

Meanwhile BSC Electronics Pty Ltd had been pursuing a parallel development of an image analysis instrument, the OFDA100 (Fig. 2.9), and this became commercially available in 1991. Both Laserscan and OFDA100 provided diameter distribution data whilst the OFDA100 also provided an estimate of mean fibre curvature. CSIRO subsequently introduced new software for the Laserscan which also provided an estimate of fibre curvature.

An IWTO Test Method for both instruments was approved in 1995. However, it was 2000 before the industry finally agreed to accept the use of these instruments for trading purposes, at which point AWTA Ltd replaced the Airflow by the Laserscan as its standard service.

Prior to this decision AWTA Ltd had also developed an electronically controlled Airflow instrument, but then decided not to proceed to commercialisation of this instrument.

2.7 Sirolan™ Laserscan.

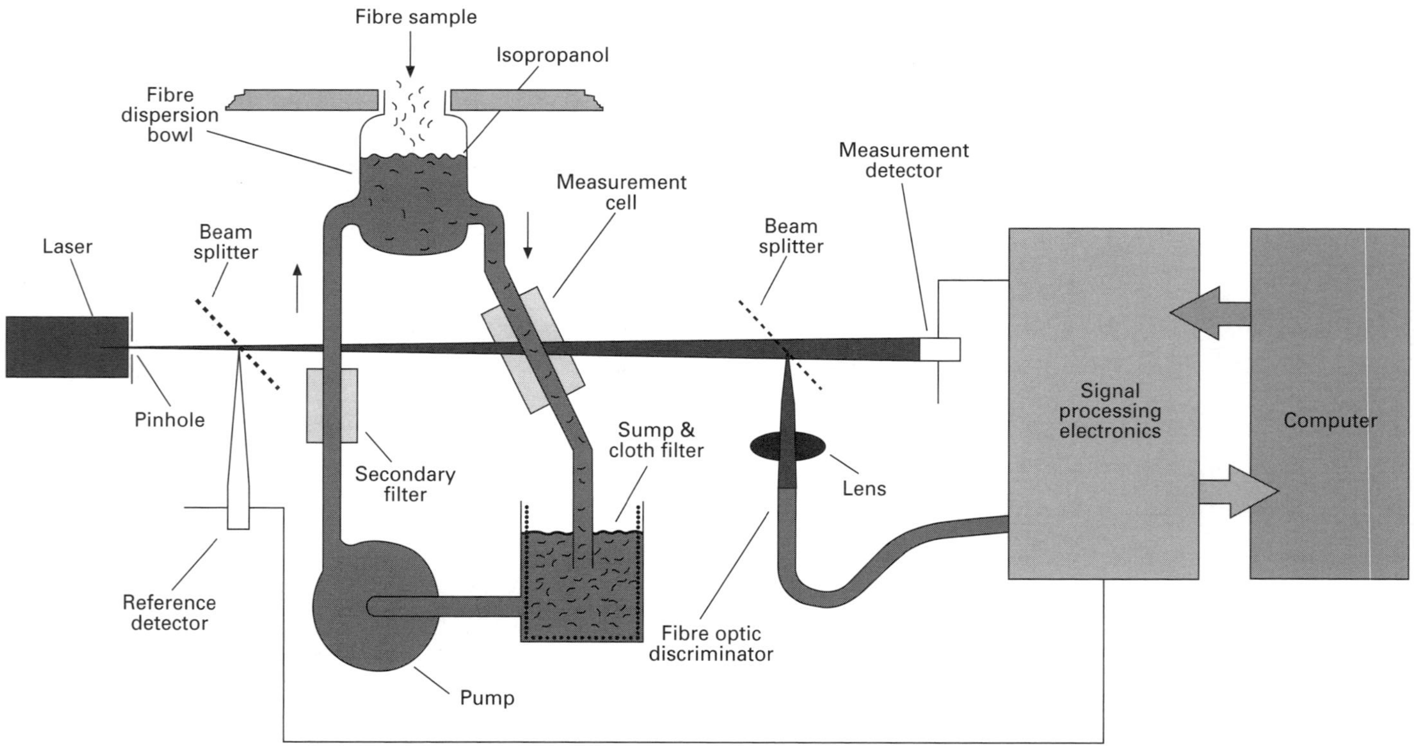

2.8 Sirolan™ Laserscan schematic.

2.9 OFDA100.

2.9.5 Staple measurements

Wool buyers have traditionally used subjective assessment of staple length, strength and position of break to estimate top length or hauteur of a mill consignment. The buyers assessed staple strength by gripping the ends of a staple and exerting a force on the staple by flicking the staple with the middle finger. There are several variables associated with this evaluation:

- the force exerted by the appraisers;
- the thickness of staples selected;
- and the number of staples evaluated.

The force exerted by this technique has been shown to range from 17 to 48 newtons. The inconsistency of the force exerted makes it harder to compare appraisals. Staples differ in thickness and therefore thicker staples require more force to break them, so appraisal requires appraisers to select staples of equal thickness, otherwise the appraised strengths will be different. In practice this is impossible to achieve. Finally, buyers typically sample only a few staples whereas 50 to 60 need to be assessed for a reasonable estimate of the breaking force to be made.

In the late 1970s CSIRO and AWTA Ltd collaborated in the development of systems that could objectively measure staple strength, staple length and the position of break.[16] Initially these systems were largely manual.[16] Staples were extracted from the grab sample using a simple grid sampling technique. The grab sample was spread over a table, and a cover pierced with an evenly spaced array of rectangular openings was lowered over it. The operator manually selected a staple from the area of wool exposed by each opening and placed this in a tray with approximately 15 parallel depressions – one for each staple. These trays were designed so that they could be stacked and still allow air to circulate between them.

The staples were weighed and their length estimated by ruler. They were then placed in a machine consisting of a pair of jaws which gripped the base and the tip. The jaws were moved apart mechanically and the peak force applied to break the staple recorded. The operator would make an estimate of the position of break and record this as tip, middle or base.

This system was used to provide the industry with an evaluation of the usefulness of the measurement via the Sale with Additional Measurements (SAM) Trial conducted by the Australian Wool Corporation in collaboration with CSIRO, University of New South Wales (School of Wool & Pastoral Science) and AWTA Ltd in 1979/80.[17] The Additional Measurements were published in sale catalogues for about 400 lots in each of six successive Adelaide sales during the first half of the 1980/81 season.

The report on these trials,[18] published in September 1981, concluded:

> The SAM Trial has demonstrated the technical feasibility of providing some further measurements as an aid to the buyer of Australian Wools. It has also shown the need to further develop sampling and testing techniques to enable them to be economically viable.

In the period 1980–85 the research effort focused on improving the efficiency of sampling and testing. AWTA Ltd concentrated upon automation of the grid sampling systems. The outcome of this research was the Mechanical Tuft Sampling (MTS) machine (Fig. 2.10). This machine consists of a long belt conveyor system, a set of three fixed sampling jaws and a sample collection system. The display sample is spread over the conveyor to a depth equivalent to one grab sample. The operator notes the length of the distributed grab along the conveyor, and enters this into the machine's programmable logic control system. The machine is initiated and the belt is moved to place the leading section of the grab under the sampling jaws. A pressure plate descends on the grab and the set of jaws is lowered with the jaws open, each descending through holes in the pressure plate to almost touch the conveyor belt. The jaws close, gripping a tuft of wool which is drawn slowly through the hole in the pressure plate to minimise the risk of breaking staples, accelerating upwards once the tuft is clear. Each tuft is then deposited upon a plastic mesh. The mesh moves incrementally and transversely to the conveyor onto a collection reel. The cycle is repeated 20 times until 60 tufts have been

2.10 Mechanical Tuft Sampling (MTS) machine.

collected. The display sample is deposited into a plastic bag at the end of the conveyor. Each mesh reel can hold samples consisting of 60 tufts for approximately 14 lots.

The reels are transported to the laboratory and a trained team of four operators further sub-samples each tuft for each lot, extracting a single staple per tuft. The staples are placed in purpose designed trays and conditioned and relaxed before measurement (Fig. 2.11).

CSIRO, the University of New South Wales (School of Wool and Pastoral Sciences) and the South African Wool Bureau pursued independent developments of prototype instruments that could automate the staple measurements. Prototypes of the CSIRO instrument (codenamed ATLAS) and the UNSW instrument (codenamed PERSEUS) were evaluated by AWTA Ltd, and the decision made to proceed to commercialisation using ATLAS (Fig. 2.12).

ATLAS measures the length of each staple, the peak force required to break the staple in half, and the weight of the two broken portions of the staple, thereby enabling a calculation of the position of break (POB). The principle of length measurement on ATLAS is to convey the staple through a light beam at a fixed speed. The length of the staple is determined by relating the speed of the conveyor to the amount of time the light beam was interrupted by the staple. Following length measurement, the staple is picked

2.11 Staple preparation.

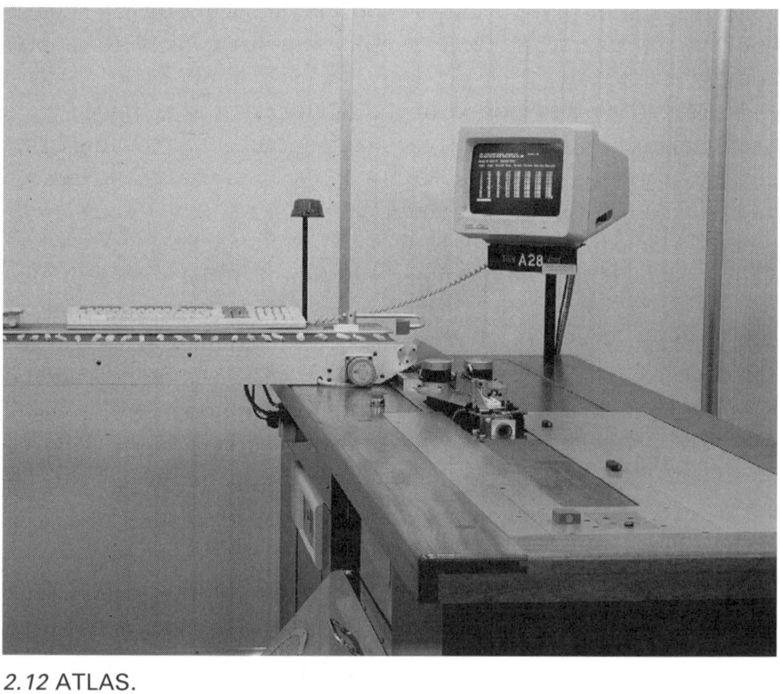

2.12 ATLAS.

up between twin rubber belts and fed to the strength section. The tip of the staple is gripped by one jaw and the base by another jaw and then the staple is broken and the peak force is recorded in newtons.

The length of staple gripped by the jaws is approximately 20 mm. Consequently any staple less than 50 mm would have only 30 mm of the staple free to break as the jaws moved apart. The peak force recorded in such a case would not be an accurate measure of the strength of that staple. Therefore a strength result is not reported for staples shorter than 50 mm. Each of the two broken portions of the staple is released from its jaw and blown down a tube to a balance where it is weighed. The sum of the two weights is used to calculate the staple mass. Broken staples are weighed and then compressed air blows the portions to a collection vessel.

The calculation of staple strength occurs in two steps. The greasy linear density is calculated by dividing the greasy staple mass (mg) by the staple length (mm) – equation 2.1:

$$\text{Greasy linear density (ktex)} = \frac{\text{greasy staple mass (mg)}}{\text{staple length (mm)}} \quad [2.1]$$

The peak force is then divided by the greasy linear density to give the uncorrected staple strength – equation 2.2.

$$\text{Uncorrected SS (N/ktex)} = \frac{\text{peak force (N)}}{\text{greasy linear density (ktex)}} \quad [2.2]$$

Finally, the uncorrected staple strength is corrected for the yield, wool type and strength factor using equation 2.3.

$$\text{Corrected SS (N/ktex)} = \left[\frac{\text{uncorrected SS (N/ktex)}}{\text{strength factor} \times \text{staple yield}}\right] \times 100 \quad [2.3]$$

Where, strength factor = 1.36704 – 0.006936 × WB + 0.06126 × Type

and

$$\text{Staple Yield} = 0.83 \times \text{WB} + 0.314 \times \text{VMB} - 6.18 \times \text{Type} + 29.6$$

Type = 1 for fleece and 0 for skirting wool
WB = wool base
VMB = vegetable matter base
SS = staple strength

The POB% is simply calculated by dividing the mass of the staple tip by the total mass of the staple (tip and base portions) – equation 2.4:

$$\text{POB\%} = \left[\frac{[\text{Tip}_{\text{mass}}]}{(\text{Tip}_{\text{mass}} + \text{Base}_{\text{mass}})}\right] \times 100 \quad [2.4]$$

A commercial service for IWTO Certification of Staple Length, Strength and POB was commenced by AWTA Ltd in 1985/86, supported by a 0.05 Australian dollar premium offered by the AWC via the Reserve Price Scheme.

Use of the measurements by wool buyers and processors was supported via the TEAM trials and later the Australian Staple Measurement Adoption Program (ASMAP).[19,20] These facilitated the buyer's and processor's understanding of the measurements by assisting them to fully test consignments and correlate the processing performance in terms of hauteur, romaine and noil with the raw wool measurements via a set of general formulae now known as the TEAM formulae. These are derived from a statistical analysis of the TEAM and ASMAP databases, and are essentially industry benchmark formulae as they represent the mean performance of all the processing mills involved in the trials.

By 2000 the adoption rate for staple measurement of sale lots in Australia was approximately 70% of all lots certified for yield and micron. The specification of these parameters in mill contracts had resulted in staple strength becoming the second most important characteristic (after micron) determining the value of Australian greasy wool.

AWTA Ltd commenced a new TEAM 3 trial to update the TEAM formulae to take account of improvements in top making that had occurred since the first TEAM formulae were developed. Improved TEAM formulae were published in 2004.[21]

2.9.6 Colour

Measurement of bulk colour is widely used in the textile industry and in other industries. In 1977 a test method, was developed in New Zealand by Wool Research of New Zealand (WRONZ) (New Zealand Standard NZ 8707). Initially this was restricted to commercial consignments of scoured wool but in 1984 this was amended and extended to include raw wool (NZ 8707). In Australia a similar standard was approved (AS 3545) in 1988.

The New Zealand and Australian Standards were merged to form a Draft Test Method, and presented to IWTO in 1986. This was later approved as IWTO-56. These methods all relied on reflectance colorimetry. However, unlike other industries which measure colour using the CIE colour space, the wool industry elected to develop its own colour space and a calibration system based on wool samples assigned values in this colour space using a spectrometer. Calibration samples were provided by WRONZ.

Wool colour changes over time, so the calibration samples were unstable and had to be regularly replaced. It was discovered that the method for assigning the values for this calibration material was also unstable, occasioning a significant shift in results when a new series of calibration samples were distributed. Consequently the wool industry has moved to the CIE colour space and calibration in this space in the area of interest for wool is provided by a set of standardised ceramic tiles. IWTO-56 was amended to affect this change in 2003.[22]

In New Zealand colour is certified for all farm lots. In Australia the adoption rate is very small, despite efforts by AWTA Ltd throughout the 1990s to encourage colour testing. In 1995 almost 1 000 000 bales were offered with a colour test available, but the measurement was not incorporated into mill contracts and hence there was little ongoing demand. To a large extent this is simply due to the difference in the types of wool produced in Australia compared with New Zealand. Cross bred types predominate in New Zealand and for these types colour, particularly yellowness, is highly variable dependent upon conditions. Australia produces predominantly merino wool, which, under the growing conditions that exist, is usually very white and bright, and much less variable.

2.9.7 Coloured and medullated fibres

In the 1980s CSIRO developed a balanced illumination method for counting dark and medullated fibres in samples of wool top. This instrument is very simple. Approximately 0.25–0.50 grams of washed and carded core sample are spread thinly between glass plates and illuminated by dual sources of light – from above and below.

When examining for dark fibres the intensity of the illumination is balanced such that white fibres tend to merge into the background, while the objectionable fibres tend to stand out. The entire illuminated specimen is examined using 2 × magnification. When a dark fibre is detected its colour can be categorised by reference to a scale, also developed by CSIRO.

When examining for medullated fibres a black background is inserted below the glass slides and the sample illuminated from above. The medullated fibres reflect this light differently and therefore can be identified and counted.

The limitation of this technology is that it relied on a painstaking examination of the illuminated sample by an observer. Specimens were small (0.25–0.50 grams) so that 20–40 such specimens needed to be examined to achieve the level of sensitivity required. This was very labour intensive and therefore very expensive.

AWTA Ltd, together with the South Australian Research & Development Institute, demonstrated that the method can also be applied to measure dark and medullated fibre contamination in wool from merino ewes that have been used as mothers for lambs derived from exotic sheep breeds such as Damaras. This contamination is caused by physical contact and as a consequence the contamination of the merino fleece by transferred dark and medullated fibres is distributed relatively uniformly across the fleece and consequently in the core samples.[23] Measurement of naturally occurring dark fibres is still a problem as these are rarely distributed uniformly throughout the fleece and consequently sampling errors are very large.

2.9.8 Fibre length in top

The measurement of fibre length in wool sliver is of fundamental importance to early stage processors as it has a significant bearing on spinning performance. The Almeter instrument, which uses capacitance to estimate fibre length and fibre length distribution, was introduced in the wool industry, as a successor to the comb sorters and in particular to the 'Schlumberger Analyser', an automated comb sorter in use since 1950.

The prototype of the Almeter was presented at the Scheveningen meeting of IWTO in May 1961. After extensive evaluation in mills, a series of inter-laboratory trials was completed in 1966. With the proving of a suitable calibration system, IWTO approved a Standard Test Method for the instrument (IWTO-17) in January 1967.

The instrument served the industry well for the next 40 years, but its continuing commercial development and manufacture were abandoned due to falling sales as the wool industry contracted during the 1990s. The Woolmark Company, through its subsidiary Wool Developments International, financed a re-engineering of the instrument to incorporate modern electronics and control systems. The new instrument, AL2000, became available in 2003.

2.9.9 Style

Considerable funds to develop systems to objectively measure 'Style' were invested by the industry from 1985 to 2000. The research was conducted by CSIRO. CSIRO nominated several characteristics that contributed to the subjective appraisal of style. The staple traits associated with these were:

- staple length;
- tip length;
- staple crimp frequency;
- crimp definition;
- wool yellowness;
- wool area;
- dust area; and
- dust colour.

Three prototype instruments using image analysis of individual staples to assign numerical values to the above traits were finally manufactured for evaluation.

Performance of the instruments was reported to IWTO in 2001.[24] There were a number of problems, particularly with respect to the repeatability between the instruments. However, it was the cost of the technology that resulted in the termination of the project. AWTA Ltd withdrew its support because it did not believe the technology would provide the measurements at a price that the industry would be prepared to accept.

2.10 Latest developments in wool metrology

2.10.1 Fibre length

A new instrument, the OFDA4000, is now available which provides estimates of length and length distribution parameters (including hauteur and barbe), estimates of diameter parameters including MFD, SD, CVD, CF, the diameter of fibre ends, curvature parameters, and an estimate of the diameter profile of fibres in top.[23] The instrument offers potential for measurements of fibre length and length/diameter relationships which have not previously been available. It is unclear how this new information will be applied in either the trading or product/process development areas. There is the potential that the extra information may improve processing prediction from greasy wool to top, and from top to yarn, and may even enable selection of fibres with particular along-fibre diameter profiles in top to spin yarns with specialised characteristics, e.g. different combinations of surface softness and stiffness.

The OFDA4000 has completed comparison trials with Almeter and Laserscan instruments[25,26] and IWTO has progressed a test specification for this instrument to a DTM status.

2.10.2 Dark and medullated fibres

The introduction of exotic sheep breeds into Australia created an impetus to developed improved instruments for measuring dark and medullated fibres in raw wool and wool sliver. With funding provided by Australian Wool Innovation and supported by AWTA Ltd an improved system based on the balanced illumination method has been developed.[27]

Wool fibres are semi-translucent, that is they both reflect and transmit incident light. In the absence of any pigmentation (colour) they absorb little of the incident white light and therefore appear white in colour.

Reflection of light occurs because the refractive index of wool fibres (1.553) is different from air (1.00). Due to phenomenon of refraction, light incident on the fibres is reflected from the exterior and interior air/fibre interface. However, if normal white wool fibres are immersed in a solvent with the same or very similar refractive index they become transparent and very little reflection occurs. If the fibres are pigmented (e.g. dark fibres) then some of the incident light will be absorbed and the pigmented fibres can be seen against a white background. Likewise, medullated fibres, due to their internal medulla continue to reflect incident light so that against a black background they appear white. Benzyl alcohol, a common industrial solvent, has a refractive index of 1.540 and white wool fibres become virtually transparent when immersed in this solvent.

The practical benefit of using this property in a test for dark and medullated fibres is that it facilitates the examination of much larger samples. CSIRO developed an efficient and safe process for immersing 5 grams of scoured and carded core samples in benzyl alcohol. These packages can be examined using a modified version of the existing dark fibre detector. This process, enhanced by AWTA Ltd's Research & Development Division, is the basis of the new test. In recent further developments image analysis software is being used to detect the dark and medullated fibres in scanned digital images of the immersed wool fibres.

2.10.3 New staple measurements

Crimp frequency

Research by CSIRO into measuring style characteristics demonstrated that crimp frequency could be measured by image analysis. When the style project was terminated in 2001, AWTA Ltd proceeded to develop an image analysis system for measuring crimp frequency only.

In 2003 AWTA Ltd advised IWTO that it had succeeded in developing a prototype instrument that could be readily incorporated into the ATLAS instrument to measure the crimp frequency of the same set of staples presented to ATLAS for measurement of staple length, strength and POB.

AWTA Ltd is in the process of re-engineering the ATLAS instrument to replace the electronics and software first engineered in the 1980s. The company is considering introducing crimp frequency, initially as non-certified information, with the roll-out of the re-engineered ATLAS instruments.

Staple extensibility

In 2003 AWTA Ltd also announced that it had designed modifications to the ATLAS instrument that allow the measurement of extension of staples before they break. Its scientists demonstrated that inclusion of this measurement in processing prediction formulae, such as TEAM, improved these predictions for certain wool types such as bellies, which tend to have a higher crimp frequency and are more stretchy than the rest of the fleece. These measurements will also become available as non-certified data with the rollout of the re-engineered ATLAS instruments.

2.11 Future trends

2.11.1 Industry trends

There are several trends in the wool industry which may influence, or even guide, future developments in wool metrology. These include:

- on-farm testing;
- development of niche markets for specialised, or particular, 'types' of wool;
- emerging Internet technologies;
- improvements in prediction equations, such as TEAM; and
- greater use of raw wool measurement in breeding strategies;

2.11.2 New measurement technologies

Spectroscopy

In recent years NIRS has been introduced to both performance monitoring (especially in scouring companies in New Zealand) and Yield Testing (IWTO-19). NIRS relies on calibration via existing measurement technologies and relates the quantity of a given non-wool contaminant (e.g. grease) to absorption patterns over a range of wavelengths. It offers a faster, simpler measurement than traditional methods and there is potential for further application. X-ray scanning, as used in security checks at airports, has also shown potential for the rapid determination of yield.

Image analysis

Over the past 30 years major gains have been made in this technology, yet its application to wool metrology has really just begun. There is considerable potential for further applications, particularly as hardware and turnkey software improves.

Refinement of existing measurement systems

Improvements are always being made to existing certifiable measurements in terms of their precision, accuracy, speed, simplicity and cost. Potential changes in current technologies could include:

- improvement of diameter calibration for measurement of ultrafine wool;
- knowledge of the relationship between fibre diameter and fibre length in tops (using the OFDA4000);
- development of an estimate of the mean fibre length within a staple to improve processing prediction;

- prediction of greasy wool softness based on current diameter and curvature technologies;
- development of inexpensive measurements of fibre entanglement in a staple and the shape of staple tips; and
- derivation of a set of parameters from fibre diameter profiles to predict staple strength.

2.12 Sources of further information and advice

IWTO test methods concerned with raw wool testing

Code number	*Full title*
IWTO-2-96	Method for the Determination of the pH Value of a Water Extract of Wool
IWTO-3-86(E)	Method of Test for the Acid Content of Wool
IWTO-6-98	Method of Test for the Determination of the Mean Diameter of Wool Fibres in Combed Sliver using the Airflow Apparatus
IWTO-7-00	Sub-sampling Staples from Grab Samples
IWTO-8-97	Method of Determining Fibre Diameter Distribution Parameters and Percentage Medullated Fibres in Wool and Other Animal Fibres by Projection Microscope
IWTO-10-03	Method of Determination of the Dichloromethane Soluble Matter in Combed Wool and Commercially Scoured Carbonised Wool
IWTO-12-00	Measurement of the Mean and Distribution of Fibre Diameter Using the Sirolan LASERSCAN Fibre Diameter Analyser
IWTO-17-03	Determination of Fibre Length Distribution Parameters by means of the Almeter
IWTO-19-98	Determination of Wool Base and Vegetable Matter Base of Core Samples of Raw Wool
IWTO-20-69(E)	Method for the Determination of the Felting Properties of Loose Wool and Top
IWTO-28-00	Determination by the Airflow Method of the Mean Fibre Diameter of Core Samples of Raw Wool
IWTO-30-98	Determination of Staple Length and Staple Strength
IWTO-31-02	Calculation of IWTO Combined Certificates for Deliveries of Raw Wool
IWTO-32-82	Determination of Bundle Strength of Wool Fibres
IWTO-33-98	Determination of Oven-dry Mass and Calculated Invoice Mass of Scoured and Carbonised Wool
IWTO-34-85(E)	Determination of Oven-dry Mass and Calculated Invoice Mass, and Calculated Merchantable Mass of Wool Tops
IWTO-35-03	Method for the Measurement for the Colour of Sliver
IWTO-38-91(E)	Method for Grab Sampling Greasy Wool from Bales
IWTO-41-92(E)	Determination of the Invoice Mass of Scoured or Carbonised Wool or Tops or Noils by Capacitance Method

Code number	*Full title*
IWTO-47-00	Measurement of the Mean and Distribution of Fibre Diameter of Wool using an Optical Fibre Diameter Analyser (OFDA)
IWTO-52-96	Conditioning Procedures for Testing Textiles
IWTO-55-99	Method of Automatic Counting and Classifying Cleanliness Faults in Tops using the Optalyser Instrument
IWTO-56-03	Method for the Measurement of Colour of Raw Wool
IWTO-57-98	Determination of Medullated Fibre Content of Wool and Mohair Samples by Opacity Measurements using OFDA
IWTO-58-00	Scanning Electron Microscopic Analysis of Specialty Fibres and Sheep's Wool and their Blends

2.13 References

Most of the reference material cited here is available on www.awta.com.au.

1. Testing the WoolClip, Australian Wool Testing Authority Ltd
2. IWTO-0-01, Procedures for the Development, Review, Regression or Relegation of IWTO Test Methods or Draft Test Methods
3. IWTO Core Test Regulations
4. IWTO Staple Test Regulations
5. IWTO Conditioning Test Regulations for Scoured and Carbonised Wool
6. IWTO Conditioning Test Regulations for Wool Tops
7. IWTO Regulations for Testing of Wool Slivers for Mean Fibre Diameter and Mean Fibre Length
8. IWTO Colour Regulations for Raw Wool
9. IWTO Arbitration Agreement and Other International Agreements (IWTO Blue Book)
10. P.J. SOMMERVILLE, The Projection Microscope, *AWTA Ltd Newsletter*, May 2002, www.awta.com.au
11. IWTO Specifications (Red Book): Edition 2007/08
12. J.L. WEAR, *The Use of NIR to Predict Residual Ash in the IWTO-19 Yield Test*, Report 02, Raw Wool Group, IWTO Nice Meeting, November 2001.
13. M LIPSON, A Method for the Estimation of Vegetable Material in Scoured Wool, *J. & Proc. Royal Society of NSW*, Vol. LXXXVI, pp 225–228
14. A. SPIES, A. BOGUSLAVSKY and L. HUNTER, *Rapid Determination of Wool Base by Means of Microwave Digestion and NIR Spectroscopy*, Report 02, Raw Wool Group, IWTO Nice Meeting, November 2000
15. P.J. SOMMERVILLE, *Fundamental Principles of Fibre Fineness Measurement*, www.awta.com.au, Publications, Reviews
16. CSIRO, *Proceedings of a Seminar of Staple Length & Strength of Greasy Wool*, Division of Textile Physics, Ryde, 7 December 1979
17. Australian Wool Corporation, *Sale with Additional Measurements Pilot Trial Explanatory Notes*, January 1980
18. Australian Wool Corporations Advisory Committee on Objective Measurement, *Sale With Additional Measurement*, September 1981
19. '*TEAM Final Report 1985*', Report to the Raw Wool Measurement Research Advisory Committee of the Australian Wool Corporation, January 1985

20. '*Report on Trials Evaluating Additional Measurements 1981–1988*', Report to the Raw Wool Measurement Research Advisory Committee of the Australian Wool Corporation, December 1988
21. *Report on Trials Evaluating Additional Measurements*, Final Report, AWTA Ltd, March 2004
22. IWTO Colour Working Group, Final Report of the IWTO Colour Working Group, Report 06, Raw Wool Group, IWTO Istanbul (Dubrovnik) Meeting, November 2003.
23. M.R. FLEET, T.J. MAHAR and S.I. DENNEY, *The Detection of Pigmented and Medullated Fibre in Core Samples of Commercial Sale Lots from Merino Ewes Mated to Damara Fat Tail Rams*, Report CTF 02, Commercial Technology Forum, IWTO, Nice Meeting, November 2001
24. M BRIMS, *Introducing OFDA4000: A New Instrument for Simultaneous Measurement of Fibre Length and Diameter in Tops*, Sliver Group, Report 02, IWTO Barcelona Meeting, May 2002.
25. V.E. FISH, A. BALASINGAM, J.W. MARLER and T.J. MAHAR, *Comparisons Between OFDA4000 and IWTO Standardised Methods for Measurements Made on Wool Tops: Part 1 Diameter Characteristics*, Report SG 02, Technology & Standards Committee, Sliver Group, Biella Meeting, November 2005
26. V.E. FISH, A. BALASINGAM, J.W. MARLER and T.J. MAHAR, *Comparisons Between OFDA4000 and IWTO Standardised Methods for Measurement Made on Wool Tops: Part 2 Length Characteristics*, Report SG 03, Technology & Standards Committee, Sliver Group, Biella Meeting, November 2005
27. AWTA Ltd, CSIRO and AWI, *Development of an Improved Test for Detection of Dark and Medullated Fibres in Presale Core Samples*, Report RWG 05, Technology & Standards Committee, Raw Wool Group, IWTO, Shanghai Meeting, November 2004

3 Wool as a natural renewable fibre

I. M. RUSSELL, CSIRO Materials Science and Engineering, Australia

Abstract: This chapter investigates the increasing demand for natural and renewable textile goods by consumers, 'brands' and retailers. 'Environmentally friendly' is not sufficient in its own right and garments must still meet the basic consumer requirements of quality, fashion, functionality, performance and value. The merits of different environmental claims such as 'eco-' and 'organic' are discussed, noting the value of Type 1 eco-labels such as the EU eco-label in validating environmental claims and assisting consumers to discriminate against 'greenwash'. The increasingly powerful role of European legislation in shaping global environmental practices and in increasing information flows in textile processing is acknowledged. The International Wool Textile Organisation has added some clarity for wool supply chains by adopting definitions for eco-wool and organic wool.

Key words: EU eco-label, eco-wool, greenwash, REACh, wool processing.

3.1 Introduction

Wool is a natural protein fibre produced by sheep. The wool fibre has unique physical and chemical properties that allow it to be extremely versatile. Wool can be warm or cool, it can be casual or formal and it has moisture-absorbing properties that allow it to be comfortable to wear in active and passive situations. In addition, because it is a natural fibre, wool is perceived by consumers to be renewable, 'sustainable' and 'environmentally friendly'.

Wool is therefore well placed to take advantage of the current consumer and retail demand for all things natural and sustainable. However, the sustainable textile space is becoming very crowded as other natural, bio-derived and synthetic fibres compete to demonstrate their environmental credentials.

The terminology used in the various environmental claims is creating confusion in the mind of consumers (and in the trade) as terms like 'organic', 'natural', 'sustainable' and 'eco-' clamour for attention and various eco-labelling schemes compete for market share. After a period of relative calm, new standards for organic, eco- and sustainable textiles are being written and endorsed by competing fibres, countries, retailers and brands. Are the environmental claims made by organic/eco- fibres/yarns/fabrics just marketing and 'greenwash', or are there real environmental benefits? Are organic or natural fibres more environmentally friendly than eco-fibres? The concern with textiles is not surprising given that textiles is collectively one of the world's largest industries and is a significant producer of high volumes of highly contaminated effluents (Cooper, 2007).

The textile industry led global manufacturing into the industrial revolution and the textile industry was an early and enthusiastic adopter of globalisation. This was partly to identify low-cost manufacturing countries, but partly also to escape rapidly expanding environmental legislation in developed countries that threatened to add significant costs. This relocation of the textile industry has meant that the pollution loads from the textile industry have also moved into low-cost manufacturing countries, many of which have not been prepared, physically or legislatively, to deal with the environmental impacts (Reuters, 2008).

Wool's overall market share has declined in real terms as well as in percentage terms and wool is now a minority fibre representing only 1.9% of world fibre production (IWTO, 2007) in a very large textile fibre market. Wool does, however, retain a significant market share in men's suiting, in knitwear and in carpets, areas where the unique properties of wool provide excellent appearance, comfort and durability that are unmatched by other fibres. In this context, it is important to recognise that 'green credentials' are insufficient in their own right. Consumers make purchasing decisions based mainly on quality, fashion, functionality and price, but for an increasing number of consumers, sustainability is becoming a critical new element.

It is essential that wool acknowledges this consumer trend. In 2004 Juan Casanovas, then President of the International Wool Textile Organisation (IWTO), reminded the global wool industry of the central importance of the consumer by stating that 'the wool demand chain starts in the shop, and not on the sheep' (Casanovas, 2004).

3.2 The environmentally aware retailer and consumer

For several years, consumers in Northern Hemisphere countries have become increasingly environmentally aware. Initially environmental purchasing was restricted to a small 'hard green' minority; however, as environmental consciousness has increased, the proportion of consumers prepared to pay more for environmentally friendly or 'ethical' goods has increased (Anon, 2007; GfK NOP, 2007; PriceGrabber.com, 2007).

An AWI/Woolmark/Queensland Department of Primary Industries marketing study (Pattinson *et al.*, 2006) notes that the drivers of increased consumer and retail interest in 'ethical' apparel products largely relate to social and processing issues and that this focus will continue in the future. The document also notes that the presence of harmful substances in textiles is regarded equally with social responsibility, product functionality, comfort and price.

Only product quality rated higher. A recent US study ranked energy efficiency as the most important factor (PriceGrabber.com, 2007).

In 2007, major retailers and garment 'brands' in the United Kingdom and United States moved from rhetoric to announcing major sustainability initiatives backed by large expenditure. High-street stores and even supermarket chains have announced designer ranges of organic fashions. 'Organic' cotton textile lines (some containing small percentages of organic cotton) are common in brand name products. Usage of organic cotton is expected to increase by at least 30% per year (Soth, 2006).

This trend to sustainable textiles is extremely important for wool. Wool is a minority fibre compared with polyester and cotton which dominate the apparel industry. Garments manufactured from these fibres are central to the business of most retailers. There is much less dependence from most retailers on wool. In the harsh business environment, wool retains space on the showroom floor only while it remains attractive to consumers and returns a comparable profit to retailers per square metre of floor space.

3.3 The growing importance of European legislation

The powerful influence of European Union (EU) environmental and product standards on International textile processing and production is beginning to be noted. Kanwar (2008) notes that

> European environmental and safety rules conceived in Brussels are increasingly becoming de facto Asian standards in the factories that churn out the televisions, clothing and furniture that fill most homes. The weight and size of the European Union's expanded market – about 450 million people, compared with 300 million in the United States – has lifted the profile and application of EU rules.
>
> … If you manufacture globally, it is clearly simpler to be bound by the toughest regulatory standard in the supply chain. As China and other sourcing countries lean towards the European approach, US companies are also beginning to work to EU rules.

One of the major tools will be the recently introduced REACh legislation (Registration, Evaluation, Authorisation and Restriction of Chemicals, EU Regulation, 2006). This complex legislation replaces 40 existing laws. REACh was adopted into EU legislation on 1 June 2007 despite extensive lobbying by chemical interest groups. A special agency (European Chemicals Agency, ECHA) has been established in Helsinki to manage the procedures and was operational from 1 June 2008.

A pre-registration period, from June to December 2008, allowed the continued production and import of chemicals during the transitional phase.

Because of the amount of work involved, the ECHA will prioritise the assessment of chemicals of greatest concern and chemicals produced in the greatest volumes.

November 2010 is the registration deadline for:

- high-volume chemicals (quantities of 1000 tonnes and above, chemicals in peak production);
- carcinogens, mutagens and substances toxic to reproduction above 1 tonne/year; and
- substances classified as very toxic to aquatic organisms (R50/53) above 100 tonnes.

Medium production chemicals (produced in quantities of 100 tonnes and more) must be registered by May 2013, while May 2018 is the registration deadline for quantities of 1 tonne and more (low production). The extent of toxicology testing increases progressively as the tonnages increase from 1 tonne to 1000 tonnes. Certain 'low-risk' chemical substances will be excluded from registration under REACh.

Substances of very high concern (SVHC) are:

- carcinogens;
- mutagens;
- substances which are toxic to reproduction;
- persistent, bio-accumulative, and toxic substances (PBTs);
- very persistent and very bio-accumulative substances (vPvBs); and
- substances causing equivalent concerns (such as endocrine disruptors).

Currently 900 SVHCs are listed and another 600 SVHCs are expected over the next 11 years.

REACh raises a number of supply chain reporting issues, even for products manufactured outside Europe. All importers of goods into the EU are required to register all substances in their imports that exceed 1 tonne per year. They must notify ECHA if the imports contain substances that are dangerous and the concentration of the substance is higher than 0.1% of the weight (Weyler, 2007).

Additional concerns for the textile industry are that specialty chemicals, such as wool-specific dyeing assistants, will need additional toxicity testing. Around 90% of the 1500 textile specialty chemicals are produced in volumes below 100 tonne a year. Examples of textile materials used in Europe in higher quantities include certain carriers (250 tonnes per year) and reactive black cotton dyestuff (more than 1000 tonnes per year). Manufacturers will either need to increase prices to cover toxicology testing, possible reformulation costs and licensing fees or withdraw the chemicals from the market. There are additional reporting obligations on users of the chemicals (Sedlak and Pellizari, 2005; Nuthall, 2007; Weyler, 2007.

REACh is currently causing concerns through the global textile chains. Briefings to the trade have been made in China, and several articles have appeared in recent months in the international textile trade magazines. Concerns have also been raised at World Trade Organization (WTO) meetings that REACh could restrict access to EU markets. Issues have been raised concerning the complexity and ambiguity of REACh, issues of inconsistent implementation across member states and problems for SMEs. Twelve of the major trading countries, including the United States, Argentina, Australia, Japan, Korea, China, Mexico and Thailand, have threatened to launch a disputes case based on WTO technical barriers to trade, asking the EU to liberalise REACh (Hogue, 2007).

3.4 Environmental claims

As 'green becomes the new black' consumers are more and more likely to be assaulted with an even more bewildering array of environmental claims. Some claims will be supported by legitimate, evidence-based arguments; however, it is probable that many environmental claims will be based on statements by overly enthusiastic marketers.

Consumer advocacy groups have long argued for truth in advertising and ISO 14021 provides clear guidance on environmental claims that are able to be made. In general, vague and non-specific claims should be avoided. This includes terms such as; 'environmentally friendly', 'green' and 'non-polluting' which are too general in nature to be of any use to consumers. Claims must be accurate, verifiable and in context. This information should be easily accessible. The UK and Australian government agencies (and others) are providing further interpretation and checklists for marketers, with warnings of the penalties under law that might be applicable for deceptive environmental claims (Defra, 2000; ACCC, 2008).

3.4.1 Type 1, Type 2 and Type 3 ecolabels and greenwash

Modern consumers are beginning to understand that there is an environmental impact associated with the manufacture of all products. As they become more environmentally aware they are seeking evidence to distinguish valid ecological claims from 'greenwash', defined by *The Concise Oxford English Dictionary* as 'Disinformation disseminated by an organisation so as to present an environmentally responsible public image. Origin from green on the pattern of whitewash' (Mohr, 2005). Wool is widely recognised by consumers as a natural and renewable fibre, but that does not automatically mean that specific wool products are sustainable or environmentally friendly.

In 2007/08, many of the major wool-producing countries initiated projects to investigate the 'environmental footprint' of wool production and processing by conducting life-cycle assessments (LCAs), not only as a defensive tool but also as a base for identifying and rectifying problem areas. The carbon footprint of commodities is the subject of a draft British Standard, PAS 2050, and this initiative is being supported by major retail chains. Many commodities are viewing this development very seriously and it is anticipated that carbon footprints of commodities will begin to have an impact on market access (Mello, 2007).

Environmental claims on products may be made through the use of eco-labels. The most common eco-labels are classified as Type I or Type II eco-labels. Both are based on the ISO 14000 series of international standards. Claims based on Type I eco-labels (as described by ISO 14024) are by far the most credible. These labels are based on criteria independently established across the life cycle of a product and certified or audited by an independent third party. Type II claims are based on ISO 14021 and are essentially self-declarations by manufacturers or retailers. Unfortunately Type II claims often present difficulties in terms of verifiability and credibility, and this is the province of greenwash.

Some research has shown that consumers often do not distinguish between Type I and Type II claims, or between third-party verified and self-declared claims. They often assume that all environmental claims have some kind of official backing (OECD, 2001).

Type III labels are little used. They are voluntary programmes that provide quantified environmental data of a product, under pre-set categories of parameters. Unlike Type I labels, a Type III label makes no claim of environmental superiority. That decision is left up to the consumer and user of the label information.

3.4.2 What differentiates organic, biodynamic, ecological, sustainable and ethical wools?

Eco-wool and the EU eco-label

The term 'eco-wool' in this chapter will be used to define wools (at various stages of transformation) that meet the requirements of a Type 1 ecolabel. Examples are the EU Ecolabel for Textiles and the Australian Good Environmental Choice label. Various forms of the term 'eco-wool' have been trademarked and used for marketing purposes but these wool products may not comply with the requirements of the above Type 1 eco-labels.

Type 1 eco-labels provide the greatest degree of transparency and the greatest assurance that the product has been manufactured with minimum environmental impact. The criteria in Type 1 eco-labels are set and assessed independently, they are readily available, they cover the total processing sequence and they apply to the processing stages with the greatest environmental impact.

The EU eco-label is arguably the most important standard for good environmental processing of textile fibres. It contains criteria that exceed 'environmental best practice' for textile processing as defined in the EU BREF document (EIPPC, 2002) and the Integrated Pollution Prevention and Control (IPPC) legislation (EC, 1996). Other international Type 1 textile eco-labels have usually taken their criteria from the EU eco-label.

The EU eco-label includes three groups of criteria:

1. All fibres have a clean fibre requirement. For the natural fibres, the clean fibre criteria are based on pesticide content.
2. The fibres must be processed using low environmental impact processing conditions (all processing agents must be biodegradable and there are strict limits on discharges of high organic load effluents and toxic materials).
3. There are 'fitness for use' criteria that guarantee that the product will be durable. This provides an assurance to consumers, but it also reduces the environmental impacts associated with processing of a replacement garment.

Some of the derived Type 1 standards (but not the EU eco-label itself) may add some social elements (fair pay, safety, child labour), but this is often accompanied by removal of some of the processing compliance stages.

Because of the international proliferation of the term 'eco-wool' and to provide clarity and guidance to the wool processing trade, the IWTO established an international working group from many of the grower and user countries to establish definitions for the term 'eco-wool'. The following definition was endorsed at the 2008 IWTO Conference in Beijing:

> *Eco-wool:*
> Wool in which residual pesticides on the greasy wool fibre do not exceed the limits set by EU Eco-label.
>
> *Eco-wool product:*
> A product manufactured from Eco-wool (defined above), which has been processed and/or transformed in accordance with the specifications of the EU Eco-label. The processor must meet the trade waste requirements of the local environmental control authority where these are more restrictive than those defined for the EU Eco-label.

In the 2002 revision, the EU eco-label recognised the value of the IWTO Draft Test Method 59 (DTM 59) as the method for measurement of pesticide residues in greasy wool. DTM 59 has a number of 'hidden' benefits for wool as it not only ensures that the analytical laboratory is accredited to ISO/IEC

17025 and that the laboratory is currently proficient, but it also requires that the wool is sampled according to IWTO core test regulations. This not only ensures that representative samples are analysed, but it also provides traceability that relates the chemical test result to the wool that is offered for sale. DTM 59 also incorporates all of the EU eco-label pesticides.

The EU eco-label requirements differ from the main organic standards in five main ways:

1. The EU eco-label applies good environmental practice criteria for most textile fibres including synthetic fibres. All fibres have their own 'clean fibre' criteria.
2. The EU eco-label includes strict and auditable processing requirements that have been established by experts familiar with textile processing. These criteria have been established by examining the life-cycle impacts for production, processing and use of the product, and limits are imposed on the key discharges with potential to damage the environment. There are many areas where the EU eco-label criteria go well beyond the requirements in organic processing. As an example, there are wool scours that have been certified for organic processing but which would not meet EU eco-label standards.
3. Organic standards focus on 'accreditation' of a farm or a textile processing mill whereas the EU eco-label focuses attention on specific certification of each processing batch. The differences are most obvious on-farm; in most (but not all) organic productions systems, any use of synthetic pesticides disqualifies the property from its organic status for some period of time. With the EU eco-label, farmers may offer specific batches of wool depending on its treatment history within the current growing season.
4. Strong weighting is placed on the durability of the final product. An area where this is very clear in wool processing is shrink resistance. The EU eco-label allows chlorine-based shrink resist treatments to be used on loose wool and slivers. The argument is that the environmental benefit conferred by manufacturing a garment that is robustly 'wash and wear' outweighs the small environmental effects from the shrink-resist treatment conducted under conditions where plant discharges are strictly controlled.
5. The EU eco-label is silent regarding on-farm issues such as animal welfare and environmental impacts, whereas this is a major thrust of the traditional organic standards. Given that the main focus in the EU eco-label is on reduction of environmental impact in processing, and that equivalent criteria apply to all fibres, it is appropriate that on-farm elements are not applied to wool and to cotton. However on-farm issues are relevant to some consumers and 'brands', especially in view of the attacks on the wool industry by People for the Ethical Treatment of Animals (PETA). IWTO has also recognised this gap and has encouraged grower countries

to develop programmes to identify good-management farming practices. Australia is developing its 'LandLeader' programme and other wool-growing countries such as South Africa are taking similar steps. However, development of auditable and consistent guidelines for land management across different climates and land types is difficult. In addition, a reporting system that allows direct 'line of sight' from the purchaser of the wool product back to the originating properties is needed. Currently, purchasers requiring on-farm assurances have established personal contacts with large wool-producing properties from where they source all of their wool.

The EU eco-label is based strongly on a supply chain approach where input of toxic materials is limited at all stages. Downstream processors are assured that only low toxicity and biodegradable components have been used on the fibre that they purchase. This information flow is often missing in textile supply chains and especially in wool supply chains.

There is increasing awareness that environmental discharges from a wet textile process such as dyeing arise from two sources:

1. the compounds deliberately added to the process by the dyer; and
2. the materials already present on the fibre when it is received. The identity of these materials is often outside the control or knowledge of the processor receiving the fibre and they may contribute a greater pollution load than the process additives. They may be difficult to remove and they may be toxic or poorly biodegradable.

The general issue of supply chain communication has been raised as an important element of good environmental practice in the EU BREF document (EIPPC, 2002) and by Cooper (2007). The problems are most severe at dyeing since dye liquors are difficult and costly to treat because of their volume and temperature. The European IPPC legislation will require EU dyehouses to better understand the levels of contaminants on the scoured wool and yarns that they process.

Finished goods processed in compliance with the EU eco-label requirements will be eligible to apply to a Competent Body in the EU to obtain the EU eco-label award, a sign to consumers that this particular textile product has been processed under good environmental conditions and with minimum environmental impact.

Most of the major manufacturers of textile processing dyestuffs and auxiliaries now manufacture specific product lines that meet the EU eco-label requirements. It is relatively simple for most textile mills to purchase processing agents that already have the necessary EU eco-label paperwork.

An important recent development has been the opening of a database in the EU eco-label website that allows late-stage manufacturers to find partially processed fibres, tops or yarns that have been manufactured in compliance with the EU eco-label. This is particularly important for wool, which has a longer supply chain than most other fibres. Because the EU eco-label requirements start with greasy wool, the task of identifying and then commissioning the processing of compliant batches of greasy wool has simply been too difficult for spinners and weavers. It is also possible for wool processors to obtain EU eco-label certification for their intermediate tops and yarns and at least one supply chain is moving in this direction.

Unfortunately the EU eco-label, at least in textiles, remains a poor marketing tool, with only around 70 textile products registered (EU, 2007a). A number of factors contribute to this poor uptake:

- lack of knowledge of the EU eco-label in the processing and retail trade;
- the perceived rigour of the processing requirements, partly due to the complexity of the paperwork at mill level;
- after-chrome dyeing cannot be used and sections of the trade believe that reactive dyes cannot produce adequate black shades;
- profit margins in the early stages of wool processing are too low to allow more than minimum compliance with environmental legislation; and
- wool garments are manufactured across several countries in extended supply chains.

Conversely the great strength of the EU eco-label is that it is a powerful standard for sustainable textile production with its links to ISO 14024 and the fact that it can support a claim that products manufactured in accordance with its criteria have a reduced environmental impact. Currently there is very wide recognition of the EU eco-label in at least the Australian wool growing industry, and it is expected that the recognition by the IWTO of the EU eco-label will increase awareness in the wool supply chains.

Other types of textile eco-labels

While there are around 40 environmental standards applicable to textile processing, many, including the EU eco-label, have small market share. By far the largest and best recognised textile eco-label is the privately owned Oeko-tex 100 label, but its philosophy and intent are quite different from the Type 1 or supply chain eco-labels. Oeko-Tex 100 is a 'human ecology' eco-label that seeks to demonstrate that manufactured articles will contain less than certain arbitrary concentrations of designated toxic agents. The Oeko-Tex 100 label is granted on the basis of chemical analysis of completed garments. This eco-label takes no consideration of environmental discharges in processing and thus cannot be used to support a claim that a specific product has been manufactured with low environmental impact. It is much simpler to obtain than the EU eco-label; however, there is less transparency (many test methods are not available) and its claims are (or at least should be) correspondingly limited.

Oeko-tex 100 should be easily achievable for finished or partly finished wool goods given the extent of removal of sheep ectoparasiticides from the wool during scouring and then in dyeing. Unfortunately Oeko-tex 100 cannot be used for identification of low-residue wools for environmental accreditation as it does not include the hydrophobic insect growth regulators, diflubenzuron and triflumuron, the main chemicals used on sheep in Australia and New Zealand. It is entirely appropriate that these pesticides are not included in Oeko-tex 100 as they are of low toxicity to humans, the main target of the Oeko-tex 100 label. However, these agents are toxic to aquatic organisms, particularly aquatic insects, and it is essential that both are considered in the EU eco-label which is concerned with environmental discharges.

There are additional eco-labels on the Oeko-tex series, Oeko-tex 1000 and Oeko-tex 100+, which require good environmental practices in textile processing mills; however, adoption of Oeko-tex 1000 is poorer than the EU eco-label (ICLEI, 2007).

Organic wool

In the minds of consumers the organic 'brand' probably represents the highest degree of environmental credibility. Most consumers carry an image of pesticide-free fruit and vegetables produced only with natural materials on small local farms, and many are prepared to pay a premium price and to accept less than perfect goods for this assurance. Most organic standards, with their focus on farming activities, deal well with simple farm produce.

Unfortunately, organic standards vary significantly, not only among accreditation agencies in the one country, but among countries, based on the needs of the regional organic farming groups that first developed their own organic standards. The International Federation of Organic Agriculture Movements (IFOAM) notes that there are hundreds of different private organic standards internationally, as well as more than 60 national government-codified organic regulations.

Government codification allows the use of the term 'organic' to be strictly controlled by legislation. Where the term 'organic' is controlled on a product label, an important part of the regulation is usually a requirement for third party auditing. In Australia the term organic is only controlled for export products; however, the organic associations are seeking to establish an Australian Standard for Organic and Biodynamic Produce that will reduce fraud and misrepresentation in the domestic market.

IFOAM has developed an Organic Guarantee System that seeks to assure organic integrity internationally. They are seeking to develop a common system of standards, verification and market identity. They encourage international organic certifiers to become 'IFOAM Accredited', based on the certifier's compliance with IFOAM Norms (basic standards). This initiative is to be strongly encouraged.

Organic wool farming

Organic standards differ significantly in the two major global markets of Europe and North America. In Europe the organic rules stem from European Commission Regulation No. 2092/91. This Regulation creates the framework for the production, labelling and inspection of organic farm products and foodstuffs. EU Regulation 1804/1999 amended Regulation No. 2092/91 to also include rules for production, labelling and inspection of organic livestock products.

Regulation No. 2092/91 allowed for the development of 'competent agencies' in each member state, and allows the competent agencies to take the rules further in their region. Article 11 of Regulation 2092/91/EEC opens the EU organic food market to products from third countries, based on the concept of equivalence. There are two main ways to export organic products to the European Union, depending on the country of origin:

1. Paragraph 1 establishes a list of countries (the 'third-country' list) where equivalence has been established. Wool producer countries on the list include Argentina, Australia and New Zealand.
2. For countries not on the 'third-country' list, import Regulation (EC) No. 1991/2006 came into force on 1 January 2007 to replace the 'importer derogation' paragraphs. Approval now largely resides with the EU Commission and requires certification to be conducted directly to the EU Regulation, or to an equivalent standard (IMO, 2007). Various transition arrangements are in place.

The focus of EEC 2092/91 is on food. Wool is mentioned only as a fertiliser and cotton is mentioned only as feed seed. EEC 2092/91 will be replaced from 1 January 2009 when the newly approved Council Regulation (EC) No. 834/2007 will apply.

In the United Kingdom the competent agency DEFRA (Department for Environment Food and Rural Affairs) took advantage of the opportunity to extend EC 2092/91 and developed the United Kingdom Register of Organic Food Standards (UKROFS). This document has a strong emphasis on organic livestock and expands on the rules for the use of allopathic (chemically synthesised, as opposed to homeopathic) veterinary medicines.

In the United States the main Institution (or set of rules) that regulates everything 'organic' under federal law is the National Organic Program (NOP). These organic standards are implemented by the USDA (United States Department of Agriculture). The NOP covers all organic producers, handlers and processors in the United States. However non-edible fibre products are not covered in the 'final rule', which therefore provides no details on the processing of organic fibres.

The USDA approves the national certification agencies that meet NOP requirements and that can therefore supply organic wool into the USA. Of the wool producing countries, three agencies in Australia are approved, as well as four in Argentina. The list is updated frequently (USDA, 2008). It is essential that potential supply chains for organic fibre identify accrediting agencies that allow them to supply wool to their intended destinations.

As a specific example relevant to wool, the UKROFS allow use of allopathic veterinary medicines (i.e. non-homeopathic medicines or pesticides) to treat internal and external animal health problems under veterinarian responsibility. Use of allopathic chemicals as a preventative is not permitted. The NOP does not allow use of allopathic medicines and the Australian organic standards follow this approach. The Australian organic standards generally allow the use of 'natural' pesticides such as magnesium fluorosilicate and spinosad for control of external pests; however, the main emphasis is on land, animal and farm management practices rather than on residues of internal and external pest control agents.

For this reason, production of organic wool in Australia is difficult and relatively limited, perhaps up to 300 tonnes per year or less than 0.1% of total production. However, there is potential for it to expand, especially in the pastoral regions where few pesticides are reported to be used. Some brokers are conducting training courses for farmers to increase understanding of the issues.

Biodynamic

A biodynamic farming system has additional requirements to an organic system. The main additional focus beyond organic is on maintenance of a healthy, well-structured soil, rich in humus and high in biological activity. Biodynamic farming systems are founded on the application of preparations originally developed by Rudolf Steiner and they have their own Biodynamic Certification Standards (NASAA, 2005).

Organic textile processing

For organic food, post-farm processing is minimal and can usually be handled through provision of lists of banned substances and requirements for strict segregation of organic and non-organic goods. This approach becomes much more difficult for textiles where more elaborate processing is required.

In 1989, the International Association Natural Textile Industry (former AKN) in Germany defined a first set of standards to process eco-friendly textiles with a practicable industrial approach. In the United Kingdom in 2001, the Soil Association published a set of 'Standards for processing and manufacture of Organic Textiles' and in the USA, around the same time, the Organic Trade Association (OTA) began to develop its set of 'North American Organic Fiber Standards'. It soon became clear there were difficult decisions to be made in developing organic textile processing standards for wool and for cotton, but most especially for wool, which undergoes more elaborate processing. In the simplest case for cotton, garments may be spun from naturally coloured cottons, albeit in restricted coloured ranges, with little wet processing and with little input of synthetic materials.

In general, for organic textile goods, there are two sets of largely incompatible consumer expectations that must be managed:

1. Consumers expect that wool goods will be reasonably durable, able to withstand (at least) hand washing without shrinkage and with colours that will not fade or leach. These properties are difficult to provide without the use of some synthetic chemicals
2. Consumers expect that organic goods will be manufactured using only natural materials, without use of synthetic chemicals or dyes.

Should organic wool be dyed only with natural dyes or can some synthetic dyes be used to improve the range of colours and colour fastness? There has been criticism of the performance and environmental impacts of production of natural dyes from plant and animal sources, especially if large quantities of these dyes were to be needed and if metal mordants are required (Holme, 2007). In general the early UK and US organic textile processing standards established a clear preference for 'natural' over the improved performance offered by synthetic dyes.

In 2002, representatives of organic cotton producers, the textile industry, consumers and standard organisations/certifiers began to develop a harmonised world wide organic textile processing standard. The first version of the Global Organic Textile Standard (GOTS) was presented in 2005 at an international cotton conference. Four important accreditation agencies have already agreed to adopt the Global Organic Textile Standard (IVN, Germany; Soil Association, UK; OTA, USA; and JOCA, Japan). IFOAM representatives attended all meetings of the International Working Group and the GOTS International Working Group aims to have the GOTS be exclusively recognised by IFOAM.

The GOTS is likely to be the most comprehensive and powerful standard for organic textile processing for the foreseeable future. Unfortunately the main focus in the GOTS is on cotton and wool is poorly considered. The best strategy for wool will be for representatives of the industry to work with the GOTS to help make it into a meaningful standard for all natural fibres.

Despite its name, the GOTS significantly blurs the boundaries between conventional processing and organic processing by allowing extensive use of synthetic dyes and processing agents. Major textile suppliers have developed extensive ranges of synthetic dyes, textile pre-treatment auxiliaries, dyeing and printing assistants and finishing agents for organic cotton (Clariant, 2007; Huntsman, 2007). Timberland has developed a range of organic fashion colours for its organic cotton range that are based on the GOTS-approved synthetic dyes, arguing that they are more environmentally friendly than natural dyes, as 'quite often pesticides and other nasty chemicals are used to grow the crop that produces the dyestuff' (Mowbray, 2007).

By allowing limited access to synthetic processing additives and dyes, the GOTS allows preparation of garments, at least in cotton, that carry an organic label and will meet modern consumer expectations for garment durability. Many of these same marketing opportunities could also become available to wool, with the exception that chlorine-based shrink resist treatments are not permitted. Currently the chlorine-free shrink resist treatments for wool are extremely expensive and/or insufficiently robust. Research is progressing to develop alternative treatments, but it is not yet clear if the new treatments will meet other requirements of toxicity and biodegradability in the GOTS (Duffield, 2007).

The IWTO has been concerned by the differences between national standards for organic wool. The following definition of organic wool was adopted at the IWTO Congress in Beijing in May 2008:

> Organically-grown wool:
> Wool grown by sheep raised on certified organic farms and harvested according to appropriate organic practices. The farm must be certified 'organic' by an IFOAM-accredited or internationally recognised (according to ISO Guide 65) certifier. The organic wool and farming practices:
>
> - must meet the standards enshrined in the legislation of the country in which the ultimate products are sold, where they exist, OR
> - where the country in which the ultimate product will be sold is not known, must meet a recognised international standard or the national standard from the country of origin.
>
> Notification of the standard to which the farm is certified must accompany the 'organic descriptor' of the wool.
>
> > Note: Recognised standards include, but are not limited to USDA National Organic Program and EEC Regulation 2092. IWTO reserves the right to exclude from this definition, standards that are deemed to be not sufficiently rigorous.
>
> Organic wool product.
> A product manufactured from organically-grown wool, which has been processed and/or transformed in accordance with the Global Organic Textile Standard (GOTS-Version 1.1 or subsequent). The product must be pure wool as defined by GOTS or by local legislation. All processors must be

certified by any IFOAM-accredited or internationally recognised (according to ISO Guide 65) certifier and must meet the trade waste requirements of the local environmental control authority.

The IWTO members were still concerned with the variations in wool production systems allowed under the foregoing definitions and noted that such variability was causing uncertainty in the market. The organisation requested that an additional code of practice be drawn up for the production and processing of organic wool 'which will augment and add guidance for the application of definitions of Organic Wool and its Products that may be applied in all world markets'. Unfortunately the right to use the term 'organic' rests with organic accrediting organisations and it remains to be seen whether the IWTO national organisations will be able to influence their national organic accreditation organisations to adopt common farming practices.

The IWTO set of definitions recognises the potential importance of the GOTS. However, it is important that certain criteria in the GOTS are amended in areas where the specific criteria do not adequately address the special processing needs of wool. There are two priority areas in the GOTS that need amendment:

1. Because some of the signatories to the GOTS allow use of allopathic chemicals for protection of animals from diseases and pests, the GOTS imposes a limit of total pesticide content of 0.5 mg/kg on 'Shorn wool certified organic'. While the GOTS outlines an analytical method, the sampling and subsampling procedures are not adequately described. The GOTS cites (Section 4.2 Residue Testing):

   > *The licensed operator is expected to undertake residue testing according to a risk assessment of contamination. ... Samples for residue testing may also be taken by the inspector during the inspection visit required for certification, either as back-up to the inspection process or in case of suspicion of contamination or non-compliance.*

   Unfortunately, because of the uneven distribution of pesticides on greasy wool, it is impossible to obtain representative samples of wool for residue testing from a processing consignment of wool during an inspection visit. A systematic risk-based sampling programme is required, linked to representative core samples taken from the supply chain in the same manner as the sampling regime in the EU eco-label.
2. The GOTS shares a common specification with the EU eco-label for textiles (2002–2007) which requires that all textile processing plants discharge effluents with a chemical oxygen demand (COD) of less than 25 g/kg of textile output. Unfortunately the GOTS exempts greasy wool scouring sites from this requirement and there are no further effluent treatment requirements for wool scours. This is presumably an oversight

as wool scours without effluent treatment produce large quantities of oxygen-depleting substances in waterways. Other organic standards are similarly silent or non-specific on the need for treatment of wool scouring effluents (NASAA, 2004). In contrast the EU eco-label includes very strict effluent treatment requirements for wool scours.

There would be benefit for wool if there was further alignment of the GOTS with the EU eco-label criteria. Certainly some of the current uncertainty in the processing trade would be reduced.

3.5 EU eco-label processing in practice

3.5.1 What are the residue targets for EU eco-label wool?

In the 2002 revision of the EU eco-label, there was alignment of the eco-label criteria with the environmental risk assessment processes that were being conducted in Australia on pesticide residues in raw wool. In 1998 Savage reviewed the available knowledge on the occurrence of pesticide residues on Australian wool and on the downstream environmental consequences that resulted when that wool was scoured internationally (Savage, 1998; Russell, 2000). A generic wool scouring risk assessment model was developed based on a single wool scouring line located on a small river system. The effluent was treated on-site and then in a municipal sewage treatment system. The model was used to back-calculate from environmental 'no-effect' pesticide concentrations in the river (based on the most sensitive aquatic organism in the ecosystem) to determine the pesticide residues on the greasy wool that could be scoured without causing environmental harm.

The EU eco-label technical groups generally adopted the Australian risk assessment process (with some rounding and grouping of data) to establish the following wool residue targets; dicyclanil was included in the 2008 revision of the EU eco-label:

- Total synthetic pyrethroids – 0.5 mg/kg.
- Total organophosphates – 2 mg/kg.
- Diflubenzuron, triflumuron, dicyclanil – 2 mg/kg.
- Total organochlorines – 0.5 mg/kg.

The EU eco-label offers a generic low-residue target/standard for wool. There are two practical applications of these concentrations for wool:

1. Wool that meets these low-pesticide requirements will be able to enter eco-sustainable supply chains and, if the EU eco-label processing requirements are followed through to final product, be converted into

products that can be awarded the EU eco-label (or other Type 1 eco-label) and be identified to consumers as environmentally preferred.

2. International scourers who purchase compliant wools will be able to demonstrate that pesticides released in their scouring operation will have no impact on the environment, even if they have only a minimal on-site effluent treatment system (such as chemical coagulation/flocculation) and discharge to a central effluent treatment system.

   This process will become important in Europe after October 2007 and when the IPPC legislation is operational. Under this legislation, there are no defined national or EU-wide pesticide residue standards. Individual scour operators must prove on an ongoing basis that they have not contaminated the environment (i.e. prove that they have not exceeded applicable environmental quality standards), and must submit processing and discharge data to public access databases. The value of a generic low residue standard is that it removes the need for detailed residue data to be provided on each lot of wool purchased.

   The same IPPC legislation will apply to wool dye houses and again they must supply information on the residue status of the wool that they process. This is a supply chain issue, as the concentration on the scoured wool that enters dyeing will be directly related to the concentration on the original raw wool. Even if a scour is able to process high-residue greasy wools because it has a total containment effluent treatment process, the high-residue scoured wool will remain as a potential problem downstream.

3.5.2 How to produce EU eco-label low-residue wool

This discussion will be based on Australian sheep farming practices, since this is where the most research has been conducted, although similar general trends will be applicable in all other wool-producing countries. Pesticides are applied externally to sheep to protect them from attack by insects. In Australia, the main pests are sheep blowfly and sheep lice. Blowfly attack has the potential to kill sheep rapidly and painfully as the flies lay their eggs in moist areas or in dung-coated skin. Lice cause the sheep to lose condition and to entangle the wool as the sheep rub against fences to reduce the itch.

The main factors that control the concentration of residual pesticides on sheep at shearing are

- the mass of pesticide applied;
- the rate of degradation of the pesticide; and
- the time between treatment and shearing to allow degradation and weathering to occur (Campbell *et al.*, 1998).

Two situations produce wool with high-pesticide residues:

1. poorly degradable pesticides applied at any stage in the growing season; and
2. readily degradable pesticides applied late in the growing season to control pests.

Specific advice is available in Australia from the various State Departments of Agriculture and from Australian Wool Innovation on methods to control pests without leaving high residues on wool (LiceBoss, 2008; Newell, 2007). However, it is important to note that compliance with the EU eco-label low-residue requirements is voluntary, and, unlike organic programmes, there are no penalties associated with leaving the programme for a year or more to attack the pests with the full armoury of available chemicals. Growers may choose to declare that their wool meets EU eco-label conditions on a mob-by-mob basis or year-by-year basis based on treatment records. A key requirement is that aspiring EU eco-label participants maintain good treatment records and understand the residue consequences of their chemical use.

3.5.3 Preparing processing consignments of EU eco-label wool

Australian Wool Innovation conducts an annual survey of pesticides on 600 randomly selected Australian fleece wools offered for sale. This survey currently shows that around 40% of wools meet the EU eco-label criteria. However, it is important to note that this does not mean that 40% of wool processing consignments will meet the EU eco-label requirements. Australian wool is predominantly sold at open-cry auctions in quantities of less than 1 tonne (on average). This system allows buyers to purchase wools from different properties in order to compile a processing lot of 20–50 tonnes that will meet a given clean fibre specification at sliver stage (micron, yield, strength, colour) at minimum cost. In fact the probability of such a consignment of wools chosen at random meeting the EU eco-label residue requirement is less than 1 in 1000. If even a single line of highly contaminated wool is present in the consignment, the average residue will potentially exceed the residue target.

Australia is therefore developing systems to reliably identify supplies of low-residue wools for the EU eco-label market at minimum cost. If all the lines of wool in a single consignment are individually tested for chemical residues, the cost is expensive (upwards of US$3000) per consignment. The problem is made difficult because the individual lines of wool for a processing consignment may reside in different wool stores across the country. They are not shipped to the scour until a scouring date is available, and the whole consignment is only assembled immediately prior to scouring. Even if all bales of wool could be sampled at that point, there is no time to perform the analysis before scouring begins.

Two schemes are currently in operation in Australia to identify the residue status of wool before sale. These schemes allow the wool buyers to make a purchase based on the residue status of the wool as well as on the other traditional specifications. Both schemes operate on the ability to test several sale lots of wool as a 'composite' sample.

The original scheme operates in Tasmania through the dominant broker. Growers make a declaration that they have not used certain chemical treatments on their sheep, and a single line of wool is chosen at random from that property annually for residue testing. If the test shows that the wool is compliant, all of the wool from the property is identified in the catalogue (Russell *et al.*, 2003). Growers who make a false declaration are visited and their wool is not identified in the catalogue. The problem remains that occasional high-residue wools may slip through and the new owner of the wool is still required to test the wool they have purchased. This can be done by combining pre-sale core samples of the wool if they are still present at the wool testing centre, or, alternatively, because all of the wool is present in a single wool store in Tasmania, the remainder of the wool can be resampled and again tested as a composite sample.

The more recent scheme operates on mainland Australia (and increasingly in Tasmania) where a farmer arranges a composite sample to be prepared from all of the wool from his property that is delivered to the wool store in a single shipment. This wool is likely to be broken up into several different sales lines representing wool of different qualities; however, the residue test certificate can apply to all of the separate lines of wool. Linkages between the residue test certificates and the lots of wool offered for sale ('chain of custody') are provided through the Australian Wool Testing Authority (AWTA) (Billing, 2008). While testing of composite samples costs more than analysis of single wool lines at present, there are still significant savings.

In other countries, wool selling systems are quite different. In the United Kingdom, wool is sold in large 8000 kg offerings based on the wool style. Single fleeces with common fleece characteristics are assigned to these large consignments and each fleece may have a different pesticide treatment history. The large consignments may contain wool from perhaps a thousand farms so that residue contents will probably represent an average for that region.

3.5.4 Wool scouring requirements

Implementation of the EU eco-label in processing chains can appear daunting for processors, with a multitude of complex compliance forms available, only a few of which are essential for each processing stage. The CSIRO strategy for introduction of the EU eco-label in wool supply chains has been to issue each processor with a checklist and to provide only the minimum set of forms to cover the specific operations of the mill.

For wool scouring, it is the responsibility of the mill management to sign off on the pesticide residue content of the wool consignment that is being processed. In practice, however, it is the responsibility of the owner of the consignment of the wool to ensure that an analysis certificate is available for each line of wool, in the same way that they ensure that other details of the wool are provided such as wool types, number and weights of bales of wool from each property, micron and yield.

Other requirements for EU eco-label wool scouring are common to other wet processing stages. The following documentation is required:

- A list of all chemicals that have been used on the fibre or in the process, along with statements of compliance from the chemical manufacturer that each of the chemicals meets all of the requirements of biodegradability and absence of toxic additives. Most mainstream suppliers of textile chemicals now have lists of compliant formulations and will supply the necessary documentation.
- A statement on the fate of the effluent leaving the factory, whether to the sewer or to the environment, is required.
- A measurement of COD concentration in the effluent leaving the factory is required, to be conducted by an independent laboratory accredited to ISO 17025. If the effluent is discharged directly to surface waters, temperature and pH measurements are required, as well as calculations to show that the COD discharge was reduced below the target mass of COD/kg of fibre processed. If the effluent is discharged to sewer, additional data on the efficiency of operation of the central sewage treatment plant must be provided. Similar information is required for other wet processing stages.
- A statement is required that no transportation preservatives based on chlorophenols, polychlorinated biphenyls (PCBs) or organotins have been added to the fibre leaving the mill.

Although this is initially complex, experience has shown that mills are easily able to manage the process after the first batch or two of eco-label processing is conducted. Some changes to established processing chemicals may be needed and it may be necessary to check with downstream customers to verify that any change, for example, to softeners, does not cause problems in subsequent dyeing operations.

A significant change for many international wool scours is that alkylphenol ethoxylate (APEO) detergents may not be used for scouring of greasy wools. The alkylphenol ethoxylate scouring detergents are cheaper than the so-called biodegradable detergents and are used extensively internationally, but they are restricted in Europe. In the environment, the APEOs degrade under anaerobic conditions to the parent alkylphenol and these act as potent oestrogen mimics. The GOTS also bans use of APEOs, and additionally requires that all processing agents meet certain aquatic toxicity and biodegradability requirements.

3.6 Future trends

Many of the building blocks for future developments to establish wool as an ethical, natural and renewable fibre are now in place. Australian Wool Innovation and other grower country R&D organisations are undertaking LCAs to provide an objective basis for establishing wool's environmental credentials and also to identify pressure points (Lyons, 2008). Already R&D activities are in progress to address some of the known on-farm issues (mulesing, pesticide use and land management) and processing issues (such as chlorine in shrink-resist treatments and chromium in dyeing).

With its recognition of the EU eco-label textile processing standard, the IWTO has established a baseline set of criteria for good environmental practices, supply chain management and improved information flow in wool textile processing. An important development will be for supply chains to make available standard lines of eco-label tops and yarns for spinners, knitters and weavers to purchase. This will shorten wool's current long supply chain that has been a formidable hurdle for garment manufacturers. The EU eco-label is also establishing a database for compliant raw materials and intermediate products (EU, 2007b).

Similarly, initial recognition by IWTO of the GOTS has been an important step for organic processing. However, there are still important issues of harmonisation of international on-farm organic standards and of amendments to the GOTS for wool processing that need to be addressed to remove some of the uncertainties associated with the organic wool 'brand'. It is essential that wool moves rapidly to meet the needs of environmentally aware consumers, retailers, brands and NGOs who are seeking assurances about on-farm practices, principally that the land has been managed well and that the sheep have been managed ethically.

As international concerns with water availability and global warming accelerate, it is likely that the trend for all things natural and sustainable will continue. Legislative leadership will continue to come from Europe as it continues to develop standards such as REACh and IPPC that will impact on all global textile supply chains. Both pieces of legislation require increased communication through supply chains. Carbon footprint labelling will become prominent and textile recycling will become an important issue as land fill becomes less available in the major textile consuming countries.

Environmental pressures in China will continue to grow as regulators and consumers become less tolerant of poor environmental practices associated with textile processing. This will lead to increased costs which may then reduce overall consumption of textiles. Already, articles have begun to appear in newspapers and fashion/trade magazines noting that the age of conspicuous consumerism in textiles is over and that a new sign of fashion status is to have a much smaller wardrobe of quality garments from designers with strong environmental credentials. This can only benefit wool, provided that it is ready to meet these developing consumer needs.

3.7 Sources of further information and advice

The monthly trade magazine *Eco-textile News* is an important source of environmental information on trends, issues and news releases. However, most general trade magazines such as *Wool Record* and *International Dyer* increasingly include environmental articles.

The AWI website (www.wool.com.au) and the AWI magazine *Beyond the Bale* provides regular articles (as well as occasional special supplements) on environmental issues. It is likely that progress on the LandLeader program will be initially available from that source.

The EU BREF document (EIPPC, 2002) is a major resource that defines good environmental practices for textile processing. It was written using European data that was obtained between 1999 and 2001 and reflects that viewpoint; however, it remains a valuable benchmarking document.

The EU eco-label textile-specific website is found at http://ec.europa.eu/environment/ecolabel/product/pg_clothing_textiles_en.htm. The site links to many other useful articles, including to the current suppliers of eco-label textiles. The background material describing the reasons for setting the specific criteria are valuable.

3.8 References

ACCC (2008), *Green marketing and the Trade Practices Act*, Australian Competition and Consumer Commission, http://www.accc.gov.au/content/index.phtml/tag/greenmarketing/, accessed 24 April 2008.

ANON (2007), 'Is there a future for ethical wool', *ecotextile news,* Issue 2, March, 10.

BILLING J (2008), 'AWTA Ltd introduces new pesticide residue analysis service', *AWTA Newsletter*, 9, April, http://www.awta.com.au/Documents/News%20Letters/2008-04-pdf/Newsletter%202008%2001%2009.pdf, accessed 30 April 2008.

CAMPBELL NJ, HANRAHAN PD, RUSSELL IM, ROBERTS GS and HORTON BJ (1998), 'Modelling pesticide residues on greasy wool: experimental studies', *Australian Journal of Experimental Agriculture*, **38**, 441–9.

CASANOVAS J (2004), Presidential Address, IWTO Congress, 13 May, Evian.

CLARIANT (2007), 'Approved Dyestuffs and Textile Auxiliaries for the Global Organic Textile Standard (GOTS), Version 1.1. Clariant Products Ltd', http://www.textiles.clariant.com/C12571C400485A16/vwLookupDownloads/

ClariantdyesauxiliariesforGOTS-issue10-14April2008.pdf/$FILE/ClariantdyesauxiliariesforGOTS-issue10-14April2008.pdf, accessed 30 April 2008.

COOPER P (2007), 'Clearer communication', *ecotextile news,* Issue 4, May, 30–33.

DEFRA (2000), 'Consumer products; green claims code', http://www.defra.gov.uk/environment/consumerprod/gcc/index.htm, accessed 23 April 2008.

DUFFIELD P (2007), 'Which standard for wool?', *ecotextile news*, Issue **6**, 27–9.

EC (1996), 'Council Directive 96/61/EC of 24 September 1996 concerning integrated pollution prevention and control', http://eur-lex.europa.eu/LexUriServ/LexUriServ.do?uri=CELEX:31996L0061:EN:HTML, accessed 28 April 2008.

EIPPC (2002), 'Integrated Pollution Prevention and Control (IPPC) Reference Document on Best Available techniques for the Textiles Industry', VD/EIPPCB/TXT BREF FINAL, November.

EU (2006), 'Registration, Evaluation and Authorisation and Restriction of Chemicals', Regulation No1907/2006; http://eur-lex.europa.eu/JOHtml.do?uri=OJ:L:2007:136:SOM:EN:HTML.

EU (2007a), 'Facts and figures', *The Flower News*, Issue 2, 3, http://ec.europa.eu/environment/ecolabel/pdf/flower_news/2007_oct/flower_0207_en.pdf, accessed 30 April 2008.

EU (2007b), 'A database of green suppliers', *The Flower News*, Issue 2, 4, http://ec.europa.eu/environment/ecolabel/pdf/flower_news/2007_oct/flower_0207_en.pdf, accessed 30 April 2008.

GfK NOP (2007), 'Ethics go mainstream', *ecotextile news*, Issue 5, June, 25.

HOGUE C (2007), 'REACH Reflux. Countries raise concerns about EU law at WTO committee meeting', *Chemical and Engineering News*, 15 Nov, http://pubs.acs.org/cen/news/85/i47/8547news5.html, accessed 30 April 2008.

HOLME I (2007), 'Getting madder', *ecotextile news*, Issue 3, April, 27.

HUNTSMAN (2007), 'Approved Chemicals and Dyestuffs for the Global Organic Textile Standard (GOTS) Version 1.1. Huntsman Textile Effects', http://www.huntsman.com/textile_effects/Media/XX4004e_organic_textiles_IMO_scan__2_.pdf, accessed 30 April 2008.

ICLEI (2007), 'GPP Training Toolkit – Background product report – Textiles, Draft 26/10/07', http://www.envir.ee/orb.aw/class=file/action=preview/id=871986/Textiles+Product+Background+Report+–+Draft+26+10+07.pdf, accessed 28 April 2008.

IMO (2007), 'New Import Regulation (EC) No. 1991/2006 commented by IMO', http://www.imo.ch/imo_services_export_import_news_en.html, accessed 28 April 2008.

IWTO (2007), 'IWTO Market Information', Edition 2007, Table 5.

KANWAR P (2008), 'The EU sets the standard', *SGS Consumer Compact*, III, (4), 2, http://newsletter.sgs.com/eNewsletterPro/uploadedimages/000006/ConsumerCompactNewsletterJanuary2008.Pdf, accessed 30 April 2008.

LICEBOSS (2008), 'Welcome to LiceBoss', http://www.liceboss.com.au/, accessed 5 May 2008.

LYONS B (2008), 'Positioning Australian Merino as the environmentally and ethically assured fibre solution', IWTO Congress, May 13, Beijing.

MELLO F (2007), 'Top labels spy a stylish edge in carbon labelling', *The Age*, September 17, http://www.theage.com.au/news/business/top-labels-spy-stylish-edge-in-carbon-labelling/2007/09/16/1189881340189.html, accessed 30 April 2008.

MOHR T (2005), 'Reputation or Reality? A Discussion Paper on Greenwash & Corporate Sustainability', www.greencapital.org.au/index.php?option=com_docman&task=doc_download&gid=12, accessed 23 April 2008.

MOWBRAY J (2007), 'Timberland makes a big organic colour splash', *ecotextile news*, Issue 1, February, 16–17.

NASAA (2004), 'The NASAA Organic Standard', December, http://www.nasaa.com.au/data/pdfs/DRAFT NASAA Organic Standard April 2006.pdf.

NASAA (2005), 'Steps to certification – primary production', http://www.nasaa.com.au/steps1a.html, accessed 30 April, 2008.

NEWELL G (2007), 'Low chemical residue wool', http://www.dpi.qld.gov.au/cps/rde/dpi/hs.xsl/30_7209_ENA_Print.htm, accessed 30 April 2008.

NUTHALL K (2007), 'Clock ticking on REACH deadlines', *International Dyer*, July 22.

OECD (2001), 'OECD Environment Directorate, Programme on sustainable consumption. Experts workshop on information and consumer decision-making for sustainable consumption, 16–17 January, 2001. Background paper',
http://www.oecd.org/dataoecd/46/19/1895757.pdf, accessed 24 April 2008.

PATTINSON R, PAHL L and HAN M (2006), 'Potential customer requirements and demand for 'Ethical Wool', *Farm Policy Journal*, **3** (4), 47–53.

PriceGrabber.com (2007), US shoppers more eco aware', *ecotextile news,* Issue 9, November, 14.

REUTERS (2008), 'Lake pollution could hit deliveries', *ecotextile news*, Issue 6, July, 10.

RUSSELL I M (2000), 'Meeting the IPPC needs of European processors, an Australian perspective', International Wool Textile Organisation, Technology and Standards Committee, Commercial Technology Forum, CTF 01, Nice Meeting.

RUSSELL I M, NUNN C R, HORTON B J and HUTCHINSON E (2003), 'Evaluation of a Tasmanian vendor declaration scheme to identify wools to meet EU eco-label criteria', International Wool Textile Organisation, Environment Committee, Dubrovnic.

SAVAGE G (1998), 'The residue implications of sheep ectoparasiticides: A report for The Woolmark Company', National Registration Authority, Quality Assurance and Compliance Section, Canberra.

SEDLAK D, PELLIZARI F (2005), 'Final Report: Impact of Reach on Textile Supply Chains', http://ec.europa.eu/enterprise/reach/docs/reach/text_final_report_051216.pdf, accessed 28 April 2008.

SOTH J (2006), 'EEB Information on the Organic Cotton Situation', http://www.eeb.org/activities/ecolabel/EEB-organic-cotton-facts-sheet-for-EUEB-Dec06.pdf, accessed 28 April 2008.

USDA (2008), http://www.ams.usda.gov/nop/CertifyingAgents/Accredited.html, accessed 28 April 2008.

WEYLER Å (2007), 'You can't get out of REACh', *Textile Asia*, July, 20.

4

Advances in wool spinning technology

M. W. PRINS, CSIRO Materials Science and Engineering, Australia

Abstract: Technologies have been developed, and are continuing to be developed, along the worsted processing pipeline to improve process prediction and control, product quality and production rates whilst maintaining or improving product quality. This chapter reviews a selection of advances made, concentrating mainly on those developed or released commercially since the turn of the 21st century. The advances include the introduction of the updated topmaking prediction formulae (TEAM-3) that is still viewed as an important tool in converting greasy wool to top, new worsted combing machines improving both production rate and quality, air-jet spinning for wool and a new generation self-twist spinning machine.

Key words: scouring, topmaking, spinning, winding, yarn clearing, twisting.

4.1 Introduction

A comparison of the shorn fleece on the shed floor with the attractive, highly desirable garments that are created from it can easily excite a sense of wonder. In the shorn fleece the quite ordered array of the staples on the sheep has already been disturbed and the variable length and state of the wool from different parts of the body are obvious. Compare the wool shorn from the neck, head, belly and crutch with that from the sides and back. There will be differences in the vegetable matter in these components, and there will also be differences in fibre diameter and colour. Add to all this a presence of a significant amount of wool wax or grease, together with some dirt and the unseen sweat salts. All this variability from one sheep, yet a consignment of wool may contain fibres from many sheep, flocks and environments.

How is it possible to profitably re-order and reassemble the fibres in shorn wool to create the fine smooth suiting fabrics, the soft pliable knitwear and the cool lightweight shirting, and to do it in such a way that there is the required uniformity of technical properties and quality in the consignments, batches and deliveries through the textile pipeline? Major issues are costs, blending or homogeneity, the removal and disposal of contaminants, the management of fibre entanglement, the degree and limitations of control of fibre position and number within slivers and yarns, and the uniformity of treatment within and between batches through the pipeline. And there is the major issue of specification and predictability of processing and product performance to meet quality and price demands.

4.2 Brief review of advances in topmaking

4.2.1 Conversion of greasy wool to top

Science takes a higher profile than art in topmaking, particularly since the development of objective measurement for raw wool. The aim is to convert an assembly of greasy wool lots in bales into combed wool or top to meet required specifications at a price. Those specifications will ultimately be governed by the yarn into which the top, often blended with other fibres, is to be spun. Blend engineering with these two factors, specification and price in mind, is really the name of the game.

One of the most important developments in topmaking has been the development of methods of predicting the processing performance of wool from objective measurements made on the raw wool. The greasy wool characteristics and the processing parameters from scouring to topmaking all determine the yield and the properties of the top. The TEAM formulae, developed as a result of the Trials Evaluating Additional Measurements project collaboration between CSIRO, the Australian Wool Testing Authority (AWTA) and the Australian Wool Corporation, is a simple regression formulae prediction tool (TEAM, 1988). It can be used to predict the fibre length characteristics in the top and the combing waste. The predictions are made from the following measurements of the greasy wool:

- fibre diameter (micrometres – commonly referred to as micron);
- staple length (mm);
- staple strength (N/kTex);
- mid-breaks (%);
- vegetable matter base (%).

Since 1988 the formulae referred to as TEAM-2 have been widely used in the industry. With the introduction of new measurements such as coefficient of variation (CV) of fibre diameter, CV of staple length and mean fibre curvature, AWTA Ltd began a review of the TEAM formulae in 2001 to improve the prediction performance. As a result of this review the TEAM-3 equation (AWTA Ltd Research Papers, 2004; TEAM-3 Steering Committee, 2004), was released in 2006. Along with readjustments of some of the coefficients, the review recommended the addition of CV of fibre diameter and CV of staple length to improve the prediction capability. The TEAM-3 formulae are as follows:

$$\mathrm{H} = 0.43L + 0.35S + 1.38D - 0.15M - 0.45V - 0.59\mathrm{CVD} - 0.32\mathrm{CVL} + 21.8 + \mathrm{MA}$$

$$\mathrm{CV\text{-}H} = 0.30L - 0.37S - 0.88D + 0.17M + 0.38\mathrm{CVL} + 35.6 + \mathrm{MA}$$

$$\mathrm{Romaine} = -0.13L - 0.18S - 0.63D + 0.78V + 38.6 + \mathrm{MA}$$

where H = hauteur or mean fibre length in the top
CV-H = coefficient of variation (CV) of hauteur
romaine = combing waste
L = staple length
S = staple strength
D = fibre diameter
M = percentage mid-breaks
V = vegetable matter content
CVD = CV-fibre diameter
CVL = CV of staple fibre length
MA = mill adjustment factor

If all the weighted objective measurements of a consignment of wool lots are known, the important benchmark processing performance parameters and top properties can be calculated. Topmakers, whether associated with exporters or combing plants, are able to buy and assemble consignments by trading off the price against the objective measurements of sale lots. A good source of information is the Australian Wool Testing Authority Ltd, http://www.awta.com.au/. Alternative prediction formulae are available in software packages such as Sirolan TOPSpec™, and Topmaker™ and a description of these is given in Simpson and Crawshaw (2002).

4.2.2 Scouring

The efficient scouring (i.e. washing) of raw wool is an essential first step in the early stage processing of wool. The aim of the scouring process is to remove the contaminants from the wool that would otherwise impede further processing of the wool while inducing the minimum amount of felting/entanglement. Fibre entanglement leads to a loss in fibre length and an increase in combing waste (noil or romaine).

The contaminants on the fibre can be divided into four groups: water-soluble material (commonly referred to as suint), wool wax (which can be both oxidised and unoxidised), dirt (of organic and inorganic origin) and vegetable matter. The aim of the wool scouring process is to remove as much as possible of the first three of these contaminant groupings. Vegetable matter, in general, is not removed in the scouring process. The amount of each of these contaminants varies widely depending on the breed of sheep, age, location, etc. Some typical figures for wax, suint and dirt for Australian greasy wool can be seen in Table 4.1.

Because wool consignments may consist of wool lots from various sources, blending is an essential part of topmaking, starting with the scouring operation. Blending may occur through bale lay down whereby the bales of wool of similar type making up the consignment are chosen in a predetermined order,

Table 4.1 Some typical wax suint and dirt levels for Australian greasy wool

Wool type	Total wax (%)	Suint (%)	Dirt (%)
Merino fleece	13.6	3.1	12.9
Merino lambs	20.3	3.1	7.4
Merino pieces	9.7	10.3	15.3
Crossbred pieces	9.3	9.8	15.7
Pieces & bellies	10.7	11.6	25.0

or through more sophisticated blending by in-line or weigh belt systems, or both. Opening and dusting of the wool generally follow.

The mechanism of wool scouring has changed little over 100 years or so and most of the wool scoured around the world is still scoured in hot aqueous detergent systems using traditional rake and harrow machines. Those developments that have occurred have centred on technology such as bowl design where hopper bottoms have now become the norm to better enable dirt recovery and continuous operation. Developments such as the WRONZ Comprehensive Scouring System and CSIRO's Lo-Flo and Siroscour systems have greatly advanced scouring technology. A description of the various opening, blending, scouring and drying technologies can be found in Simpson and Crawshaw (2002).

4.2.3 Topmaking

The sequence in topmaking typically involves the addition of moisture and lubricant (to adjust the regain of the wool after post-scour drying and to lubricate the wool in preparation for carding), blending, carding, gilling (typically three passages), combing, followed by two more gilling passages, with the tops being produced at the second post-comb gilling passage. Multiple slivers are creeled at the input of each topmaking stage, continuing the fibre blending from start to finish. Descriptions of the equipment used in topmaking can be found in Simpson and Crawshaw (2002).

Since 2002, two new combs were introduced to the topmaking industry at the 2003 ITMA exhibition held in Birmingham, namely the N. Schlumberger & CIE (NSC) ERA™ comb and the Sant'Andrea Millennium™ comb. NSC has redesigned some key elements of the comb (Schenek, 2003). The circular comb has been reduced in diameter (103 mm) and is now fully pinned around the circumference. The comb rotates at a constant speed and there are 2.86 revolutions per cycle. This provides a staggered starting point for each cycle, spreading the wear of the pins more evenly. Only fine pin segments are used on the circular comb in place of the progressive pinning on previous models. The stroke has been dramatically reduced to 45 mm. Comb speed is

260 cycles/min and through a combination of increase in speed and comb width, the claimed production rate has been increased by about 25% over NSC's previous model (Strehlé, 2004; NSC website). The Sant'Andrea Millennium comb is an evolution of the P100 which on its introduction was radically different from previous Sant'Andrea combs. Claimed comb speeds are up to 280 cycles/min through its revolutionary drawing-off carriage working cycle with differential motion and continuous feeding. By increasing the circular comb pinned arc to 290 mm improvements in cleaning action while reducing the stress on the fibres are claimed (Schenek, 2003; Sant'Andrea/ Finlane website).

4.3 Yarnspec™: predicting spinning performance

The spinner demands that a top meets what can be a long list of technical specifications as well as a price. These specifications are designed to ensure that the top will perform as expected in the spinning mill and also downstream in weaving, knitting and fabric. Some top attributes such as quality of scouring and combing, and suitable lubricant and antistatic agents, are difficult to determine from measurements of the top. Therefore, mills will usually liaise closely with a few selected suppliers who reliably supply to the mill's requirements. A few attributes such as dark fibre and contaminant levels can affect the suitability of the top downstream of spinning, though not directly impact on spinning performance. Finally, there are the fibre properties, particularly diameter and length, which are so important to spinning performance that allowed limits are chosen according to the yarn being spun.

Spinning performance and subsequent yarn performance are critical because both spinning and weaving cost, typically, three to four times as much as all of topmaking. Yarn breaks are so expensive in terms of labour cost and lost productivity that it is common for less than one break every 40 km to be allowed (Lamb and Yang, 1996). For a fine yarn being produced at 1 km/ hour/spindle this equates to a maximum of 50 ends-down per thousand spindle hours (50 EDMSH).

In order to improve the predictability of wool processing performance CSIRO has introduced a yarn and spinning prediction program – Sirolan-Yarnspec™ (Lamb and Yang, 1996). The program aims to predict what a good modern mill can expect to achieve using a particular wool top for a given yarn under the specified spinning conditions. This is a powerful and necessary tool for a closed quality control system that enables ongoing improvement and reduces error margins on cost and performance.

Sirolan-Yarnspec™ incorporates theory and algorithms derived from fits to experimental data. However, theory shows that all mills run up against the same limits so that expected yarn properties can be predicted without the need for a mill-dependent correction factor (once measurement systems are uniformly calibrated).

4.4 Worsted spinning

4.4.1 Spinning technology

At the spinning frame twin rovings are pulled off freely suspended packages, separated, and enter the drafting zone at each spindle position. Drafting control is via two synthetic aprons (top and bottom) driven by rollers of which the top rollers are recessed so that a light pressure is exerted on the drafting strand. Drafts around 20 are typical. Upon emerging from the delivery nip, twist is inserted by the rotation of a spindle. Because worsted yarns are generally fine, the ring diameters are of the order of 45 to 55 mm in diameter, and the spindle speeds are of the order of 8000 to 12 000 rpm. The twist inserted at the worsted spinning frame is usually in the Z-twist direction and depending whether knitting or weaving yarns are being spun, the delivery speeds are of the order of 12 to 20 metres per minute.

Because of the random positioning of fibres along the yarn inherent to the process, it is impractical to spin yarns having much fewer than 35 to 40 fibres on average in the yarn cross-section. Worsted spinning is principally a balance between the maximum mean fibre diameter that can be used to spin a yarn of a designated count while achieving satisfactory spinning performance and yarn quality. Beyond this, the next most important parameters in order of importance are hauteur, fibre strength, CV of diameter and sometimes crimp.

Apart from automation and electronic control, there have been a number of advances in long staple spinning technology based around ring spinning. These advances include the development of weavable singles yarn directly from the spinning frame namely Sirospun (Plate and Emmanuel, 1982a,b; Plate and Lappage, 1982; Plate, 1983; Plate and Feehan, 1983), Solospun (Prins, *et al.*, 2001) and Compact (Krifa and Ethridge, 2006; Salhotra *et al.*, 2003) spinning.

Taylor (1988) noted that in the manufacture of worsted weaving yarns, the reason for the two-folding stage is that single worsted yarns will not weave. They are too hairy, and their surface abrasion resistance is not high enough to enable them to survive the abrasion forces they experience on the weaving machine. The poor abrasion resistance arises because fibre on the surface of the yarn has a high probability of lying on the surface for a substantial part of its length, and thus can be readily abraded away. Two-folding overcomes the problem by twisting two yarns together in such a way that the surface fibres are trapped between the two yarns. The twofold yarn then has sufficient abrasion resistance to enable it to be woven. An important point to note here is that, to achieve trapping of the surface fibres, the two component yarns must not only be twisted about each other, they must also individually contain twist. Two-folding of two untwisted yarns gives a structure in which the surface fibres are not trapped, but are on the surface of the twofold yarn for its entire length.

CSIRO researchers (Plate and Emmanuel 1982a,b; Plate and Lappage, 1982) recognised the requirements outlined above for the structure of such a yarn and undertook a study of the mechanisms needed to impart individual strand-twist while concurrently folding two strands together. By feeding two separated strands together on to one spindle, they showed that if the twist equilibrium at the point at which the two strands converge was disturbed, it was possible to trap small amounts of alternating twist in each of the strands while the strands were twisted together. This produced in one operation a yarn, Sirospun™, with some characteristics of a single yarn but one in which surface fibres are adequately trapped to withstand the abrasive forces on a weaving machine.

Sirospun™ is a technology that can be retrofitted to existing ring spinning frames to produce a two-strand weavable yarn. Sirospun yarns are spun from two drafted strands of roving, spaced about 14 mm apart in the drafting zone, that are allowed to combine in the twisting zone just below the front draft rollers. To avoid spinning a single strand, each Sirospun yarn passes through a breakout device mounted above the ring rail. Because Sirospun yarns do not require any further two-folding or twisting, the splices made in these yarns needed to have sufficient strength and abrasion resistance to survive the weaving process. This requirement lead to the development at CSIRO of the Twinsplicer™ and Thermosplicer™ technologies.

During 1998 a new spinning technology, Solospun™, was released and subsequently displayed at the 1999, Paris ITMA. This technology was developed in collaboration between CSIRO, the Woolmark Company and WRONZ, based on an initial clip-on, roller attachment developed at CSIRO. As the name suggests, Solospun is a spinning technology that produces a weavable singles yarn in a single step from a single roving. The Solospun technology (Solospun Technical Manual; Anon., 1998) is a simple, inexpensive clip-on attachment to standard long-staple (worsted) spinning frames. The hardware consists of a bracket that holds a friction pad and a pair of Solospun rollers (Fig. 4.1). The bracket clips on to the shaft of each pair of top front draft rollers of the spinning frame, with each Solospun roller being positioned just below and parallel to, but not in contact with, its corresponding top front draft roller. The Solospun rollers are rotated by being in contact with the bottom front draft rollers. Unlike Sirospun, Solospun is spun from a single roving strand; therefore there is no longer a need for a double roving creel or breakout devices. The benefit of producing a fine, weavable singles yarn is the ability to manufacture lightweight pure wool and wool blend fabrics that

(a) (b)

4.1 Solospun rollers.

would not ordinarily be achievable via the conventional singles followed by folding (twisting) route.

In an interesting combination, Shaikhzadeh Najar *et al.* (2006) investigated the benefits of combining the Sirospun and Solospun spinning technologies and dubbed it the Solo-Siro spun process. Wools with two different fibre diameters were spun to a single yarn count of 40 Tex (1/25 N m) over a range of twist levels. The authors found that in comparison with a conventional singles and a Sirospun yarn, the yarn hairiness of the Solo-Siro spun yarn was significantly less. The Solo-Siro and Sirospun yarns recorded similar yarn strengths with both being stronger than the conventional singles yarn. The Solo-Siro spun yarn did not exhibit any advantage in yarn evenness. Although at the time of writing the combination may be of academic interest, it remains to be seen whether this has any practical application.

The Compact spinning system has been recognised by some authors as a revolution in ring spinning. The benefits claimed for Compact spun yarns are increased strength (at the same twist), increased elongation and reduced hairiness. This technology was primarily developed for the short staple system, but companies such as Cognetex and Zinser offer Compact systems for the long staple sector. Using an SKF-developed system, Hechtl (1996) compared conventional and Compact spun long staple yarns. This study showed that in comparison with conventional ring spun yarns both yarn tenacity and elongation for Compact yarns were significantly increased at the same twist level and that yarn hairiness was significantly reduced. It was also noted that greater production for Compact yarns could also be achieved by reducing the yarn twist level while retaining yarn strength, therefore eliminating the need to increase spindle and traveller speeds. In a study by Basal and Oxenham (2006), using 100% Pima cotton yarns and 50/50 cotton/polyester, the indication was that the rate of fibre migration as well as the amplitude of migration is higher in Compact spun yarns. These findings were attributed respectively to the minimised spinning triangle and the resultant greater fibre density associated with Compact yarns. Further discussions of these spinning technologies can be found in Simpson and Crawshaw (2002).

It is quite often desirable or necessary to impart particular yarn and hence fabric characteristics at the spinning frame because of the impracticality of blending the components prior to spinning. The most common example today is the production of stretch wool knitting and weaving yarns by the introduction of elastane filament in the core of the yarn during spinning. The introduction of a filament core is known as core spinning. For core spinning with elastane, the filament must be positively driven to provide a known tension (Invista, 2006). Typically, the core component is guided through a series of rollers and stationary guides and introduced into the drafted fibre strand immediately behind the top front draft rollers. Core spinning may involve the introduction of a wide range of filaments and even pre-spun staple yarns to impart aesthetic or technical attributes. The core component can provide strength and integrity through components such as high-strength aramid filaments. Staple fibres can also be used as the core by wrapping filament around them, providing strength and cohesion. This system is generally referred to as wrap spinning, although for many, wrap spinning may refer to the hollow spindle system whereby a twistless drafted fibre strand passes through a hollow spindle on which is mounted a filament package. As the spindle rotates the filament is unwound and wrapped around the staple fibres, imparting strength to the resulting yarn. Other methods for achieving filament wrapping are the Selfil™ and Sirofil™ systems. Selfil is based on the Self-Twist principle and Sirofil is based on the Sirospun system. Self-Twist yarn is spun by inserting alternating twist into each of two fibre strands and immediately bringing them together so that, in trying to untwist, they twist about each other. For Selfil spinning, one strand of staple fibre feed usually used in Self-Twist spinning is replaced by two continuous synthetic filaments. These are then self-twisted using two consecutive twisting systems and produce a fine, strong, torque-balanced single strand yarn at 300 m/min. In Sirofil spinning, one of the two roving strands is replaced by a filament, with the filament being introduced immediately behind the top front draft roller in much the same way as that used in core spinning. However, the spacing between the filament and single roving strand is maintained at the standard Sirospun spacing of about 14 mm prior to emerging from the front draft rollers. The drafted fibre strand and filament are allowed to combine in the twisting zone; because the filament is typically the lighter of the two components, it wraps around the staple fibre strand.

Robinson and Marsland (1984) briefly describe five variations of core spun and wrapped core spun based on the Sirospun principle using wool rovings and synthetic filaments. The five variations can be spun on conventional spinning frames and were given the following names:

1. double rove wrapped core spun;
2. double rove double core spun;
3. double rove wrapped spun;
4. single rove wrapped spun;
5. single rove wrapped core spun.

The authors concluded that these yarns showed considerable potential for knitting and weaving and that it would also be possible to spin finer yarns.

Open-end (OE) spinning, air-jet spinning and friction spinning systems are available for wool but have not to date found wide adoption. Short wool (40 to 45 mm) has been spun on the OE system but the speeds achievable are not as high as for cotton. Contaminant build-up in rotors is cited as a problem.

4.4.2 Winding and clearing

The yarn on the spinning bobbin is 'twist lively' which means that, if wound off the bobbin under moderate to low tension, it will tend to wind around itself and snarl. The bobbins are therefore steamed to impart temporary set.

Once spinning has been completed, the yarns are wound at high speed from their spinning bobbins onto larger packages for further processing. During this procedure the faulty sections of yarns are removed and the fault-free yarns are rejoined, either by knotting or splicing. If knots are used, they may fail in subsequent processing, may cause other faults in processing, or require labour for their removal during mending of the final fabric. The ultimate solution would be a yarn joint completely indistinguishable from the parent yarn, and knotting has been generally superseded by splicing. CSIRO has been involved in the development of splicing technology suitable for wool yarns, partly motivated by earlier work on Sirospun.

Splicing involves the untwisting of the fibre ends at the two yarn ends to be joined, then bringing the two yarn ends together and inserting 'twist' into the join. The splice must have the same appearance as the parent yarn (i.e. be inconspicuous) and have almost the same strength. Two splicing systems have been developed by CSIRO for worsted yarns, mechanical and pneumatic. CSIRO has licensed the mechanical Twinsplicer™ technology to Savio (Italy) and the pneumatic Thermosplicer™ technology to Schlafhorst (Germany).

In the Twinsplicer (Fig. 4.2), the yarn ends to be joined are sandwiched

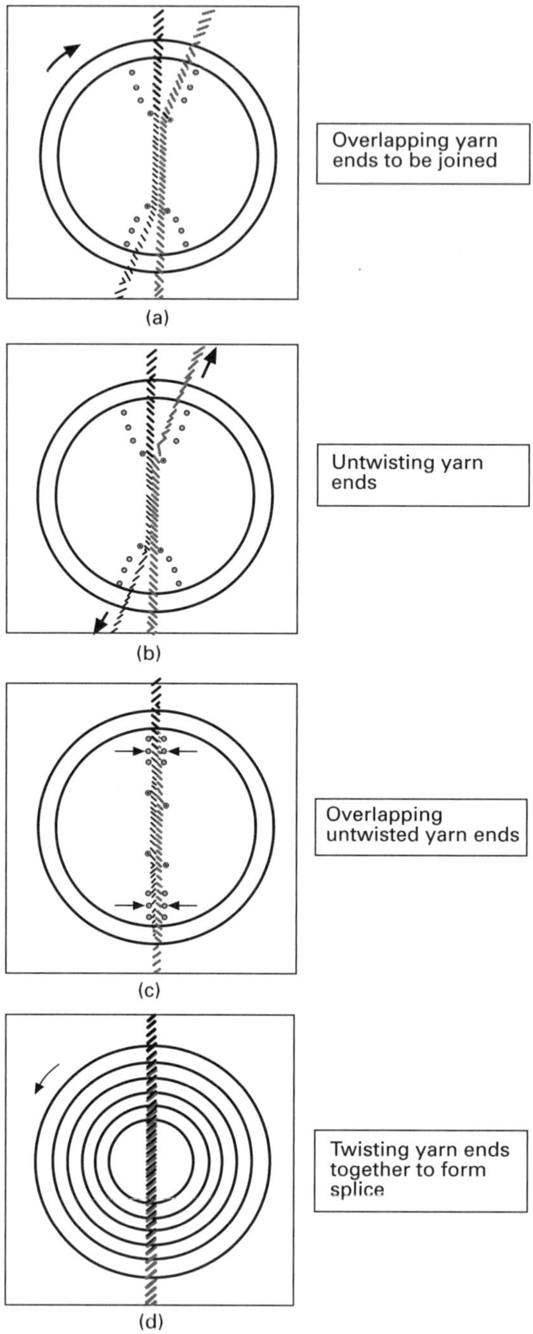

4.2 Mechanical splicer operation.

between two annular discs, which are geared together in such a way that they rotate in opposite directions around their central axes. To produce the yarn splice, the discs are first rotated to remove the twist over a short length of the two yarn ends to be joined. The untwisted ends are then overlapped and 'twist' is inserted into the join by rotation of the discs in the opposite direction. Although initially developed for wool, the Twinsplicer is primarily used for cotton yarns.

The Thermosplicer for worsted yarns (Fig. 4.3) was developed after the observation that heating wool fibres increased their flexibility. The Thermosplicer works by rapidly heating the wool fibres above their glass transition temperature during the yarn joining phase of the splicing operation. This is the temperature at which memory of past stresses is lost. The fibres become more pliable and consequently are easier to bind into the splice. The result is a stronger, inconspicuous splice. Investigation has shown that hot-air splices in wool yarns, irrespective of yarn type or state, are far more abrasion resistant than cold air splices. In weaving, cold air splices recorded the highest failure rate. During fabric inspection, hot air splices were judged to require the least levels of mender attention.

During the winding operation, the opportunity is taken to monitor the yarns for faults. Traditionally, the yarns were monitored for thick and thin

4.3 Thermosplicer.

faults. It has now also become common practice to monitor ecru yarns for coloured contaminants such as vegetable matter, dark and medullated fibres, non-wool coloured fibres and grease contamination. Siroclear™ (licensed to Loepfe) is an optical sensor incorporated into the thick and thin fault sensor to monitor the colour of the ecru yarn being wound; (Fig. 4.4). Both Loepfe Brothers Ltd (http://www.loepfe.com) and Uster Technologies AG (http://www.uster.com) incorporated sensing technology for the detection of polypropylene (undyed) in ecru yarn. The Loepfe technology is based on a triboelectric detection principle whereas Uster appears to have combined a capacitance detector with an optical detector. Keisokki Kogyo Co. Ltd (http://www.tmgoogle.com/en/Keisokki-Kogyo-Co.-Ltd.html) has also introduced an optical foreign fibre detector into their yarn clearing technology. Any coloured contaminant or foreign fibre that is detected and falls outside preset limits is automatically removed and the yarn spliced as described earlier.

The preference of course is to minimise the presence of coloured and non-wool fibres. Once blended with the wool fibres, fibre-like contaminants are almost impossible to remove. Hence, CSIRO developed a system to detect and remove coloured contaminants early in the wool processing pipeline to prevent the contaminants being blended in to the wool. This system (Dark Lock Sorter™ licensed to Loptex S.r.l, (http://www.loptex.it)) is typically incorporated in the fibre opening line after scouring. Recently, Loptex introduced polypropylene detection into their sorter by incorporating an acoustic reflection measurement system. Another contaminant detection system developed by Jossi Systems AG (http://www.jossisystems.ch) uses an ultraviolet light/fluorescence detection system in their sorter for the same application.

4.4 Siroclear.

Just as for the spinning of cotton and synthetic fibres, there has been a big move to automation in worsted spinning. Automatic doffing of full spinning bobbins has become standard where the full bobbins are removed from the spindles and replaced by empty bobbins. The empty bobbins are presented to the spinning frame on a conveyor and the full bobbins are taken away by the same conveyor. Using this conveyor system, the spinning frames can be directly linked to winders. However, one problem that has had to be overcome in worsted spinning is that wool singles yarns are normally steamed before winding to reduce twist liveliness. Several companies have introduced in-line steamers where the bobbins are transported from the spinning frame through the in-line steamer on a conveyor before being presented to the winder. At the same time, winder manufacturers have also improved their machines to allow winding of twist-lively yarns by maintaining the yarn ends under tension.

There is strong demand to bring quality control in spinning on-line but at the moment it seems that it is too expensive to be introduced on the spinning frame apart from the detection of 'ends-down'. However, on-line quality control remains an important part of the winding process. Although coloured fault detection was first developed to remove vegetable matter contamination in ecru wool, the technology has achieved large penetration in both the worsted and cotton sectors. Yarn hairiness can also be measured on-line during winding. Moreover, it is now possible, with electronic tagging of bobbins, to measure yarn quality in winding and to generate a list of individual spinning frame spindles that need attention. In general, the demand for automation is increasing in high labour-cost countries while there has been a very marked trend for spinning to move to the low labour-cost countries in Asia and Eastern Europe.

4.4.3 Twisting

Yarns for weaving, particularly warp yarns, are usually twisted or plied, although it is not uncommon to use singles yarns in the weft. Knitting yarns are almost invariably plied; however, there is a trend now for lightweight knitwear to use singles yarns. The purpose of plying is twofold. Plied yarns are much more resistant to abrasion than a singles yarn of the same count, so they will more easily resist the torture test of weaving. Knitting yarns are plied to create a balanced yarn which is not twist-lively and which will not cause spirality in the resulting knitwear.

Twisting (Simpson and Crawshaw, 2002) is now almost universally carried out using two-for-one twisters, which can take either two packages of singles yarn, or an 'assembly-wound' package which is formed by winding two yarns together. Twist is inserted by continuously looping the pair of yarns together around the package thus inserting two turns of twist for each rotation of the loop.

Plying twist is usually in the opposite direction to the singles twist, so two singles yarns of Z-twist will be plied in the S-direction. This has the effect of trapping the singles yarns' surface fibres in the structure while increasing the yarn bulk and rendering the fibres in the singles components parallel to the yarn direction. For knitting yarns the ply twist in turns per metre will be between a half and two-thirds that of the singles twist. There are variants on this for weaving yarns and, for some fabrics, even twist-on-twist yarns may be made where the ply twist is in the same direction as the singles. These are very hard, lean yarns of high density and are typically used in crepe fabrics. The two-fold yarns again require steaming to give set to the new fibre configurations and eliminate twist liveliness.

4.5 Future trends

4.5.1 Air-jet spinning

The air-jet spinning principle involves the use of high-speed rotating air in one or more chambers to impart differential twist into a fibre stream. The air jets mainly operate on the surface fibres, creating a fasciated yarn. As a result the yarns have limited application and are mainly suited to the short staple sector. The Murata Machinery Company of Japan has continued the development of the air-jet spinning technology and in 1997 introduced the Murata Vortex Spinning™ (MVS) system (Fig. 4.5). This system imparts a greater degree of fasciation in the fibre stream, giving the resulting yarn an appearance similar to ring spun yarn. The MVS system was designed for spinning 100% cotton and MVS yarns have a smooth, low hairiness finish and are consequently low pilling. Production rates of up to 400 m/min are possible with the MVS system.

The MVS system has three zones: *drafting*; *spinning*; and *winding*. The drafting zone uses a high draft ratio (up to 220) in order to reduce the high numbers of fibres in the input sliver cross-section to the number required to spin the yarn (typically between 70 and 100 for fine yarns). In the spinning zone, the drafted fibre stream is passed through an air-jet nozzle and hollow spindle to make a yarn strand. The rotating fibre balloon and resulting upward twist motion is controlled by an air vortex created around a needle connecting the path of the drafted fibre ribbon at the front rollers to the shaft of the hollow delivery spindle. Twist insertion relies upon the upper portion of some of the fibres separating from the false twist created by the air vortex and expanding outwards in the high velocity air stream as they trail the fibre assembly entering the spindle shaft. The separated fibre ends expand due to the whirling force of the air-jet stream and twist around the entrance of the stationary spindle. The fibre ends are then twisted around the parallel fibre core as they are pulled into the spindle shaft 'A' as shown in Figure 4.6(a).

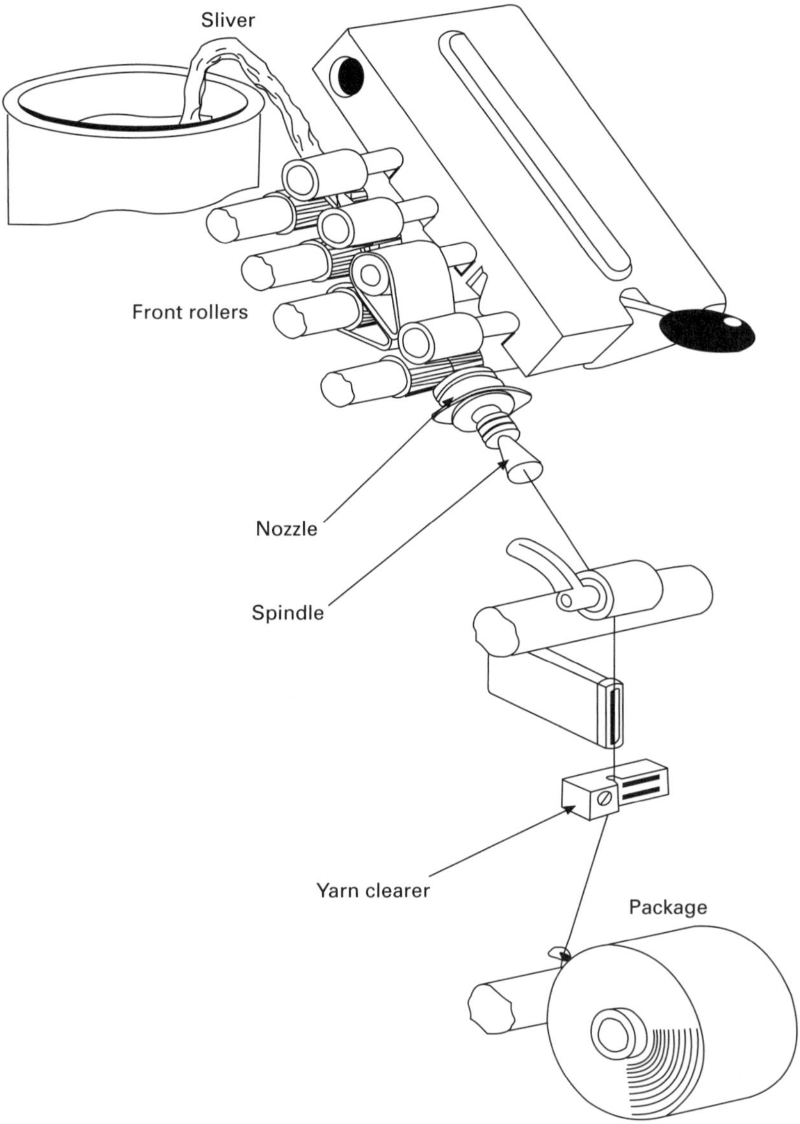

4.5 Murata Vortex Spinning system.

The yarn appearance and production speed have obvious appeal for the worsted sector and interest has been shown worldwide to adapt the MVS system to spin wool (Gordon *et al.*, 2005).

Little work has been undertaken to characterise wool MVS yarns, but a study undertaken on cotton MVS yarns (Soe *et al.*, 2004) has shown that the core fibres in MVS yarns are highly ordered and parallel and have almost no

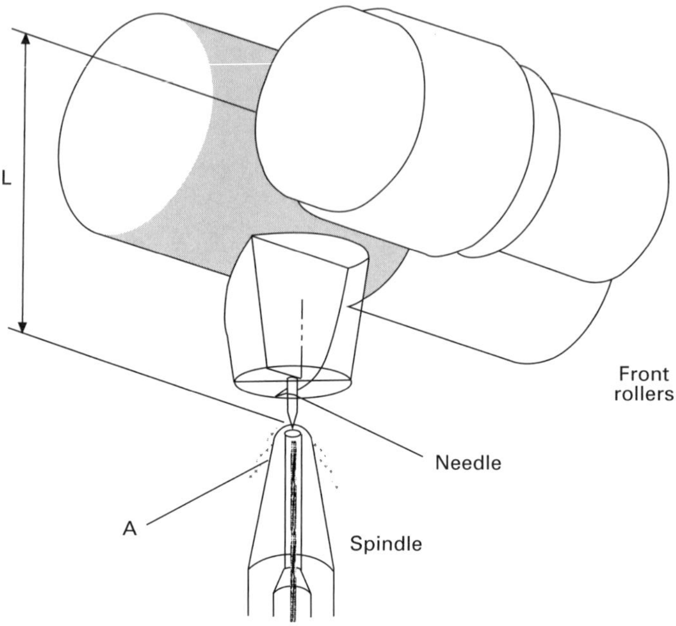

(a)

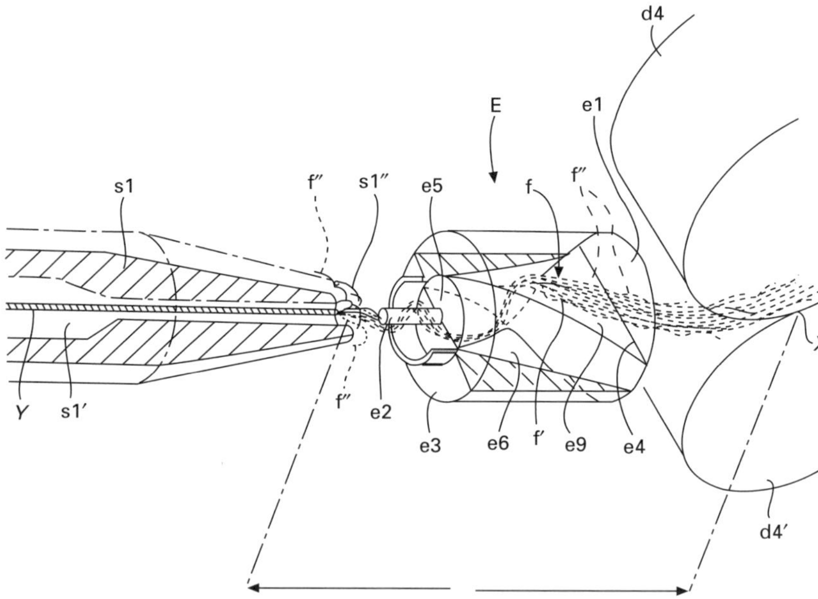

(b)

4.6 MVS twist insertion principle.

twist. In comparison with ring spun and rotor spun yarns, MVS yarns have the greatest proportion of wrapper fibres and these encircle the core fibres. In comparison to ring spun yarns, MVS yarns appear to be more uniform, bulkier and stiffer, while having a lower tenacity. Whether the properties of short staple (cotton) MVS yarns will transfer to wool MVS yarns is still to be determined, but in general MVS yarns have been shown to have good pill resistance and good dimensional stability. The yarn count range for MVS yarns is typically 1/20 N m to 1/66 N m and the fabric handle, particularly for coarse count yarns, tends to be stiffer and harsher than equivalent fabrics produced from ring spun yarns. The high production rate, however, can be quite attractive with MVS yarns, being up to 30% cheaper than equivalent short staple ring spun yarns.

4.5.2 New generation self-twist spinning

To address the knitting industry's demands for greater yarn production rates and lower input costs, the Oerlikon, Allma Volkman Company (Oerlikon) recently released a long staple yarn manufacturing technology called WinPro™ based on the self-twist principle. The WinPro™ system produces four-ply knitting yarns from conventional rovings in the yarn count range of 4/10 to 4/120 N m. WinPro™ is a two-stage system with the first being WinSpin™ which is a high-speed spinning system based on the self-twist principle and the second stage is WinTwist™ which is high-speed, two-for-one twisting. Each WinSpin™ spinning position consists of a three roller drafting system:

1. a pair of oscillating rollers to impart alternate S and Z twist;
2. a set of six idler rollers to combine the four individual yarns; and
3. yarn wind up onto a parallel-sided package.

The draft system is of sufficient width to accommodate the four roving strands which are kept separated while alternating S and Z twist is inserted into each strand. The alternating twist direction interval is about 110 mm. At the immediate output of the oscillating rollers, the fibre strands are initially combined into two-ply yarns followed by the four-ply combination by three pairs of idler rollers. It is claimed that production speeds of 250 m/min are possible. Each pair of idler rollers is offset such that the twist reversals do not coincide, improving yarn strength and appearance. The WinSpin™ operation is followed by a tailored two-for-one operation with claimed production speeds of up to 120 m/min. It is claimed the higher work to break of the 4-ply WinPro™ yarn makes it suitable for whole garment knitting due to improved knitting efficiency. With up to 50% less twist than conventional ring-spun yarn, WinPro™ is said to result in softer handle knitwear.

4.6 Sources of further information and advice

Australian Wool Testing Authority Ltd, http://www.awta.com.au/
Australian Wool Innovation Ltd, http://www.woolontheweb.com.au/LivePage.aspx?pageId=2
CSIRO Textile & Fibre Technology, http://www.csiro.au/science/Textiles.html

4.7 Acknowledgements

The author would like to thank the Australian Wool Textile Training Centre (http://www.awttc.com.au), and in particular Dr Barry Harrowfield and Mr Barry White, for allowing the use of some of the material presented at the textile training courses held at CSIRO Textile & Fibre Technology, Geelong, Australia, September/October 2006.

4.8 References

ANON (1998) Weavable singles yarn – A technological breakthrough in worsted spinning, *Wool Technology and Sheep Breeding*, Vol. 46, No. 1, 46

AWTA Ltd Research Papers (2004) Trials evaluating additional measurements – combined reports for TEAM-1, TEAM-2 and TEAM-3 1981-2004, http://www.awta.com.au/Publications/Research_Papers/Staple_Measurment.htm

BASAL, G. and OXENHAM, W. (2006) Comparison of properties and structures of compact and conventional spun yarns, *Textile Research Journal*, Vol. 76 No. 7, 567–575

GORDON, S. G., HORNE, S. L. and HORNE, F. A. (2005) The properties and structure of Murata vortex spun wool and wool-blend yarns, *Proceedings 11th International Wool Textile Research Conference*, Leeds, CD-ROM, 153YF

HECHTL, R. (1996) Compact spinning system – an opportunity for improving the ring spinning process, *Melliand Textilberichte*, April, 188–189 (E37–E38)

Invista (2006) *Producing Core-Spun Yarns Containing Lycra® Elastane Fibre*, Technical Document #120, February 10.

KRIFA, M. and ETHRIDGE, M. D. (2006) Compact spinning effect on cotton yarn quality: interactions with fibre characteristics, *Textile Research Journal*, Vol 76, No 5, 388–399

LAMB, P. R. and YANG, Y. (1996) Choosing the right top for spinning, Top Tech Papers, Presented at CSIRO Geelong Australia, 11–14 November, 258–276

NSC http://www.nsc.fr/us/nscfibretoyarn/index2.asp

Oerlikon, http://www.oerlikon.com/textile/ (Allma Volkman or fiber spinning)

PLATE, D. E. A. (1983) An alternative approach to two-fold weaving yarn. Part V: the properties of two-strand yarns, *Journal of the Textile Institute*, Vol. 74, No. 6, 320–328

PLATE, D. E. A. and EMMANUEL, A. (1982a) An alternative approach to two-fold weaving yarn. Part II: the theoretical model, *Journal of the Textile Institute*, Vol. 73, No. 3, 107–116

PLATE, D. E. A. and EMMANUEL, A. (1982b) An alternative approach to two-fold weaving yarn. Part III: testing of the theoretical model, *Journal of the Textile Institute*, Vol. 73, No. 3, 117–123

PLATE, D. E. A. and FEEHAN, J. (1983) An alternative approach to two-fold weaving yarn.

Part IV: factors affecting strand-twist, *Journal of the Textile Institute*, Vol. 74, No. 4, 204–212

PLATE, D. E. A. and LAPPAGE, J. (1982) An alternative approach to two-fold weaving yarn. Part I: control of surface fibres, *Journal of the Textile Institute*, Vol. 73, No. 3, 99–106

PRINS, M. W., LAMB, P. R. and FINN, N. (2001) Solospun – the long-staple weavable singles yarn, Proceedings of the Textile Institute, 81st World Conference, Melbourne, Australia, April, CD-ROM, WS17B

ROBINSON, G. A. and MARSLAND, S. G. (1984) Core-spun and wrapped core-spun ring yarns, *SAWTRI Bulletin*, Vol. 18, issue 4, 5–11

SALHOTRA, K. R., ISHTIAQUE, S. M. and KUMAR, A. (2003) Compact spinning, a comprehensive survey, *Textile Asia*, November, 34–41

SANT'ANDREA, http://www.finlane.com

SCHENEK, A. (2003) Long staple spinning, *ITB International Textile Bulletin*, June, 10–19

SHAIKHZADEH NAJAR, S., KHAN, Z. A. and WANG, X. G. (2006) The new Solo-Siro spun process for worsted yarns, *Journal of the Textile Institute*, Vol. 97, No. 3, 205–210

SIMPSON, W. S. and CRAWSHAW, G. H. (eds) (2002) *Wool: Science and Technology*, The Textile Institute, Woodhead Publishing, 2002

SOE, A. K., TAKAHASHI, M., NAKAJIMA, M., MATSUO, T. and MATSUMOTO, T. (2004) Structure and properties of MVS yarns in comparison with ring yarns and open-end rotor spun yarns, *Textile Research Journal*, **74**(9), 819–826

Solospun™ Technical Manual, http://www.solospun.com/tech/pop.htm

STREHLÉ, P. (2004) 50 years of combing and the worsted drawing lines, *ITB International Textile Bulletin*, March, 22–24

TAYLOR, D. S. (1988) Australian innovation in textile technology, Chapter 5 in *Technology in Australia 1788–1988*, Australian Academy of Technological Sciences and Engineering, 253–309

TEAM (1998) Report on Trials Evaluating Additional Measurements 1981–88, Raw Wool Measurement Research Advisory Committee of the Australian Wool Corporation, December

TEAM-3 Steering Committee (2004) TEAM-3 Processing trials report, International Wool Textile Organisation, Technology and Standards Committee – Raw Wool Group, Evian Meeting, May 2004, Report No: RWG02

5 Advances in wool weaving and knitting

S. GUPTA, National Institute of Fashion Technology, India

Abstract: This chapter reviews the machinery developments in the preparation, weaving and knitting of wool. Weaving preparation comprises operations such as sectional warping, creel and yarn feeding, leasing, drawing in, tying, dropper pinning and quick style changing which, though considered a subsidiary process, plays the important role of minimising the machine down time. The chapter then discusses the advances, benefits and applications of new weaving technologies in air-jet and rapier weaving. Recent radical transformations in jacquard weaving with the introduction of modularity and modern technology are also discussed. The chapter goes on to discuss advances and applications to wool knitting using flat knitting machines, computerised complete garment machines, circular knitting machines and associated computer-aided design (CAD) solutions.

Key words: yarn preparation to weaving, air-jet weaving, rapier weaving, modular jacquard shedding, computerised complete garment knitting.

5.1 Introduction

The single most important development that has influenced technology trends in the textile sector is the globalisation of textile trade. Globalisation, led by the phase-out of the Multi-Fibre Arrangement (MFA), the global movements towards the liberalisation of trade, and the entry of China into the World Trade Organization (WTO), has completely changed the way textile companies do business. The traditional producers are hence facing over-capacity in domestic production as well as low-priced imports of commodity products from low-cost countries. Their technology focus is thus shifting towards production of high value added product lines not presently threatened by imports. Low-cost countries on the other hand are moving towards increasing productivity so as to offset the price pressures and become more competitive in view of the emphasis on right first-time, right-on-time, right every-time deliveries and higher levels of service being demanded by large retailers.

Furthermore, Asia has emerged as a new consumer base of the world owing to its high population density and growth rate. Sixty per cent of all Australian wool is now processed in China. Growth in information technology has introduced this large population to changing fashion and lifestyle trends, thereby fuelling a retail boom and consumer demand for wider variety of colours, designs, styles, aesthetics, improved handle, comfort and easy-care garments. Changing lifestyles have driven the growth in casual wear, sportswear, footwear and technical textiles. This in turn has led to growth in use of a wide variety of new fabrics.

In weaving and knitting, the existing generation of technology is proving more than sufficient for most manufacturers' needs of commodity products. Since most machinery manufacturers are based in developed countries and buyers of technology are probably now in the low-cost countries, there has been a spate of mergers and acquisitions among European textile machinery makers and alliances/joint ventures with Asian machinery makers. It is also important to note that machines are no longer made specifically for wool. Furthermore, there has been an enhanced focus on environment protection, water/energy saving, versatility and flexibility in the new generation of machines.

5.2 Advances in yarn preparation

The yarn preparation-to-weaving stage is evolving all the time. By eliminating abnormalities, defects and irregularities in yarns, and by monitoring every single thread during the preparation of the warp and wefts, the yarn preparation step creates the conditions for optimal use, on the loom, of the wide range of yarns used in woollen and worsted weaving.

Furthermore, one of the most important demands in weaving today is complete utilisation of the production capacity of the weaving room with least time being lost in changing when a weaving machine runs out of warp. Each warp change results inevitably in machine down-time and the associated loss of production. Weaving preparation, positioned between warp productions and weaving mill, plays an important role in minimising the machine down-time. It includes operations such as leasing, drawing in, tying, dropper pinning and quick style changing.

5.3 Warp preparation

High-quality warp preparation prevents problems from arising not only during the weaving stage, but also later on during the downstream processing of the fabric (i.e. during dyeing, finishing and making up). Over the years, warping technology has had to evolve considerably in order to adapt to the rapid technological evolution and growing production flexibility of weaving devices and weaving machines, as the latter strive to win over new markets and new consumers to more sophisticated, higher added value textiles. Nearly all the manufacturers have developed (to varying degrees) their preparation machines, which not only match the exacting demands of the new generation weaving machines but also help to bring down running costs. A number of technologically advanced machines have emerged that have solved the problems of the past and, at the same time, introduced certain innovations for specific processes.

Ever-increasing production capacities of the weaving machines imply that the warps have to be changed at ever shorter intervals and that even mills producing only a few different articles will be forced to make frequent warp changes. Even with the use of larger warp beam flanges and increased warp lengths, warp change frequencies cannot be reduced further. Conditions in mills with a very wide range of articles are even more extreme. A wide range of articles often means that a different fabric is being produced on each individual weaving machine, and also that the piece lengths are becoming shorter and shorter. Add to this the increased weft insertion rates and the result is an increasing number of warps that have to be made ready for laying-in in the weaving room at ever shorter intervals.

It is no longer acceptable practice to use the expensive weaving machine as a knotting machine stand, and in consequence lose up to 10% absolute efficiency as a result of last pick/first pick downtime of 8 hours. A dramatic reduction in down-time is achieved with the introduction of 'divided frame' looms and the quick style change (QSC) systems. The style change is accomplished by splitting off the warp beam and associated components (such as warp-stop motion, warp tension sensor, harness frames and the reed) and replacing them with a pre-prepared new warp beam with its ends drawn in and heat sealed to a plastic sheet for quick and simple insertion into the loom take-up rollers, without causing tension problems. The removed warp beam is taken to the warp preparatory room on a special mobile module carrier and reloaded. From the moment the machine stops until the new style begins to weave takes less than 30 minutes.

5.3.1 Sectional warping

The latest generation of automatic sectional warping machines brings together solutions to all the most urgent demands of the modern weaving mill, particularly in terms of flexible solutions that will allow it to produce small quantities at reasonable costs. This is very important because the ability to process small batches is a key requirement of weaving in today's global market, where one needs to be able to operate effectively in the more profitable niche markets targeted by countries, such as Italy, which, has a highly developed wool weaving culture.

The Benninger's Versomat, for example, is an automatic sectional warper that is fast and versatile, with the ability to handle small as well as larger production batches, using all types of yarn. Such machines usually have a small footprint with an automatic leasing device that does not need permanent floor fixtures and works without an operator. Other features include a leasing device with double splittable reed, automatic control of the section tension, automatic measurement of the width of the section and of the feed pitch, the laser system for detecting threads that are out of the selection and the possibility of fitting the machine both with electrical beaming that allows extremely light pulls and reversal of the beaming direction. Most manufacturers are focusing on:

- the possibility of processing sections made up of a minimum of 16–30 threads, depending on the count;
- complete automation of option selection with a wide range of choice;
- shorter management times;
- small and large warp sections to be tied, depending on the weaving requirements;
- possibility of using the system with the machine idle, that is, with the leasing device distributing the threads on the rotating drum without horizontal motions, which results in a considerable reduction of machine dimensions;
- innovative predisposition for striped warping, with reed centring system; and
- quick machine stops in the event of thread breaks.

All these elements, newly developed, make it possible to achieve remarkable production targets that meet the needs of the modern weaving mill, which is constantly under enormous pressure to deliver quality, process flexibility and lower running costs.

5.3.2 Creel and yarn feeding

When it comes to creels, the challenge is always about improving package unwinding and tension control. The problem is made complicated by the fact that creels come in different models, according to the field of application, the size of the packages, the textile materials to be processed, the space available, and the processing methods. In general, it is possible to identify four basic models: fixed creels, creels with rotating package-holding panels, store-type creels (ideal when working with wool) and carriage-type creels. A few significant advances that have led to an increase in the productivity of these systems are detailed below.

Ionisation of the warp-stop motions

Static charge is a major problem which negatively affects the quality of the warps. To eliminate the static charge, ionisation bars are used. In a new approach, the vertical ionisation bars previously on both sides at the exit from the creel have been applied directly to the warp stop-motions. The result is that horizontal ionisation bars now cover the entire width of the thread path at every creel level. The powerful discharge electrodes emit positive and negative ions that cross the bundle of threads through the air and allow better unwinding control and complete elimination of the static charge.

Automation

Adequate automation reduces idle times considerably, thereby rendering the system more cost effective. The new modular structure allows a level of automation to be attained to suit specific requirements. The mill can opt for an automatic cutting device to reduce down-times, automatic carriage insertion to save time and effort, or an automatic tying and cutting device for optimal batch change.

Choice of thread tensioning device

There are several of these devices available on the market. Isotens devices provide optimal and even tension control, suitable for the processing of all yarn types, both filament and non-filament yarns, centrally operated and regulated, with dustproof protection, and soft thread handling. For technical yarns Dynatens-type tensioning devices allow perfect unwinding of technical yarns without damaging the fibrils, they are centrally operated (moved and adjusted) and feature long roller life. For yarns that need high unwinding tensions Megatens-type tensioning devices are available.

Waxing devices

The waxing device is positioned between the warper and the beamer. As an example, in the Versomat, provision is made for a waxing group that applies waxes and special products before the beaming stage. The device is made up of a grooved roller supported by the motorised waxing roller with a waxing tank which can be lowered. This device is useful for dampening and lubricating the threads and, above all, for binding the lateral fibrils of the yarns with special additives so as to ensure optimal weaving on the loom. For single wool worsted yarns, a size bath is used at warping to apply either hot or cold size.

5.3.3 Drawing-in, tying-in and dropper pinning

In modern drawing-in machines the process of taking the warp ends from a warp sheet and drawing-in into healds, drop wires and reed is fully automated. Drawing-in is done using robot-like machines on a special type of heddle. Heddles are separated from the stack and brought to the drawing-in position. A plastic knife opens a gap in the reed, and a hook draws-in the warp end through the heddle and reed in one step. A drawing rate of 50 000 warp ends per 8 hours (200 ends per minute) on these machines provides independence from availability of trained staff, consistent quality of drawing-in and perfect draw-in independent of drawing-in pattern, number of frames and type of yarn.

Stäubli is a leading manufacturer of these machines. The Delta series of drawing-in machine, which was launched on to the market in 1991, continues to have the highest degree of application flexibility and drawing-in capacity. Delta works without any restrictions on the harnesses which could be drawn in, that is to say the healds, drop wires, reeds and frames. No separating aids such as angled parts or keyholes are now required for separating the healds and drop wires from the feed stacks. The user is not forced to renew the harness park in the event of buying a drawing-in machine. The new Opal automatic leasing machine by Stäubli can combine several warp sheets (from one to eight sheets) of multi-colours or single colour into one lease to prepare such a warp for automatic drawing-in or tying-in. The machine is equipped with a camera system to ensure the correct yarn colour order as programmed by preventing doubles and distinguishing between threads of different colours. Depending on the application, the leasing speed of the Opal is up to 100 000 threads per 8 hour shift. Some Japanese manufacturers are offering an air-jet automatic drawing-in machine, which uses compressed air instead of needle hooks or grippers to draw in the warp yarns through the harness (droppers, healds and the reed), giving improved fabric quality by eliminating damage to the harness.

On a loom equipped with quick style change, the Delta product range is supplemented by Warplink equipment in order to rationalise and automate the processes between warp beam storage and the weaving room. The drawing-in process is followed by the process of welding a piece of plastic foil to the warp using the Warplink. Thanks to Warplink there is no longer any need for tying the warp ends to the tying cloth when laying the warp into the weaving machine. This is normally a time-consuming operation, which has to be carried out by specialists. In contrast, the laying of the plastic foil into the drawing-off rollers takes only a couple of minutes. Another advantage is that the plastic foil guarantees an even yarn tension over the whole width of the warp, making weaving start-up much simpler and less time-consuming. In this way, time savings of between half an hour and an hour can be achieved at each article change.

Tying-in

A small portable robot is used on or off the weaving machine for tying-in. A typical warp-tying machine can knot single or ply yarns of varying fineness and content at a usual rate of 60 to 600 knots per minute. The number of warp ends to be tied together can be preprogrammed; once this number is reached, the knotter stops automatically. A dual knotting system is used on a double-beam weaving machine. The knotters work from left to right and from right to left simultaneously. In recent years, knotting machines have been improved not only in mechanical reliability but also in ergonomic aspects. All machines now provide the four essential knotting alternatives, lease-to-lease, sheet-to-sheet, sheet to lease and lease to sheet. In recent years the free length of yarn between the clams has been extended by some 40% in some machines so that the same extensions in knot formation can reduce the strain and stress by the same percentage. The machine is highly flexible with a capability to tie single and double knots, to tie yarns in a count range from 0.8 to 500 tex and the reference thread tension is recorded independently and continuously for each individual warp sheet.

Dropper-pinning

The use of the open type of drop wire has increased steadily in recent years and it is estimated that 50–60% of all drop wires in current use on weaving machines and looms are of this type. Closed drop wires must be drawn automatically, semi-automatically or by hand, but open-ended drop wires can be either drawn or pinned. The pinning operation is normally carried out in the loom on drawn warps, as knotted warps will already be drawn through the existing drop wires. Some mills with a wide variety of heald frames and different types and lengths of heald would find it extremely difficult to effectively integrate a fully automatic drawing installation. Under these conditions semi-automatic drawing, automatic entering of the reed and dropper pinning of open-ended drop wires, either in a frame or the loom, provide an economical alternative. Fischer-Poege is making dropper-pinning machines that allow dropper pinning away from the loom.

5.3.4 Transportation and loading

The multi-width weaving machines have greatly increased the complexity of the materials handling problem. Two 190 cm beams, frequently 10 or more frames, six banks of drop wires and bars and commonly two or three reeds are a handling package requiring special equipment. There are significant problems associated with handling a two tonne, inconvenient and vulnerable two-part load. The yarn must not be damaged or subjected to tension during transportation from the preparatory department to the weaving machine or in the transfer operation from truck to weaving machine. Appreciating the nature of the problem and the desirability to minimise its frequency and associated lost machine time, beam flanges have been steadily increased by loom designers up to 1000 mm, but this further complicates the problem for the preparation department. An electro-hydraulic loading device is essential to transfer a heavy set of frames, drop wires and reeds into the correct location for making the appropriate connection of the machine. The time for complete transfer of a multi-width heavy dobby set using the electro-hydraulic loading device is about 20–25 minutes.

5.4 Advances, benefits and applications of new weaving technologies

5.4.1 Structures and constructions in woven wool fabrics

Because wool is a natural protein fibre, it is difficult to weave with the consistency associated with a wholly synthetic product, whose composition can be fully controlled throughout the manufacturing process. Woven wool fabrics are defined by the three categories of manufacturing machinery for production of wool yarns: woollen, worsted and semi-worsted. The category used is determined by the fabric's intended end use. Worsted is a tightly woven fabric made by using only long staple, combed wool or wool-blend yarns. Wool-like fabrics are made from worsted blends. Mohair and other animal fibres were processed on standard woollen, worsted and semi-worsted yarn manufacturing machines with occasional modifications to adjust for variations in fibre length.

Production of worsted (long fibre) yarns for weaving is expensive, due to the need to two-fold, or ply together, two singles yarns, to create yarns that survive the stresses of weaving. Typical single yarns break too easily during weaving. An alternative is to use sized ring spun yarn or use Solospun yarn made using CSIRO's Solospun™ attachment. Solospun yarns are robust, single-ply worsted yarns that withstand the stresses of weaving (or knitting). These yarns are finer, resulting in lighter fabrics and opening up new possibilities in product design.

Earlier, most woollen and other animal fibre fabrics were produced either on projectile weaving machines or on rigid/ flexible rapier weaving machines. Air-jet weaving machines are not suitable for production of heavyweight woollen fabrics, but are increasingly being used for making lightweight worsted products. Air-jet weaving technology is a good choice for weaving long runs with little variation. In all other cases rapier technology is advisable. Water-jet weaving machines cannot be used to produce fabrics of wool and similar animal fibres.

In the early 1990s, only Japan's Tsudakoma Corp. and Toyoda Automatic Loom Works Ltd manufactured air-jet weaving machines for making worsted fabrics. With the introduction of new generation, more flexible air-jet machines, the choice has widened. Sulzer continues to be the leading manufacturer of projectile machines that were widely used to produce fabrics of wool and similar animal fibres. Flexible rapier weaving machines are available from a small number of Italian and Belgian companies, while rigid rapier weaving machines are available through several European woollen fabric makers.

5.4.2 Rapier machines (rigid and flexible)

All modern rapier weaving machines are modular in design so that each machine is based on a common platform designed to accommodate future extensions or conversions. This modularity safeguards the mill's investment: no other weaving machine is so easy to adapt in order to take advantage of new market opportunities. Designed for universal use, a very wide range of fabrics, from 20 to 850 g/m^2 can be woven on these machines for uses ranging from household textiles to industrial goods. One unique advantage of rapier machines is the minimal stress on the filling yarns. Up to 16 different filling yarns, with yarn counts ranging from 5 to 1000 tex can be inserted at rates of 1400 m/min. Many other features have been added recently.

Shed geometry has been an area of attention. The short stroke of the sley and the frames and the redesigned rapier heads allow the machine to weave with smaller shed opening. Less stress is thereby put on the warp ends, resulting in increased warp friendliness. The optimised shed geometry leads to uniform fabric characteristics over the whole width. The location of the sley cams is below the fabric and this allows heavier beat-up forces, so that fabrics with high cover factors can be woven with ease.

Rapier tape guiding is another area that has received recent attention. Most machines now offer two interchangeable guiding systems for the rapier tape. The 'Guided Gripper' system offers higher speed while the 'Free Flight' system provides higher versatility. The rapier tape in the Guided Gripper system is guided by one-piece hooks. Industrial speeds are obtained through use of a small shed and a small rapier head. In Free Flight systems, the rapier tape is not guided by hooks, as the yarns can be damaged by hooks diving into the warp yarns. Rather the rapier is guided by supporting hooks for wool or worsted weaving.

Versatility is enhanced by a Quick Step filling presenter that is modular, allowing a loom to start with a four-colour insertion and to then add more channels depending on the need. The colour and weave pattern are controlled by microprocessor or jacquard. Such Quick Step modules are interchangeable, and there are no mechanical drives, so no maintenance or lubrication is required.

The same basic thinking applies to setting of warp tension with the unique Tension Sensor Feeler (TSF) device. A sensor incorporated in the ends of the whip roll eliminates unwanted effects from friction and vibration. In a similar way, the programmable filling tensioner (PFL) for the weft yarn can be programmed to add prescribed amounts of tension to the weft at precisely chosen moments during the insertion. In this way, not only are the weakest yarns treated gently, but also the weft tension in the fabric is under control. Fine tuning of the insertion settings is made very easy and accurate with the tension meter.

Ultimate quality is obtained with the knot extracting device, which makes sure that every metre of fabric is monitored to remove knots in the weft yarn. This fully automatic system detects and removes the knots in the weft yarn before they enter the fabric, so ensuring the best quality.

Rapier looms for woollen and worsted weaving are offered by manufacturers such as Dornier (HTV S), Somet (Super Excel), Picanol (Gamma 4R), Sulzer (G6200/250/300) Vamatex (Leonardo) and others.

5.4.3 Air-jet machines

Most of the recent advances have been witnessed on the air-jet weaving machines. These machines are emerging as the preferred alternative for high-speed weaving with low power consumption, flexibility and versatility with minimum maintenance and maximum profitability. Machinery designers have focused on three basic criteria: higher performance, immediate fabric quality and full modularity.

Speed or rates of filling insertions (RFI) of 2000 m/min on air-jet weaving used to be the limit a few years ago. This limit has been overcome by a number of manufacturers. The 2003 ITMA saw a number of manufacturers offering fast, wide-width air jet machines. Lindauer Dornier's air-jet machines AWSL 4/EasyLeno was shown weaving leno fabric at a speed of 450 picks per minute (ppm), or RFI of 2349 metres per minute (m/min). The machine featured a new main nozzle and positive weft clamp that allows the machine to hold the yarn prior to insertion. Dornier showed another air-jet weaving machine with dobby machine in a 200 cm nominal width that weaves the finest wool fabric with wool/elastane filling at 900 ppm. Picanol featured a wide range of air-jet machines of which at least four exceeded 2000 m/min RFI. By making the insertion cone shorter and more slender on its OMNIplus 800 loom, the load on the weft yarn was reduced, allowing higher speeds. At the same time, an entirely new 16-hole relay nozzle has been developed which, in combination with the short reaction time, enables the loom to weave more fabric with less usage of air. Promatech showed three air-jet machines, the fastest being the 3.4 m wide Mythos Tec, shown weaving at a speed of 2496 m/min RFI. Sultex featured the 2.1 m wide L5400 machine that wove a womenswear fabric at a speed of 2079 m/min RFI.

The higher speeds also demand that the frame and all moving parts are perfectly balanced and robustly built. The new hybrid harnesses are made of aluminium with carbon fibre reinforcements. Another innovation has been the introduction of variable speed weaving. It becomes important when weaving different filling yarns of different qualities. A weaker yarn needs to be run at low speed while stronger yarn can handle much higher speed. The variable speed function makes it possible to program the speed pattern according to the yarn quality, the number of harnesses and the pattern, so that a higher overall speed can be maintained – the maximum speed is no longer determined by the weakest yarn or the most complex part of the pattern.

Picanol was the first to introduce a powerful motor named Sumo. This allows the machine to be driven directly without belt or clutch at different speeds as desired. The speed of the motor is controlled electronically, without a frequency converter, reducing power consumption and permitting greater flexibility. The starting torque is very powerful and adjustable. The beat-up force is constant from the start throughout the entire weaving process. Since then, Dornier & Promatech have also showed variable speed weaving machines. The highest difference in speed was shown by Dornier and Picanol to be 100 picks/min.

Along with variable speeds these machines also offer variable pick densities that allow the weavers to produce fancy effects in the filling direction. The variation in pick density has been achieved through modularity and by controlling electronic let-off and take-up mechanisms via a microprocessor. The difference between low and high pick density (pick density range) is of the order of magnitude 11–200 picks/cm.

5.4.4 Jacquard shedding mechanism

With the high pace of innovation it is important that weaving machines offer a modular basic platform designed to receive all future extensions and adaptations, thereby safeguarding the investment in the machine. Modularity also allows easy adaptation of new developments to take advantage of new market opportunities. On most looms today the basic machine structure for the cam, dobby and jacquard versions is identical, making it possible to change the shed formation system at some time in the future. It is possible to change from cam to dobby or jacquard, and vice versa. The superstructure mountings are also identical, making it possible to decide at some time in the future to add features such as a fancy beam. On the latest versions, even the controller area network (CAN) bus technology allows new features to be added without complex extra wiring or printed circuit boards being required.

Recently jacquard weaving has been radically transformed by the introduction of modularity and modern technology. With the introduction of new jacquard shedding concepts by the two major jacquard manufacturers Stäubli and Grosse, jacquard machines are free of all mechanical constraints. In the new concept, the jacquard head is independent of the weaving machine drive since it has its own drive without mechanical coupling to the weaving machine. It sits much lower and vertically over the loom frame itself. The dimension of the head as well as that of the tie width is the same as the reed width. This allows the control of individual warp ends through harness cords set vertically and eliminates the need for magnets, hooks, pulleys, springs and the gantry. In turn, this has resulted in a smaller machine size and hence lower building and air-conditioning costs, elimination of mechanical parts leading to reduced maintenance in terms of lubrication/oil/filter change etc., quick style changes, faster speeds, and more freedom to designers.

Stäubli's Unival 100 electronic jacquard machine individually controls each end of a jacquard harness using stepping motors or actuators. The need for hooks, knives, magnets and pulleys is eliminated because each harness cord or heddle is attached directly to the motor. The selection and lifting of the harness is performed electronically and hence fabric design is achieved in the same way as any current electronic jacquard system. Electronic control allows unlimited possibilities for shed formation (symmetrical, asymmetrical, shifted, venturi effect shed, etc.) and frees the choice of shed opening and motion for each warp thread or groups of warp threads. This provides weavers with new opportunities that were not previously available in jacquard shedding. Further, the Unival is also modular in its construction where actuators are assembled in groups of 16 forming a module. The modules are mounted on aluminium rails enabling jacquard capacities in the range of 6144 to 20 480 warp threads. This modular concept allows the Unival to adapt to any loom width and yarn density. Various configurations are possible for the same number of hooks depending on the density, the width of the fabric and the available overhead clearance.

In Grosse's Unished, the shed formation has been achieved by leaf springs that are moved by actuators. The leaf springs control the bottom shed as well as the top shed (a positive jacquard shed type). Each leaf spring is connected to a heddle that controls one warp end. The configuration of the jacquard head and the individual control of each heddle (or warp end) allow the heddles to be set vertically. The jacquard head is mounted directly on the side frames of the weaving machine, thus making quick style change (QSC) possible in jacquard weaving. It is also easy to exchange the entire jacquard head, including the heddles.

Jakob Müller AG Frick of Switzerland has also introduced a harness-free jacquard shedding concept for narrow fabrics on its MDL/C label machine. The shedding concept is based on individual electronic selection of warp yarns using special heddle wires. Again, there is no traditional jacquard head, harness or comber board and no hooks, pulleys and returning springs.

Advances in weaving technology in conjunction with the new jacquard concepts have not only resulted in RFI values higher than 2000 m/min but also in high levels of versatility in terms of style change (QSC), variable speed and pick density. This means that intricate jacquard fabrics can now be woven at the same production speed as that of commodity fabrics.

5.4.5 Quick style changes and other advances

Beside increasing speed, loom makers have continued to innovate in making weaving more versatile. All major loom makers now offer the QSC system mentioned earlier, which is essential for versatility and productivity. QSC is an important advance for wool weaving as wool tends to be woven in short runs, is an expensive fabric, and quality must be maintained throughout. Grosse's QSC for jacquard reduces the time for a style change from 90 to 10 minutes, enabling short runs. Dornier's air-jet and rapier machines are equipped with a patented pneumatic shaft lock (PSL) system that permits locking and unlocking of shafts using a touch display, facilitating quick change of warps and styles in less than 30 minutes. Dornier also offers on-the-fly pattern change on their air-jet machines that allows change from, say, sheeting to napery fabric without stop. Today, pattern changing (within the same fabric style) has been completely withdrawn from the list of standard loom stoppages.

In addition to QSC, loom makers have continued to innovate by introducing many unique features for minimising non-productive time. All major loom makers, for example, offer automatic pick finding, fast frame connections, quick fastenings for warp beam and cloth roll, rapid width reductions on both left and right side and microprocessors for setting the shed and selvedge crossing times to increase the amount of time available for producing fabric at high speeds.

In fabric forming, control technologies today offer the possibility of monitoring and setting of control points like warp tensioning, pick density change, machine speed, shed closure setting, selvedge binding and even compressed air-jet opening and closing. Other options include control of scheduling and generation of reports. 'WeaveMaster' and 'Loomdata' from Barco connect the looms by means of a standard Ethernet network and utilise the standard transmission control protocol/Internet protocol (TCP/IP) to set up a connection with a company's enterprise resource planning (ERP) system. If the looms are equipped with on-loom inspection devices, the company can automatically download the weave schedules as well as upload work-in-progress reports, efficiency and quality data on stops and defect information. The WeaveMaster production monitoring system also allows automatic export and viewing of reports in a website environment, thereby allowing travelling managers to keep close control over the operations in their plants anywhere in the world via the company's website. They have access to any of the WeaveMaster reports used within the office. Weave supervisors/production managers using personal digital assistants (PDAs) with wireless capability can monitor their looms remotely. In case of a loom stoppage for example, the supervisor can select the one that has stopped, check the conditions for stoppage/defect, can override the error and turned off the machine error light. This allows the operator to restart the machine.

In other cases, the supervisor can make a decision on the difficulty of the stoppage condition and have the proper people notified. The on-loom inspection systems use either cameras or scanners to scan the entire fabric. The system can now be installed between the front fabric rest and take-up rolls. While this is an improvement over off-loom cloth roll and post-weaving inspection, more than 1 metre of fabric can pass before the system can detect it. The on-loom inspection systems can identify defects, capture and store their images, report and map defects and stop the weaving process. EVS (Elbit Vision System) offers an on-loom inspection system termed Loomtex, while Barco and Uster provide their own versions of on-loom inspection, Cylopes and Fabricscan.

Barco's automatic on-loom inspection system, which detects warp and filling defects, has the ability to generate defect maps and various quality reports when connected to Barco's QualiMaster. The defect maps contain the quality of the fabric rolls in terms of defect types, along with the location of those defects within the roll. The system has the ability to generate an XML document on a secure website. The customers can download these XML files and utilise them directly to optimise their cutting systems. Promatch offers another type of inspection system, with a new electronic eye. ITMA has demonstrated the electronic eye on one of their Lenardo Silver FTS Rapier machines. The distinguishing features of the system are its small dimensions, low cost and integration with the machine. The function of the electronic eye is not only to detect defects but also to interact with the machine and change the setting. While this may be true for some defects, other defects would not be repaired or avoided without operator intervention.

5.5 Advances, benefits and applications of new knitting technologies

5.5.1 Structures and constructions in knitted wool fabrics

Unlike woven fabrics, a knitted fabric shows stretch in all directions. Stitches can be worked from either side, and various patterns are created by mixing basic stitches with the purl either in columns (ribs), rows (garter, welting) or more complex patterns. Each of these fabrics has different properties: a garter stitch has much more vertical stretch, while ribs stretch much more horizontally. Using different combinations of knit and purl stitches, along with more advanced techniques, fabrics of variable consistency, from gauzy to very dense, from highly stretchable to relatively stiff, from flat to tightly curled, can be produced. Some more advanced knitting techniques create a further variety of complex textures like small eyelet holes and 'knitted lace' – a very open fabric resembling lace. Changing the order of stitches produces a variety of cables, honeycombs, ropes and Aran sweater patterning. Using entrelac techniques, rich checkerboard textures are formed.

Since knitted fabrics are much more elastic, dresses and lingerie are more form-fitting than counterparts made from a woven fabric. Knit fabrics can stretch from 0 to 500%, depending on their material and knitting pattern. Lace knitting generally produces the most flexible fabric, since it has large holes that can deform in shape; by contrast, cable knitting generally produces the least flexible fabric, since the stitches are crossed under tension, which inhibits deformation. The elasticity of knitted fabrics gives them an excellent drape, but this is opposed somewhat by their generally greater thickness compared to wovens. Thus, the turn of the cloth (i.e. the maximum curvature of a fold of the fabric) is generally finer in woven fabrics than in knitted fabrics. For this reason, knitted fabrics resist wrinkles better than wovens, but do not generally take a crease.

The appearance of a construction is also affected by the weight (or fineness) of the yarn, which is dependent on the fineness of the wool fibre. The thicker the yarn, the more visible and apparent stitches will be; the thinner the yarn, the finer the texture.

Knitted fabrics are generally warmer and more comfortable than woven fabrics, which is why they are worn closer to the body. Moreover, many knitted fabrics are made from wool, which stays warm even when wet; wool is preferred since it is more elastic than most fibres and produces more even, beautiful knits. In general, elasticity and warmth are opposing qualities in a knitted fabric, since the most elastic knitted fabrics, such as lace, have the largest holes and are thus less insulating.

5.5.2 Weft knitting

Flat knitting machines

In industrial knitting applications, the terms 'Flat' or V-bed knitting machine refer to two flat needle beds arranged in an upside-down 'V' formation. These needle beds can be up to 2.5 m wide. A carriage, also known as a cam box or head, moves back and forth across these needle beds, working the needles to selectively, knit, tuck or transfer stitches. A flat knitting machine is very flexible, allowing complex stitch designs, shaped knitting and precise width adjustment. It is, however relatively slow when compared with a circular machine.

Flat knit or V-bed machines have the ability to produce fully-fashioned sweaters and cardigans as well as three-dimensional structures and garments. Fully fashioned knitting machines are flat knitting machines that produce custom pre-shaped pieces of a knitted garment. Instead of knitting a whole rectangular sheet of fabric, instructions from a knit pattern on a punch card or computer file guide a fully fashioned knitting machine's needles to add or drop stitches to create custom two-dimensional shapes appropriate to the desired finished garment structure like parts of a cardigan. The pieces emerge from the machine ready to be sewn together.

Fully fashioned knitting cuts down on the amount of material required to make a garment by eliminating selvage, the remnants that would be left after cutting from a rectangular fabric sheet. For example, a sweater requires at least four pieces of fabric: two sleeves, the front piece and the back piece. Prior to fully fashioned machine techniques, a full sheet of material would have to be produced, each of the four pieces would be cut out, and the remaining fabric would be discarded. With full fashioning, the machine produces only the four required pieces.

The necessary techniques for changing the fabric width or diameter are achieved by changing the knit structure (e.g. rib to interlock), varying the structural elements (stitch length, weft insertion, knit, tuck, float), shaping through loop transfer, wale fashioning by 'needle parking', and segmented takedown for varying rates of takedown across the width of the fabric. These knit options above may also be used to change the structure of each piece to create limited curvature (such as convexity at the bust of a sweater) in the relatively two-dimensional output. Both Shima Seiki of Japan and Stoll of Germany produce these machines.

Computerised complete garment knitting machines

Complete garment knitting is a next-generation form of fully fashioned knitting that adds the capability of making a three-dimensional full garment. Unlike fully fashioned knitting, where the shaped pieces must still be sewn together, finished complete knitted garments do not have seams. The technology of knitting seamless garments on V-bed machines has gained significant commercial acceptance since its introduction in 1995. Besides offering higher comfort and better fit to consumers by eliminating seams, the innovative technique creates entire garments with minimal intervention of cutting and sewing processes, leading to substantial savings in cost and time, higher productivity, quick response and just-in-time production.

Structures that are most often made with the complete garment technique are clothing (sportswear to sweaters) or technical textiles (car seat covers and other three-dimensional shapes such as the therapeutic elastic supports for foot, ankle, knee, elbow, etc.). These machines can produce a variety of topologies that were difficult or impossible to create with knitting machines before, including: connected tubes, circles, open cuboids and even spheres (for helmet shells).

Complete garment knitting requires two needle beds for three-dimensional structures (such as clothing). As is the case with all fully fashioned knitting, machines require individual single needle selection (through electronic control) and presser feet (to hold down formed loops). Both Shima Seiki and Stoll offer comparable solutions for the knitting of complete garments on computerised V-bed knitting machines. Shima Seiki's WholeGarment range of machines has range of gauges from 5 to 18 needles per inch and knitting widths ranging from 50 to 80 inches (127–203 cm). Stoll's Knit & Wear machines have gauges ranging from E2.5 to E9.2 and knitting widths between 72 and 84 inches (182 and 213 cm). Both offer multiple gauge system, which permits different gauge areas to be knitted in one single course. The technology allows the combination of many different patterns within one garment, which cannot normally be accomplished on a traditional knitted garment. For example, a jersey knit can be placed side-by-side with a mesh knit, which can be placed side-by-side with a rib knit, which can be placed side-by-side with a jacquard knit, etc.

Stoll has filed a patent describing a technique for manufacturing seamless garments using the same needle space as the body width for their Knit & Wear machines. With the new technique, the sleeves are knitted subsequent to the body, using the needles previously used in knitting the body. Narrowing is performed without fashioning, using what Stoll refer to as the 'gore' technique. This technique is sometimes also referred to as 'flechage' or 'short-row' knitting and involves knitting successively shorter rows to narrow or impart shape into a fabric. Since carriage strokes are shorter compared with the conventional Knit and Wear technique, knitting time is reduced. Production time is reduced because the knit and transfer traverses required to join the sleeves to the body are not required in this case. Joining is carried out in the gore technique.

One of the most significant by-products of these technological advances has been mass customisation, i.e. the ability to make one garment of a particular design and eventually allow the consumer to give the final input regarding the features of that design. The volume of production in terms of number of garments made in a particular time will not be relevant in order to achieve economies of scale.

A further benefit is that knitting a complete garment on these new seamless machines allows the product to remain in digital form until it reaches the manufacturing/conversion plant. This ability to communicate digitally offers the potential to completely change the concept of a factory. Instead of a configuration where, say, 150 of these machines are working in one place, today it is conceivable to geographically distribute these same 150 machines in department stores across the city. And if there were enough demand to keep this single machine running, then the traditional factory concept would not be necessary.

While digitisation has assisted the production of woven products through design, pattern making, creation of markers and digital printing, at this stage the fabric must be cut into its component parts and sewn. At that time, the product assumes a physical form and maintains its physical form through the assembly process.

Circular knitting machines

The needle bed in a circular knitting machine is known as the *Cylinder*, the diameter of which can range from 4 inches (10 cm) across for sock/hosiery machines, to 4 feet (1.2 m) for large fabric machines. Circular knitting is faster, but has limitations in design scope when compared with a flat machine.

There are many types of circular knitting machines which produce long lengths of tubular fabric and quite often they are manufactured with very specific end uses in mind. Single jersey machines are equipped with a single 'cylinder' that produces plain fabrics (single thickness). Wool production on single jersey machines used to be limited to 20 gauge or coarser, as these gauges can use two-fold wool yarns which will produce spirality-free fabrics. Recent innovations have managed to control the spirality of single yarns with twist and yarn settings so they can be knitted into very fine next-to-skin garments on fine gauge machines. Retailers such as Icebreaker have launched a complete range of such garments. Terry loop machines are also based on single jersey machines, on which fleece fabrics are produced by knitting two yarns into the same stitch, one ground yarn and one loop yarn. These protruding loops are then brushed or raised during finishing to create a fleece fabric. Sliver knitting machines are single jersey machines that have been adapted to trap a sliver of staple wool fibre into the knit structure.

Single jersey machines can be fitted with an extra set of needles on a 'dial' which houses an extra set of needles positioned horizontally adjacent to the vertical cylinder needles. These double jersey machines allow the production of fabrics that are twice as thick as single jersey fabrics. Typical examples include interlock-based structures for underwear/base layer garments and 1 × 1 rib fabrics for leggings and outerwear products. Much finer yarns can be used as singles yarns do not present a problem for double jersey knitted fabrics as the 'double layer' construction works to cancel out the residual torque between the face and reverse sides, the net effect being no spirality.

Several circular knitting machine manufacturers are now offering machines with 44 needles per inch in both the cylinder and dial. These very fine-gauge machines appear positioned to compete with fine-gauge warp knitting machines. South Korea-based Keumyong Machinery Co. offers its KILM-108V interlock machine for underwear, swimwear and nightgown markets. Orizio Paolo of Italy is offering JOHN/TC in a 30 inch (76 cm) diameter, 40-gauge version. Orizio has developed a special thin needle of 0.30 mm for the purpose. Terrot Strickmaschinen GmbH, Germany, also has a 40-gauge single-jersey machine with a 34 inch (86 cm) diameter, 3.2 feeds per diametrical inch, four knitting tracks and a speed factor of 1020. Germany-based Mayer & Cie. has introduced a new concept machine, which employs a new 'hooked' needle without latch. The hooked needle has the action of a compound needle, but the sliding component of the compound needle is now a hooked element. Two hooks positively guided in a cam track relax the stress on the yarn and knitting elements. Referring to the process as 'relaxed knitting', Mayer & Cie. claims there is less wear on the needle, and loop formation is more reliable. The machine is available in models with up to 44 needles per inch.

Because of the increasing size of fabric rolls on the knitting machine, significant development has been undertaken to rotate the 'jumbo' fabric rolls on the same machine without increasing vibration in the knitting zone. Improved designs from Vanguard Supreme, a division of Monarch Knitting Machinery Corp., New York, and Terrot offer machines specifically designed for jumbo roll knitting. Vanguard's jumbo roll machine has a speed factor of 1500 for plain jersey fabric and 1000 for fleece. Mayer & Cie. also offers a high-speed knitting machine, Relanit 4.0, with 22 gauge and four feeds per inch that has a speed factor of 1500 and is equipped with a jumbo roll (500 pounds; 226 kg) take-up system.

Another trend evident in fine-gauge garment machines is for production of lightweight knit tops. Garment-length machines usually are offered in a 34 inch (86 cm) diameter and in gauges ranging from 12 to 18 needles per inch. Monarch offers V-AERGY, a 34 inch diameter, 18-gauge machine with a speed factor of 600, equipped with three-position needle selection on dial and cylinder, synchronised or delayed timing by an auto-timing system, and four-colour striping and knit-in draw thread. Terrot offers a computerised double-jersey transfer machine that has versatility combined with computerised jacquard patterns for better-quality sweaters, with or without separation thread. Coupled with the dual transfer from cylinder to dial, this machine can produce typical flat-knit structures with very high productivity.

In a recent advance, the individual electronic needle selection capability of the machinery has enabled production of complete or near complete garments on small and larger circular knitting machines that has long been believed the domain of the V-bed machine. Some circular knitting machinery manufacturers (such as Santoni) have since created 'the garment sensation of the decade' – the whole seamless bodywear concept for applications which range from underwear to sportswear, swimwear, leisurewear and outerwear. The concept exploits most of the known advantages in circular knitting. Significantly the technology produces garments in a gauge fineness, lightness, comfort and tactile character that the flat machine, for all its virtues, cannot yet match. Similar models are also offered by Orizio Paolo and Steiger of Switzerland. Steiger is collaborating with Lectra to offer KnitExpert software, comprising Lectra's Prima Vision and Steiger's Model software that allows for three-dimensional structure and jacquard and intarsia pattern design.

Circular knitting is also being employed to create pieces that are circular or tube-shaped, such as hats, socks, mittens and sleeves in apparel, upholstery for home/office furniture and buildings and vehicles, as well as a whole host of other technical products. They are easy to decorate with various patterns and can be made with various surface effects.

Computer-aided design

The interface between the knitwear designer and the knit technician is critical to the development of new samples. The designer's concept sketch must eventually be realised on the machine through the capabilities of the technician. Until now computerised machine makers have been focusing on the knit programming module where a trained technician creates a pattern on a dedicated CAD system that saves the data on a diskette. This is then transferred to the machine, and manufacturing starts. It is only in the past few years that the two major computerised knitting machinery makers, i.e. Shima Seiki and Stoll, have focused their energies on improving this interface.

Shima Seiki has enhanced the interpretative process in their SDS ONE CAD integrated knit production system that allows all phases including planning, design, evaluation and production. Knit designs are created in the Paint program and are simultaneously converted to programming data. Various combinations of knit structures and yarn selections are available in the design system, which allow virtual simulation without actual knitting. There is also a pattern structure database of knitted parts, such as collars, plackets and cuffs, through which parts can be incorporated or changed at will. Alternatively, Stoll's M1 CAD system has tried to establish communication between designer and technician by offering two different windows for designers and technicians who need different information for the same design. The technical window shows the developing design in the form of running yarn notations and technical data, while the design window presents the design as a knitted structure. Both windows are easily displayable and automatically convertible.

Shima Seiki, manufacturer of computerised Wholegarment® knitting machines has tested a fully integrated CAD concept in Japan. The company offers 'an original, one-of-a-kind sweater just for you' in Japan's knitwear market through their Knit Factory Boutique – a production factory and retail boutique combined in one facility. This revolutionary hybrid set-up offers complete on-demand production of customised items designed and fitted perfectly to each individual customer. The boutique is located in a three-story building. One floor contains a couple of knitting machines. Another floor has designs and samples to assist customers in making decisions regarding the products they want. The third floor is a customer lounge where garments can be tried on in a virtual CAD environment. The concept is for customers to select a design and see how it might look in a particular setting. Changes are made, and the digital file used to drive the in-store knitting machines. These steps take only about an hour or two, depending on the style, colours and patterns selected. The completed product can be picked up by the consumer at the agreed time. Prices range from 15 000 to 30 000 Japanese yen, which are comparable to high-quality items sold in department stores.

The project is being considered one of the most promising forms of retail fashion next to electronic shopping via the Internet. Once a customer places an order, his or her measurements and preferred designs can be input into a customer database, so the likelihood of repeat orders is very high. The project is not just about a profitable retail store however. It is about integration of cutting-edge technologies in knit designing, graphics programming, simulation, whole-garment programming and manufacturing, and value-addition through printing and/or embroidery. The recent innovations in CAD and seamless manufacturing have made it possible.

5.5.3 Warp knitting

Warp knitting has emerged as the most versatile fabric production system with a very favourable performance/cost ratio. Warp knitted fabrics can be produced to be elastic or stable, with an open or closed structure. They can be produced flat, tubular or three-dimensional. Fabric width of over 6 m is available without seams or even up to a multiple of this width if it is a net construction. The gamut of warp knitting includes Tricot, Raschel, Milanese and Crochet knitting machines. Stitch bonding machines are also included with warp knitting. These machines normally make flat fabrics. Knit-braiding machine or the circular warp-knitting machine is a lesser-known but useful member of the knitting machine family that can produce tubes of various small diameters from 2 to 7 inches (5–18 cm).

Tricot is very common in lingerie. Milanese is stronger, more stable, smoother and more expensive than tricot and hence is used in better quality lingerie. Raschel knits do not stretch significantly and are often bulky; consequently, they are often used as pile fabrics and as an unlined material for coats, jackets, straight skirts and dresses.

By using weft insertion on warp knitting machines to insert length-wise warp ends the extensibility of these fabrics in that direction can be reduced and strength increased. Biaxial fabrics can be made by choosing the correct structure and yarns of the same strength, to provide extensibility in both directions. A 'multi-axial' structure can also be achieved by laying on additional diagonal ends, rising both from right and from left. This complex structure of yarn sheets is stitched through by compound needles and fixed together by warp knitting. The combination of multi-directional fibre layers and matrices has proved capable of absorbing and distributing extraordinarily high strain forces and therefore find widespread industrial applications.

Double needle-bar Raschel machines have been used to produce double-sided fabrics, which are then slit to make warp knitted plush for upholstery. When talking about upholstery most people think of home textiles and similar applications for fabrics. However the automotive industry is also a huge market for these textiles. In 2003 the European automotive industry alone consumed nearly 45 500 tonnes of interior textiles for car seats, head-rests, door covers, headliners and parcel shelves. Additionally, technical textiles and semi-technical textiles are also used by the car industry in the form of fabrics for air bags, seat belts, molded carpets, V- and timing belts, and fabrics in the engine compartment (hoses, etc.). While woven fabrics dominate this sector with 51% share, circular knits are the second largest fabric group with a 19% share. Warp knitted fabrics hold the third largest share with a 12% of the market. This includes the Tricot family of knits (6%) brushed or structured, with Tricot pile sinker fabrics having a 4% share. Raschel knits represent just 0.79%. Brushed or structured Tricot fabrics are predominantly used as headliners or are found in the seat wings. Owing to their fine dense velour, Tricot pile sinker fabrics have a 4.73% share.

More recently, Raschel machines have been adapted to make spacer fabrics where the two fabric surfaces are linked by relatively thick synthetic monofilaments. Spacer knits are double-layered circular knits with a cushion of air and ‘spring-like’ yarns between the two sides. Today, these fabrics can reach a thickness of 60 mm and can be fashioned for diverse applications ranging from helmets, boot soles, composites, medical products, body armour and bedding. In another variant of this technique, conventional cotton or wool yarns are used for linking the two fabric surfaces. The space between the two fabrics can be filled with synthetic foam or sand to make insulators or other special composites. Many recent models by European car manufacturers (Citroen Grand, Lancia Argento, Smart Fortwo and Forfour, and VW Polo Fun) have used spacer fabrics primarily as seat covers. Germany-based Karl Mayer Textilmaschinenfabrik GmbH is the major machinery supplier in this sector.

Another end use for spacer fabrics is moulded bra cups. A single spacer fabric replaces three layers usually used in bra cups, i.e. the pretty ‘Face’ fabric, foam or fibrefill middle layer and the ‘lining’ fabric. Spacers mould well and can be produced in different densities, allowing production of matching wings and straps.

5.6 Future trends

In the 21st century, research institutes and enterprises involved in wool textiles and textile technology have begun to utilise diverse technological advancements in the areas of information technology, automation and robotics. These developments have led to more efficient, innovative production technologies and more uses of wool textiles. The trend will not only be to produce inexpensive articles, but rather articles containing wool with real value in practical uses.

5.7 Sources of further information and advice

The machinery manufacturers provide detailed information, not only on the equipment they manufacture, but also on the products that can be produced.

http://www.benningergroup.com/
http://www.staubli.com
http://www.fischer-poege.de
http://www.tsudakoma.co.jp
http://www.toyoda.com
http://www.sultex.com
http://www.lindauer-dornier.com
http://www.promatech.it
http://www.picanol.be
http://www.grosse-jac.com
http://www.mueller-frick.com
http://www.barco.com
http://www.shimaseiki.co.jp
http://www.stoll.com
http://www.mayercie.de/
http://www.santoni.com
http://www.steiger-zamark.com/
http://www.orizio.com
http://www.terrot.de
http://www.karlmayer.de

5.8 Further reading

GUPTA S, High performance air jets blowing up the weaving business, *J Asia Text Apparel*, **17**(3), 2007.

GUPTA S, Advances in jacquard weaving, *J Asia Text Apparel*, **16**(3), 2006.

GUPTA S, Knits advancing into technical textiles, *J Asia Text Apparel*, **15**(5), 2005.

GUPTA S, Mass customization: The way forward for seamless knitted garments, *J Asia Text Apparel*, **16**(2), 2006.

GUPTA S, Developments in data control systems for spinning & weaving, *J Asia Text Apparel*, **15**(1), 2005.

GUPTA S, Weaving preparation for cost-effective production, *J Asia Text Apparel*, **14**(1), 2004.

GUPTA S, The challenge of digital printing, *J Asia Text Apparel*, **13**(6), 2003.

GUPTA S and GUPTA D, Convergence – keyword for CAD/CAM, *J Asia Text Apparel*, **13**(2), 2002.

ADANUR S, *Handbook of Weaving*, Technomic Publishing Co. Inc., Lancaster, Pennsylvania, 2001.

SEYAM ABDELFATTAH M., Weaving and weaving preparation at ITMA'03, *J Text. Apparel Tech & Management*, **3**(3), 2003.

SEYAM ABDELFATTAH M., Weaving technology: advances and challenges II, *J Text. Apparel Tech & Management*, **3**(1), 2003.

6 Advances in the manufacture of nonwoven wool

N. FINN and R. WOOD, CSIRO Materials Science and Engineering, Australia

Abstract: A comparison is made between the mechanical and aesthetic properties of nonwoven fabrics and those of knitted and woven fabrics and the differences are described and explained. The wool types commonly used in nonwovens are described and the range of nonwoven manufacturing technologies that are applicable to processing wool are explained in detail. These include web formation via carding and cross-lapping, air-laid, and wet-laid processes. Web bonding methods include needle-punching, hydroentanglement, stitch bonding, and thermal and chemical bonding. Recent developments in these areas are also discussed.

Key words: nonwoven, carding, spunlace, cross-lap, air-laid, wet-laid, stitch bond.

6.1 Introduction

In nonwoven manufacturing the fibre is converted directly into fabric without passing through a yarn stage. Although significant cost savings due to the shorter process route make nonwovens attractive, nonwoven fabrics have inherent physical limitations and weaving and knitting will remain the favoured production processes for some time. These limitations can be understood through a comparison of properties of wovens and knits with nonwovens:

- *Wovens and knits:*
  - within yarns: intimate fibre contact, helical fibre paths – strength and elasticity,
  - between yarns: looser linkages, yarn crimp.
  - drape, bulk, good handle, fluidity.
- *Nonwovens:* Intimate fibre contact throughout is required to achieve the requisite strength and fibre security. This leads to:
  - stiffness,
  - poorer handle,
  - poor drape,
  - poor stretch recovery.

The strength and elasticity of woven and knitted fabrics are provided by the yarns, the yarn crimp and the yarn arrangement. This allows high fabric strength and good fibre security within the yarn while the fabric's flexibility and fluidity are provided by the looser links between the yarns. Nonwovens cannot easily imitate this effect because the strength and fibre security of the nonwoven is derived from fibre entanglement or bonding throughout the fabric. These shortcomings currently restrict nonwovens to certain applications. These include industrial fabrics and medical and safety apparel but some consumer apparel products are well suited to nonwovens and so research is continuing to overcome the limitations and widen the applicability of nonwovens.

6.2 Nonwoven processes and recent innovations

The nonwoven process can be broken down into several stages:

- web formation;
- bonding;
- finishing;
- coloration;
- converting.

6.2.1 Web formation

In synthetic fibre nonwovens the major web formation processes are:

- carding (with cross-lapping);
- air-lay;
- wet-laid;
- spunbond;
- melt-blown.

As wool is a staple fibre and not found in continuous form, web formation for wool nonwovens can be achieved only by carding, air-lay or wet-laid processes. These could be combined with non-staple fibre nonwovens, e.g. spunbond or melt-blown processes; a less common alternative is the air-laid process.

The wool inputs are usually similar to those chosen for woollen processing, i.e. shorter fibre than for worsted and usually free of vegetable matter and so carbonised wool or broken top is used. Broken top is shorter wools that have been combed to remove vegetable matter (VM), very short fibres and neps and is expensive compared with carbonised wool. As with woollen processing, it is also possible to piece-carbonise the fabric having made it with low-VM wools but this is rare and the VM may cause other problems in the nonwoven plant if it is not dedicated to the wool process. Blends of wool and synthetic fibres are also possible; a huge range of fibre blends are commonly used in nonwovens processing. Web formation for staple fibres is usually via a carding process a less common alternative is air-laid systems.

The schematic diagram of Fig. 6.1 shows carded web formation. The card rollers are covered with card clothing (Fig. 6.1 inset) which carries the fibre tufts through the machine. The workers rotate in the opposite direction to the main cylinder with the card wire points are also opposed. The main cylinder has high surface speed (1000 to 1500 m/min) while the workers have low surface speed (20 to 100 m/min). This action opens the tufts and individualises the fibres before forming them into webs on the workers. Stripper rollers pass the fibres back onto the main cylinder. This recycling intimately blends the fibres together. At the doffers a fraction of the fibres on the main cylinder is transferred to each doffer to form the output webs. In double doffer cards the two webs are combined by the take-off conveyors into a single web and delivered forward for further processing. Some cards have only one doffer. The double doffer allows higher productivity. The output speed is typically around 30 m/min but may be as high as 100 m/min.

For lightweight fabrics, the web produced by the carding machine goes directly into the bonding process. For heavier fabrics several cards can be positioned in line and their webs combined by being laid one upon the other onto a conveyor belt travelling beneath the machines. The fibre orientation leaving a card is usually predominantly in the machine direction. This leads to the strength being greater in the machine direction (MD) and weaker in

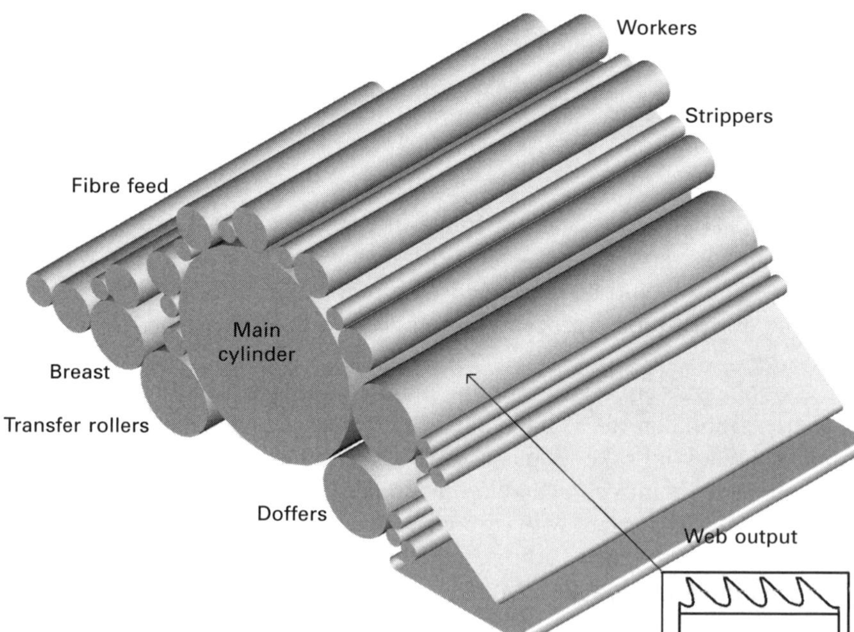

6.1 Basic non woven card. (This is the most common means of web formation).

the cross-machine direction (CMD). This results in a high MD/CMD ratio, which is undesirable for some applications such as wool apparel.

Condenser rollers such as those shown in Fig. 6.2 can improve the MD/CMD ratio by reorienting the fibres. These rollers are clothed with special wire and each one rotates at a slower speed than the prior roller so the web is passed on to the next roller with greater areal density and with more fibres oriented across the web. This also has the added effect of increasing the doffer speed while keeping the output line speed the same. The higher doffer speed takes more fibre from the card per revolution of the main cylinder and so lowers the total fibre load of the card and can allow higher production rates.

A recent development has been the random card. This card has an extra roller between the doffers and the main cylinder with special wire; see Fig. 6.3. This roller turns at high speed counter to the main cylinder and partially strips it; the transfer is said to be mostly via air-flows and a more random orientation is produced. MD/CMD is claimed to come down to 3:1 and with condenser rollers and randomiser down to 1.5:1.

The desire to increase productivity while maintaining quality has led to nonwoven cards with three doffers being developed. The three doffers further increase the rate at which fibre can be cleared from the main cylinder, allowing higher throughput because a limit is usually reached when the card overloads and excessive nep generation occurs. The three doffers also provide

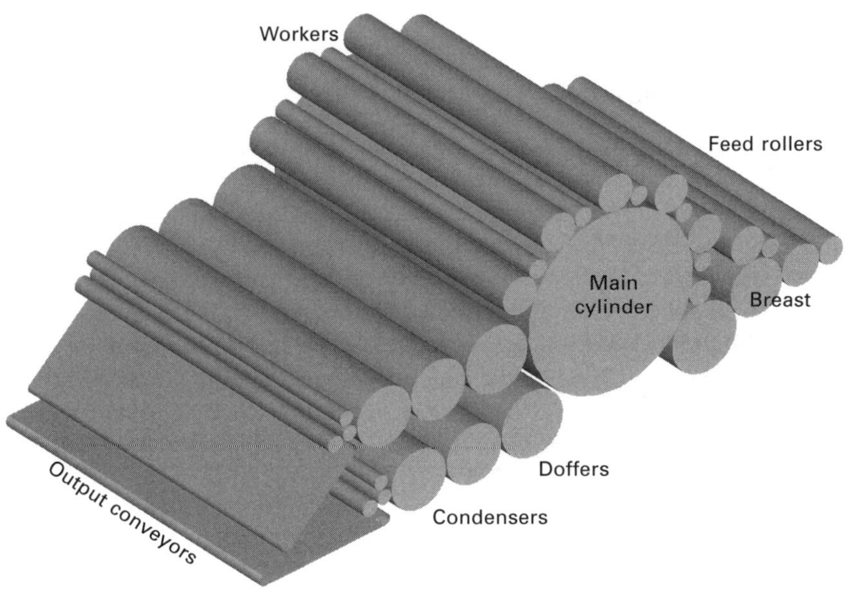

6.2 Schematic diagram of a typical nonwoven card with condenser rollers.

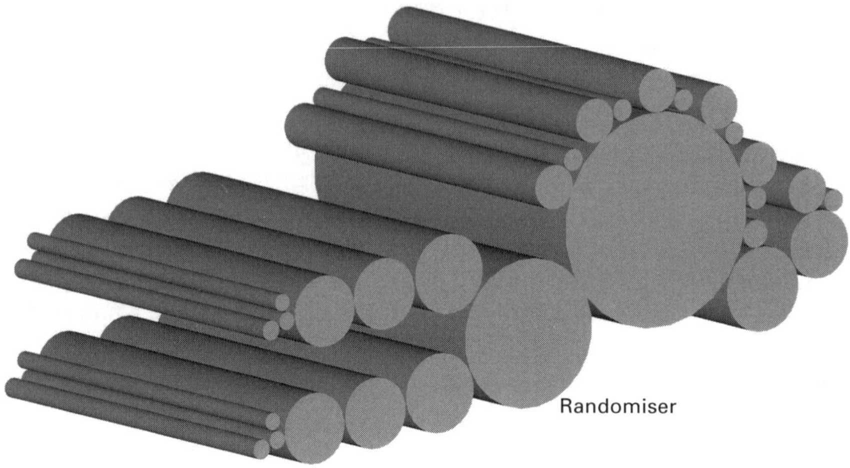

6.3 A random card. There is an extra counter-rotating roller between the main cylinder and the doffers.

three webs and should lead to further averaging of the uncorrelated random web variations. Correlated web variations, however, would be additive if they arrive at the point of combination in phase. The reduced recycling due to the three doffers means less blending and opening within the card and could lead to poorer web clarity, possibly cancelling the benefit of averaging the three webs.

6.2.2 Cross-lapping

To produce heavier fabrics from a single card, a cross-lapper as illustrated in Fig. 6.4 is often used. This device lays up the lightweight web leaving the card so that the multi-layered web leaving the cross-lapper is perpendicular to the card direction. The ratio of the card web speed to the cross-lapper output speed determines the number of layers and the weight of the cross-lapped web. In this case the predominant fibre direction is across the fabric; the MD/CMD ratio is then less than 1.

Care has to be taken to align the edges of the overlapping layers otherwise step changes in fabric density will occur at the mismatch. This means that adjustment of the cross-lapper and the card web density together is required to achieve the desired fabric weight.

6.2.3 Web drafting

The heavy cross-lapped web can be drafted, or stretched, to pull the fibre orientation towards the machine direction. Web-drafters use multiple closely

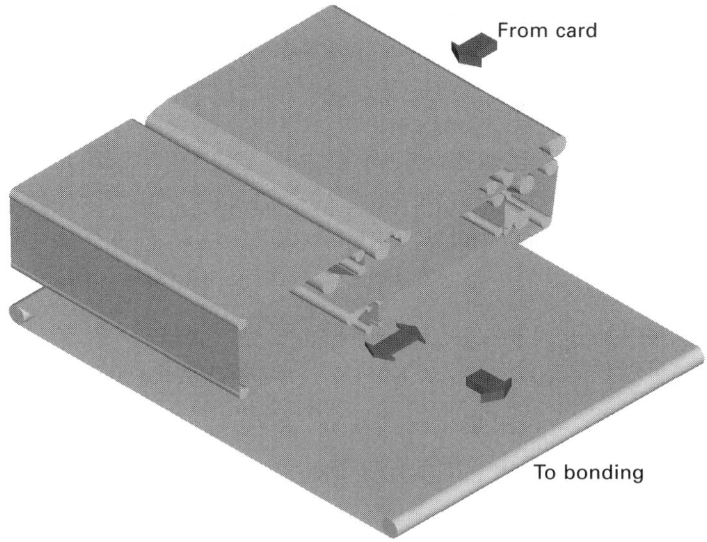

6.4 Cross-lapper.

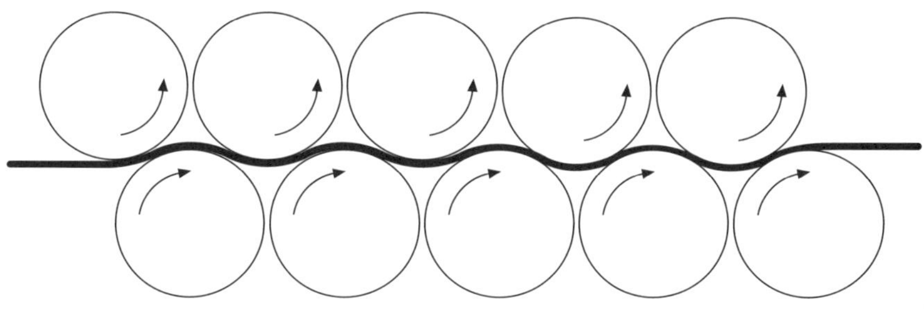

6.5 Web drafter.

spaced roller-sets moving at successively increasing speeds to draft the web in several stages (Fig. 6.5). In this way MD/CMD ratios close to 1 can be achieved but at the expense of fabric uniformity as the thinner areas tend to stretch more easily than the thicker regions. It is impossible to fully control the web because the wide rollers need to have a diameter sufficient to withstand bending under their own weight. Their diameter determines the closest distance between the roller nips and hence the degree of draft control possible. Web drafting usually occurs after some pre-needling to provide some draft control.

NSC and Spinnbau have recently developed alternatives to cross-lapping, known as the Turbo™ and Airweb™ doffing systems. These use a conventional card followed by a new system that throws the fibres into an air-stream to be collected by suction onto a drum or belt to give a randomised pattern of orientation. These devices are claimed to produce MD/CMD ratios down to

1.2 directly. The output speed can be reduced to produce quite heavy webs without cross-lapping, and high bulk is also possible with some fibre orientation in the vertical direction.

6.2.4 Air-lay systems

Air-lay processes are an alternative to carding and cross-lapping. These machines use high-speed rollers and air-flows to generate and air-lay a stream of staple fibres into a three-dimensional web with high loft. The blending action in the air-lay machine is less than a full nonwoven card as it has a small main cylinder, fewer or no workers, and little or no fibre recycling around the cylinder. However, the web properties are very different from conventional card webs, having a random orientation distribution and, in the high-loft versions, many fibres arranged in the vertical plane compared with cross-lapped webs. This can give high vertical resilience and loft with high throughput speeds. The lofty webs are usually spray bonded or thermally bonded but may be needle-punched or spunlaced. Production rates can be much higher than for the card cross-lapper or carded air-lay systems (such as Turbo or Airweb). As little blending and opening occurs in the air-lay process itself, greater attention must usually be paid to blending and opening prior to the air-lay machine.

A number of companies manufacture long fibre air-lay machines of various designs (Fig. 6.6). A process for adding short-fibre cellulose fluff-pulp (~5 mm) to nonwovens is also called air-lay but this is not relevant to wool nonwovens.

6.2.5 Wet-laid systems

Wet-laid systems are similar to air laid except for the method of distribution and transport of the fibres into a batt or web (Fig. 6.7). The mode of fibre

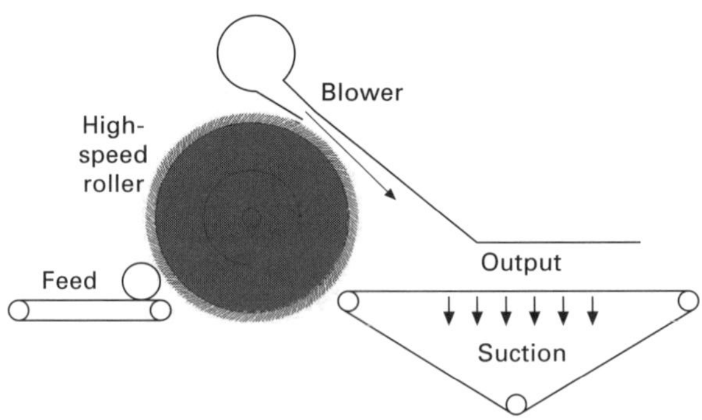

6.6 A schematic diagram of a generic air-lay machine.

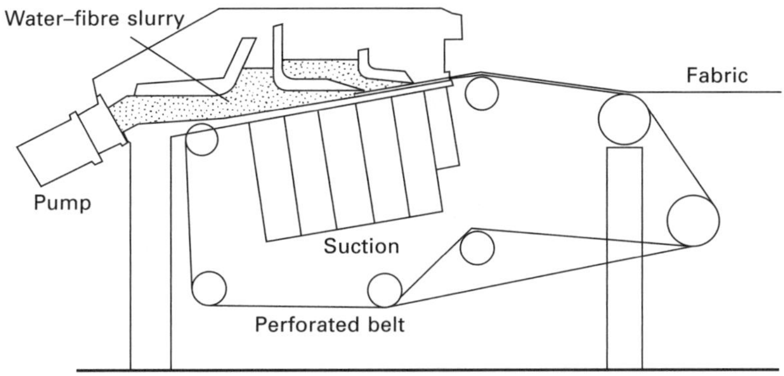

6.7 Wetlaid nonwoven process schematic.

suspension used is water rather than an air-stream. The wet-laid system is also evident in paper making. The fibres are generally very short when compared with conventional textiles and generally fall into the range of 2 to 5 mm. These fibres are suspended in water at a concentration generally below 0.005% which is much lower than for paper making systems. The web or batt is formed on a mesh or perforated belt, also known as forming fabric, where the water is removed and the short fibres are distributed randomly. The forming stage is followed by pressing and drying for bonding of cellulosic fibres. Bonding for wet laid can also be achieved using the same methods as used for other nonwovens. These are chemical bonding with the addition of a polymer binder or blending a low melt bicomponent fibre into the blend. The fabric is then passed through a bonding oven and to further finishing as required. Wet laid products have very low loft and in fact often resemble paper in their physical properties. Products that are suitable for the wet laid process include bedding, linings, medical textiles, wipes and cover stock.

Common wet-laid products are:

- medical textiles;
- interlining fabrics;
- cover stock;
- filter media;
- special papers;
- webs for printed circuit boards;
- serviettes, etc.

6.3 Bonding systems

6.3.1 Needle punch

Figure 6.8 shows the needle-punching process which is very commonly used to manufacture medium to heavyweight fabrics. The needles are used at a

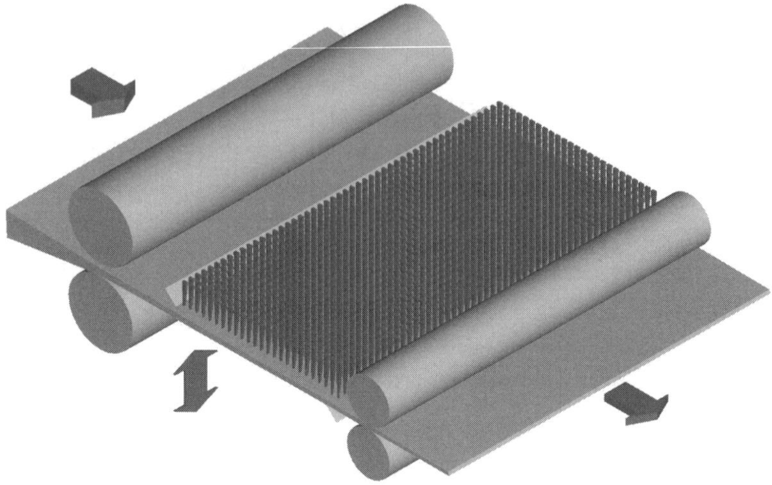

6.8 Schematic of the needle-punch process.

density of several thousand per square metre and reciprocate through the fabrics. The barbs on the reciprocating needles catch the fibres and entangle them to form a felted fabric. A stripper plate above and below the fabric has a hole for every needle and strips the fabric from the needle as it passes down or up through the plate.

Often several stages of needling from each side of the fabrics are required to give the fabric sufficient strength. Because of the reciprocating motion, the speed is limited by the 'advance per stroke', the fabric cannot be pulled forward very far while the needles are in the fabric without damaging either the fabric or the needles. Needle board frequencies have now reached 2000 rpm but are commonly lower. Needle selection is crucial to efficient felting. The shaft diameter, the barb position, the number of barbs and the needle shape are key parameters. Needle penetration depth can be adjusted on the machine.

Over-needling can lead to excessive fibre breakage and so there is an optimum level of needling with respect to fabric strength. For wool nonwoven textiles, an attractive needling process is the velour needle punch. This machine needles the fabric into a brush underneath rather than into a steel stripper-plate with a hole for each needle to strip the fibres from it. The velour needle punch pushes the fibres through the fabric to the opposite surface in a controlled way and produces a velour or velvet-like finish to the fabric. The fabric can also be patterned in this way using special arrangements of needles and brushes.

A recent development is the Hyperpunch system from Dilo (Fig. 6.9) where the needles follow an elliptical path rather than straight up and down so that they partially follow the fabric as it advances through the needle-

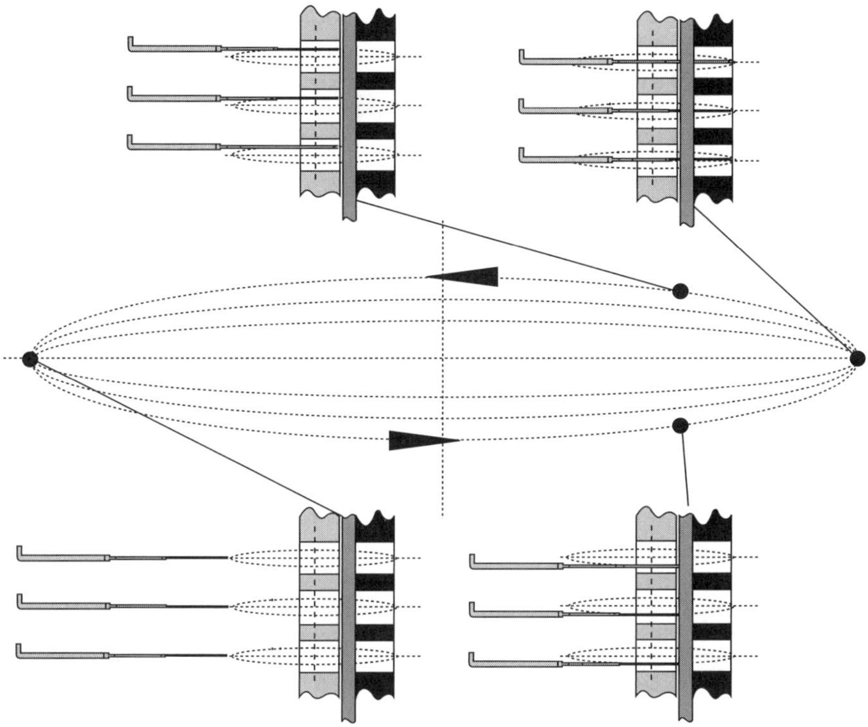

6.9 The Dilo hyperpunch system.

loom, increasing the distance advanced per stroke of the needle board. This allows higher fabric delivery speeds and hence higher productivity.

6.3.2 Stitch bonding

Because nonwoven fabrics need to derive their strength from the intimate entanglement of the fibres and this leads to greater stiffness compared with wovens and knits a compromise is reached in pure-fibre nonwovens between handle and performance. However, nonwovens can be reinforced so that the fibres can be more loosely entangled but the strength is provided by the reinforcement. One means to do this is called stitch-bonding and one example machine is the Maliwatt™ from Karl Meyer. In stitch bonding sewing threads are inserted by sewing needles aligned across a cross-lapped web. The threads provide mostly machine direction strength while the natural fibre orientation of the cross-lapped web provides the CMD strength. The sewing threads, constituting only about 5% of the fabric weight, are buried into the fabric in finishing where a pile may be raised or the fabric lightly wet-felted so that they are not visible in the final garment. Such wool and wool-blend fabrics can be used in outerwear and have a lower fabric weight limit of around 250 gsm. A fleece-like fabric can also be produced in this way.

An alternative to stitch bonding to give extra strength is the use of a 'scrim'. These are light woven fabrics that are incorporated into the nonwoven, usually by insertion between two webs before bonding. Needle punching the webs through the woven fabric can produce a strong fabric with a lower degree of entanglement and so give a softer fabric. However, the fibre security can then be low and pilling and fibre shedding can become a problem. The cost of the woven fabric often has to be low and so synthetic fibres are used and for disposables welded nets are favoured. However, for highly specified technical fabrics, such as some wool-containing paper-making felts, the scrim is a carefully designed key component rather than a cheap reinforcement.

6.3.3 Hydroentanglement

Also known as Spunlace or Jetlace, hydroentanglement uses rows of fine high-pressure water jets to entangle the fibres of the web into nonwoven fabrics as illustrated in Fig. 6.10. The water from the jets is removed by suction slots behind each injector. Because there are no reciprocating parts the production speed is not limited as needle-punching is by the 'advance per stroke'. The speed can be very high and is limited only by the energy that can be injected by the water jets into the entanglement process. Speeds of hundreds

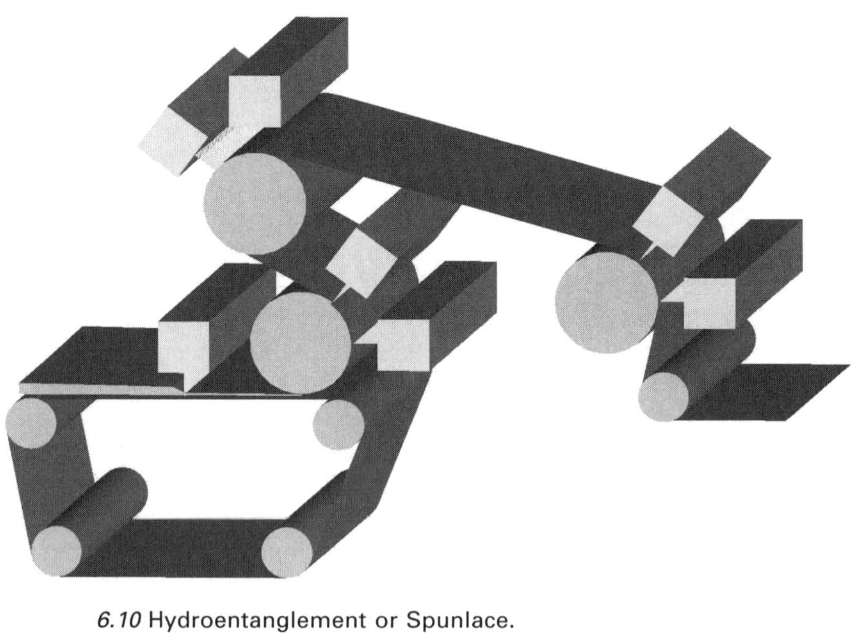

6.10 Hydroentanglement or Spunlace.

of metres per minute are used on lightweight synthetic fabrics for disposables but lower speeds are often used for heavier or more durable fabrics. Power consumption is relatively high but the energy cost per kg remains low because of the high production rates possible. Only two companies supply large-scale Spunlace lines, Rieter Perfojet and Fleissner.

The fine, high-pressure water jets are applied against the fabric backed by either a mesh belt or a drum. The drums have mesh shells or, in the case of Rieter-Perfojet, may have random perforated shells designed to improve entanglement and reduce striping by the jets. Spunlace process speeds may exceed 300 m/min but are usually much lower for wool, which is harder to entangle than some finer synthetic fibres.

Common Spunlace products are:

- wipes, towels, tissues;
- filters;
- protective apparel;
- surgical gowns and covers;
- synthetic leather;
- sanitary products;
- home furnishings;
- interlinings (some wool).

Spunlace fabric weights have an upper limit, if the fabric is to be entangled throughout its thickness, of about 400 gsm. The main advantage of the Spunlace process for wool is that lighter-weight fabrics can be produced compared with needle-punch nonwovens. Also a higher degree of entanglement can be achieved with less fibre damage than needle-punching. Reinforcing scrims can also be used in spunlace fabrics to add strength. While there is currently very little commercial production of spunlace wool fabrics, research and development is ongoing and is expected to provide commercial outcomes in the near future.

Hydroentanglement jets have very high energy density and can fibrillate fibres with weak tranverse strength compared with their longitudinal strength. This makes the process ideal for making micro-fibre fabrics from fibres deliberately designed to fibrillate. These 'splitable' microfibres are manufactured in bicomponent fibre extrusion systems where two different polymers are co-extruded into one filament with a cross-section such as that shown in Fig. 6.11.

When these fibres are formed into a web via carding the adhesion between the two polymers is sufficient to avoid fribillation during all processes up to hydroentanglement. During hydroentanglement the fibres break up into the microfibres and produce super-soft flexible fabrics with interesting technical as well as aesthetic characteristics. If blended with wool the microfibres can provide improved softness and drape compared with pure wool or normal

6.11 Citrus-style microfibres.

wool-blend nonwovens. Because of the problems of poor stretch recovery, reinforcing scrims can be used to prevent bagging of fabrics during use or wear. Such fabrics are still stiffer than woven fabrics but can be made with equivalent quality to milled wool outer-wear fabrics at lower cost. It should be noted, however, that much of the cost in wool fabrics is in the fibre itself and in dyeing and finishing processes. Thus while the full top-making and spinning processes are avoided, the savings may not be as high as expected at first glance and there is usually a quality penalty.

Many technical attributes of wool favour its use in nonwovens. These include its inherent fire resistance, odour absorption, electrostatic effects, moisture absorbency and natural image. It also does not melt, is durable and when it wears it produces a soft non-abrasive powder and so is still used in felts for protecting bearings in various machines.

The electrostatic properties can be used in electret filters where it is combined with a resin or fibre with opposite electrical affinity as defined in the triboelectric series. The charged fibres attract dust particles and so enhance the filtration efficiency but with low pressure drop and high dust-holding capacity, which are the key quality parameters for filters. A lower pressure drop means less energy is used to filter the air.

6.3.4 Chemical bonding

Latex bonding is the common chemical bonding technique used where a binder is applied to the fibre batt. The term latex is used loosely in the textiles industry for historical reasons to refer to any curable polymer binder

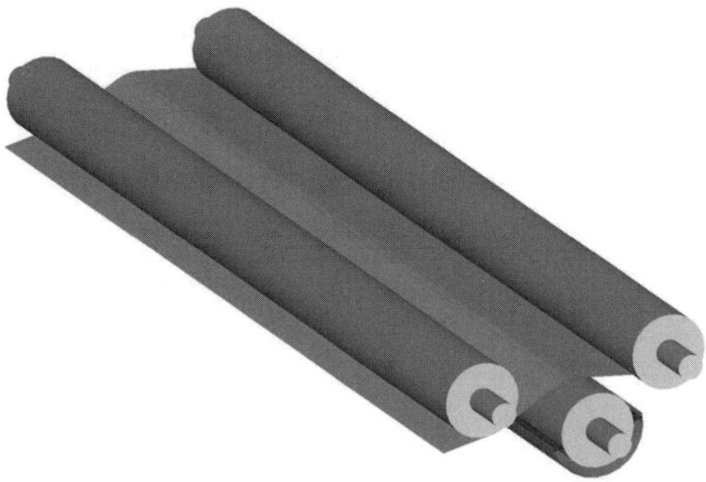

6.12 Kiss roller application (courtesy of Andritz Kusters).

rather than true latex binders. The resin application can be achieved in several ways, the most popular being kiss-roll (Fig. 6.12), saturation, foam and gravure roller application. Saturation application involves the fabric being run through a bath of the binder, the fabric wets out and excess binder is removed by passing the fabric through nip or squeeze rollers. Gravure application is by gravure rollers having a fine textured surface that picks up and transfers a layer of the resin emulsion onto the surface of the fabric and then passes the fabric through nip or squeeze rollers. Other methods include spraying of the resin as well as foam applicators. All methods require drying and curing of the binder and this is an added cost because of the high energy required to remove the water component of the binder.

6.3.5 Thermal bonding

For wool nonwovens thermal bonding is used with blends of wool and thermoplastic bonding fibres. These bonding fibres are usually sheath–core bicomponent fibres with cross-sections as shown in Fig. 6.13. The low melt-temperature component melts and wets out the other fibres during heating with hot air, usually in a through-air bonding oven as shown in Fig. 6.14. This process usually produces an open lofty structure often used for acoustic or thermal insulation. The fraction of bonding fibre used affects the stiffness, strength and resilience of the nonwoven.

In polyester bonding fibres the low melt-temperature component is usually a copolymer of the high melt-temperature polyester component. A range of melt temperatures is available from 110 to 160 °C. Polyamide (nylon) bonding

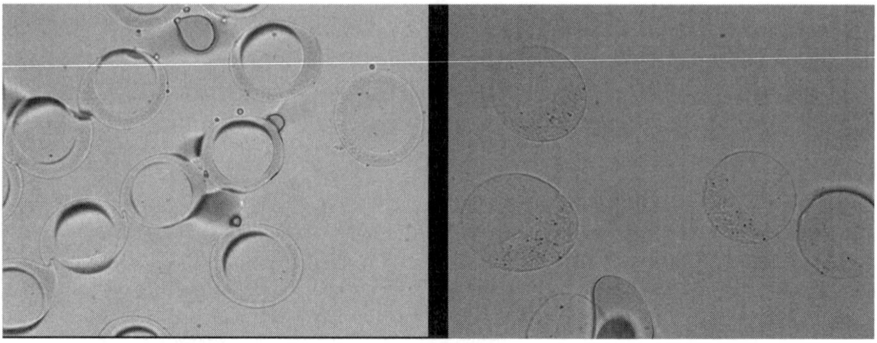

6.13 Sheath–core and side-by-side thermal bonding fibre cross-sections.

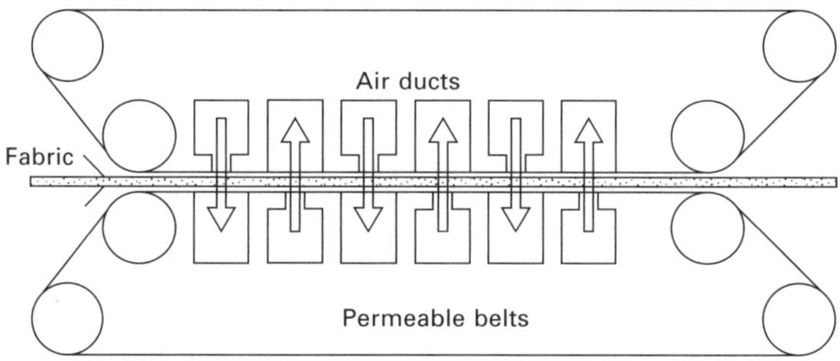

6.14 Through-air bonding oven.

fibres are available with PA6 as the low melt-temperature component and PA66 as the high melt-temperature component, the difference is only around 20 °C (about 225 versus 245 °C). Nylon is often used with wool when the fabric is to be dyed because nylon readily takes up the same dyes as wool. Polyester is cheaper and easier to thermally bond than nylon and can also be dyed, but dyeing is slightly more difficult and requires higher temperatures. Often thermal bonding is used in insulation layers such as bedding products that do not require dyeing. The wool provides good moisture management properties and some odour management effects.

Thermal bonding is also possible using heated calendar rollers whereby the fabric is passed through a high-pressure nip point and heat is transferred via conduction. A variation on calendar bonding is point bonding where a patterned or embossing roller is used (Fig. 6.15). When the raised section of the roller comes in contact with the fabric and the base roller it creates a bond only on this site, simultaneously generating a pattern in the fabric.

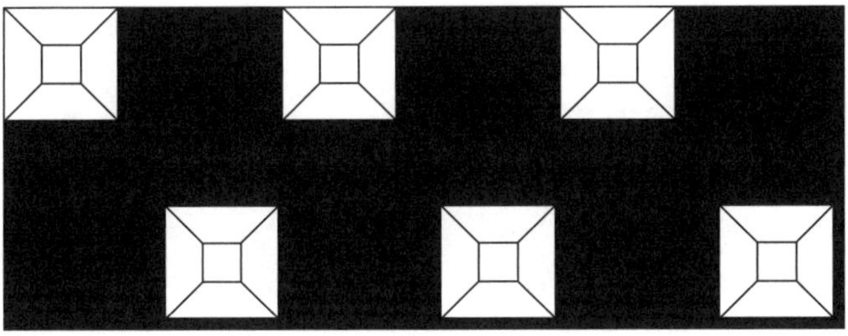

6.15 Example of a point-bonding pattern.

6.4 Finishing

Finishing of nonwovens encompasses all traditional finishing techniques as well as many more that have been adapted specifically for nonwovens and the subsequent products produced. Raising and brushing are used for the production of velour and suedes. Wet finishing is also possible just as for traditional textiles in piece form or roll goods. This enables the application of polymers to modify surface properties such as fibre to fibre friction to improve handle or to provide water repellent, stain blocking, antistatic, antimicrobial or flame-retardant properties. The range of finishing techniques is ever growing as the nonwoven industries strive to produce novel and functional textiles.

6.5 Future trends

Nonwovens continue to expand into new areas, entering new technical textile areas as well as competing in traditional textile markets. Product developers have already placed wool into new nonwoven markets, markets where wool was previously excluded because of cost. Wool has the potential to bring unique properties to nonwovens in areas as diverse as electrostatic air filters and domestic wipes. Environmental pressures are now being widely felt with industries and the public demanding environmentally sustainable solutions. This opens areas of opportunity for composite materials made of natural fibres, including wool, if price points can be met. Traditional textiles will also come under pressure from the competitive costs of nonwoven production. As research and product development improves properties such as fabric drape, stability, abrasion and pilling resistance, we will see a greater part of this sector taken up by nonwoven products.

6.6 Sources of further information and advice

The area of nonwoven textiles has been reviewed in two recent books:

- *Nonwoven Fabrics*, edited by W. Albrecht, H. Fuchs and W. Kittelmann, Wiley-VCH 2000;

- *Handbook of Nonwovens*, edited by S. Russell, University of Leeds, UK, Wiley 2006.

There are numerous bodies worldwide that can be of assistance in nonwoven research as well as industry bodies in most countries a cross-section is listed below:

- CSIRO Materials Science & Engineering (Australia)
- EDANA: European Disposable and Nonwoven Association (Europe)
- INDA: Association of the Nonwoven Fabrics Industry (US)
- TTNA: Technical Textile and Nonwoven Association (Australia)
- Leeds University, Nonwovens Research Group (UK)
- NCSU: North Carolina State University (US)
- Korea Nonwovens Industry Cooperative
- Taiwan Nonwoven Fabric Industry Association

The websites of machinery manufacturing companies provide useful background information as well as news of their latest developments:

www.fincarde.com
www.karlmayer.de
www.nsc.fr
www.rieter.com
www.spinnbau.de
www.dilo.de
www.kuesters.com
www.fleissner.de
www.neumag.oerlikontextile.com

6.7 Acknowledgement

The authors would like to acknowledge the Australian Wool Textile Training Centre (http://www.awttc.com.au), where some of this material was presented during textile training courses held at CSIRO Textile & Fibre Technology, Geelong, Australia, in October 2006.

7 Wool finishing and the development of novel finishes

J. SHEN, De Montfort University, UK

Abstract: This chapter provides an overview of recent developments in wool finishing. Since wool fabric finishing is a very extensive topic, this review focuses on functional finishes such as shrink-resist, flame-retardant, insect-resist, stain-repellent and stain-release finishing. In recent years, there has been considerable interest in the use of enzymes to achieve a variety of finishing effects on wool. Incorporation of enzymes into wool wet processes to improve wool fabric softness, reduction of prickle and shrink resistance are reviewed.

Key words: wool finishing, shrink resistance, flame retardancy, mothproofing, stain repellence and stain release, enzyme application.

7.1 Introduction

Wool fabric finishing can be categorised into two main areas: chemical finishing and mechanical finishing. The aim of finishing is to improve the serviceability of wool fabrics and/or to produce the desired properties of fabric in order to meet the specified requirements of consumers and end uses.

There are three main objectives to be achieved during the finishing of wool fabrics:

1. Removal of contaminants or impurities from the fabrics by scouring.
2. Improvement of the required properties of wool fabrics for improved serviceability and durability. This includes handle, softness, fullness, control of dimensional stability by decatising and relaxation shrinkage.
3. Application of functional finishes for specific end uses including shrink resistance, flame retardancy, waterproofing, insect resistance, stain repellence and stain release.

Wool fabric has natural warmth and a hydrophobic character which means it is an ideal choice for use in outerwear garments, as the fabric possesses water-repellent properties. Wool finishing can improve the existing properties and/or add the desired functional properties to wool fabrics to meet modern consumer requirements and the needs of specific end uses. General finishing processes for wool are scouring, fulling, antistatic treatments, relaxation shrinkage, anti-felting and functional finishing such as flame retardancy, mothproofing, stain repellence and stain release, among others. These properties are conferred in a sequence of processes carried out in a logical order. However, the order in which they are applied may vary, depending on fabric type, desired appearance or properties.

As it entered the 21st century, the textile and garment industry began to focus more on comfort, environmental protection and functionality. A wide range of wool finishing processes has being developed. The demand for environmentally friendly finishing processes has greatly increased, for example AOX (absorbable organic halide)-free shrink-resist finishing for machine washable wool and PFOS (perfluorooctane sulfonate)-free stain-repellent finishes. Advanced nanotechnology is being employed to produce wool with multi-functional properties. Micro-encapsulation sensory perception technology enables fragrance or active ingredients such as aloe vera or vitamin E skin care as well as medical care products to be incorporated into garments. These will be released from wool clothing or household textiles over a period of time.

This chapter provides an overview of recent developments in wool finishing. Wool fabric finishing is a very extensive topic, so this review will focus on functional finishes such as shrink-resist, flame-retardant, insect-resist, stain-repellent and stain-release finishing. In recent years, there has been considerable interest in the use of enzymes to achieve a variety of finishing effects on wool. Incorporation of enzymes into wool wet processes to improve wool fabric softness, reduction of prickle and shrink resistance are briefly reviewed.

7.2 Dimensional stability and shrink-resist finishing for machine-washable wool

Changes in fabric dimensions during use can be a major defect and are often the cause of unserviceability and consumer complaint. There are two types of shrinkage (Saville, 2000): relaxation shrinkage and felting shrinkage. Relaxation shrinkage is the first stage of shrinkage and occurs when the stresses or strains introduced during the processing of textile materials are relaxed in water, or water plus detergent. Felting shrinkage is non-reversible and is caused by the progressive entanglement of wool fibres through mechanical action during washing. This tendency of wool to felt prevents the use of untreated wool materials as machine-washable textiles.

The physical structure of the scaly cuticle layer of the wool fibre is considered to be the main reason for felting. The modification of the wool surface either by degradative methods and/or by application of a polymer on the wool surface is the major approach to achieve anti-felting properties of wool.

Electron scanning microscopy shows that wool fibre is covered by a thin sheath of overlapping scales like the slates on a roof (Fig. 7.1). These scales are responsible for the felting and shrinkage that occurs during laundering. This is due to the differential frictional effect (DFE) caused by the difference

7.1 Scanning electron micrograph of wool fibres.

in friction between μ_a (friction coefficient when rubbing in the against-scale direction) and μ_w (friction coefficient when rubbing in the with-scale direction). Finishers can make use of this property in a process known as milling to achieve consolidation of the fabric and make it fuller and denser. On the other hand, felting shrinkage can be detrimental to the performance of wool fabrics. Shrink-resist finishing is required to give wool fabrics and garments machine washability.

An understanding of the structure of the cuticle cells in the surface scale of fibres is necessary for the development of successful shrink-resist finishes. The wool cuticle is about 400 to 500 nm thick. The cuticle cells comprises an endocuticle (120 to 180 nm thick), an exocuticle (150 to 200 nm thick) and an outermost epicuticle layer (5 to 7 nm thick) (Fig. 7.2). The epicuticle layer contains 18-methyleicosanoic acid covalently bonded to the surface of wool fibre via a thioester bond (called the F-layer) to form a hydrophobic barrier. The exocuticle layer contains a high proportion of crosslinked disulphide and isopeptide bonds, resulting in the resistance to attack from alkaline agents and proteolytic enzymes. The endocuticle, a layer lying below the exocuticle, has a relatively low crosslink density (3% half-cystine) and thus is easily permeable and more susceptible than the exocuticle to chemical attack (Feldtman *et al.*, 1983; Mori and Inagaki, 2006).

Anti-felt or shrink-resist finish processes have been developed for decades in order to obtain easy-care (machine-washable) wool products. Commercially

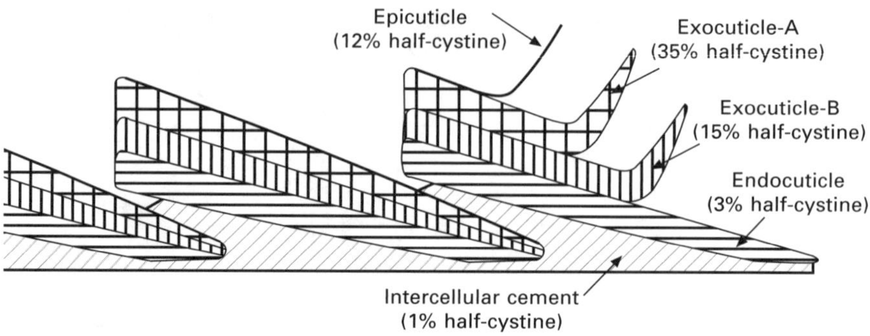

7.2 Schematic diagram of wool cuticle (Feldtman *et al.*, 1983).

successful shrink-resist processes used by the textile industry in the past, and technologies currently being developed, can be divided into four groups:

1. Combination of oxidation and additive polymer processes.
2. Additive polymer only processes.
3. Plasma treatment followed by resin polymer or softener finishing.
4. Enzymatic processes (Section 7.6).

Many useful reviews of wool shrink-resist processes are available (McPhee and Shaw, 1984; Byrne, 1996; Holme, 2000, 2007a). It is useful to distinguish between shrink-resist treatments applied at loose fibre stage to either scoured wool or to sliver, and to shrink-resist treatments applied at fabric or garment stage. The loose fibre treatments rely on reducing the differential friction effect either by degrading the scale structure or by masking the scales. The fabric treatments generally rely on at least a degree of interfibre bonding to reduce interfibre movement.

Of the loose fibre treatments, the chlorine-Hercosett shrink-resist technology has been the leading process for the continuous superwash treatment of wool tops since its early development in the 1960s. In this procedure wool top is first chlorinated using acid hypochlorite or chlorine gas (the Kroy process), anti-chlorinated with sodium sulphite, neutralised and passed through a Hercosett solution (a polyamide-epichlorohydrin polymer) followed by softener application; all these steps are accomplished continuously in a suction drum back-washer line. The chlorination treatment removes covalently bound lipids from the wool surface and also oxidises surface cystine disulphides to cysteic acid (RSO_3^- residues); these dual effects create a more hydrophilic and anionic fibre surface. The latter surface is highly receptive to the cationic polymer, Hercosett, which 'exhausts' on to the surface of each individual fibre and self-crosslinks on drying. The latter process ensures that Hercosett polymer is fixed and garments made from the treated wool show long-lasting shrink resistance in subsequent machine washing cycles (Lewis, 2005). The

AOX generated in the process arises from the use of chlorine as an oxidant and also from the resin; however, low-AOX grades of Hercosett are available. It is important to note that there are no free phenols in wool and the potential for generation of highly toxic AOX species such as dioxins is low (Shaw, 1990).

Dichlorodicyanuric acid (DCCA) is a major alternative chlorination reagent for imparting shrink resistance to wool. DCCA reacts to release hypochlorous acid which acts as an oxidising agent (Veldsman and Swanepol, 1971; Levene and Cohen, 1996; Cardamone and Yao, 2004; Cardamone *et al.*, 2004a).

Although the chlorine-Hercosett processes impart excellent and robust shrink resistance to wool tops, pressures caused by AOX legislation and general concerns about the use of chlorine remain. New approaches are highly desirable to remove the problem of AOX. Currently, effective alternative resins are being developed. The ideal resin should be effective on wool top as well as on fabric, be linked permanently to the fibre surface; its application should be water based, any AOX contamination in the effluent must be avoided, and characteristic wool properties must not be changed, e.g. avoidance of yellowing.

Alternative commercial resins that have been proposed for oxidation/resin shrink-resist finishing are Listrilan SR from Stephenson Speciality Chemicals; Polymer DP5570 from Devan-PPT Chemicals, Ambergate, UK; Lanaperm VPA from Clariant; and Beetle Resin PT763 from BIP, among others. The resin Listrilan SR is applied in continuous shrink-resist processing for wool tops from a suction drum backwash bowl immediately after rinsing following the intermediate pre-treatment stages which include chlorination, antichlorination, neutralisation and rinsing (Stephenson technical information). The polymer, DP5570, is a reactive acrylic copolymer. It is claimed to be able to compete economically with the Hercosett polymer, but offers additional benefits such as non-yellowing, natural softness and control of hairiness/facing/pilling that occurs during washing and has become a problem with the standard process. The addition of low polymer solids is all that is required to fully meet easy-care standards, with the benefit of a reduced polymerisation temperature leading to energy saving (Benisek, 2006; Devan-PPT technical data sheet). Other natural polymers such as collagen (Hesse *et al.*, 1995b) and chitosan (Julia *et al.*, 1998, 2000; Pascual and Julia, 2001; Roberts and Wood, 2001) have been suggested as alternatives to the Hercosett process.

Alternative oxidation agents to prepare the surface of the wool to allow adhesion and spreading of the polymer have been widely investigated. Potassium peroxymonosulphate (Caroat), the oxidative constituent of which is permonosulphuric acid (H_2SO_5), has received the greatest attention. However the extent of oxidation is much less and the level of shrink resistance is sometimes inadequate, especially for low-twist yarns.

An AOX-free continuous shrink-resist treatment for wool tops (the Perachem process) has recently been developed at the University of Leeds, UK (Holme, 2007a; Lewis and Hawkes, 2007). This consists of a preliminary step in which the surface energy of the wool fibres is increased by the partial removal of the hydrophobic covalently bound surface lipids. This is achieved by treatment with a nucleophilic agent (e.g. *N*-hexadecyltriammonium bromide) under alkaline conditions. This step is followed by oxidation of the cystine bonds in the A-layer of the exocuticle (see Fig. 7.2) and sulphitolysis with sulphites to generate Bunte salt groups (—SSO_3^-). These groups create a negatively charged wool fibre surface and cause cuticle swelling which also contributes to the overall shrink-resist effect of the treatment. The last step is exhaustion of a cationic low-AOX Hercosett resin. The whole process is carried out in a continuous six-bowl treatment. It is claimed that this patented process results in a wool that is whiter and softer than that achieved by conventional chlorine-Hercosett treatments while the dyeing properties remain the same.

For treatment of wool fabrics, a number of additive treatments alone using polymeric finishes for machine washable wool have been commercially applied. Currently available synthetic polymers include the surface-active Bunte salt polymer (Securlana from Cognis); silicone rubber with silicone-modified polyurethane (Dicrylan 7702 from Huntsman Textile Effects, formerly Ciba); reactive polysiloxane-based softener (Arristan 64 from CHT Group), among others. The aim in this process is to achieve adequate adhesion of the polymer to the fibre surface so that the interfibre bonds hold during the washing process, but at the same time to avoid deterioration of the fabric handle caused by the interfibre bonding.

Traditionally the most successful commercial polymer for fabric treatment has been the Sirolan BAP process, in which a water-soluble bisulphite adduct of a polyether polyisocyanate (Synthappret BAP) is mixed with a polyurethane dispersion. Synthappret BAP is a self-crosslinking polymer containing carbamoyl sulphonate groups that react readily under alkaline conditions (Fig. 7.3). This mixture is made alkaline with sodium bicarbonate, padded on to wool fabric, and the polymer crosslinked on the fabric surface by stenter curing at elevated temperatures (150 °C) (Guise and Jackson, 1973; Guise, 1977; Cook and Fleischfresser, 1985).

Bunte salt (—SSO_3Na) polymer finishing for shrink-resistant wool has been reported in a number of publications (Bell and Lewis, 1975; Lewis, 1977, 1982, 1999). Preparation of a Bunte salt polymer is based on esterification of the relevant alcohol with chloroacetic acid, followed by reaction with sodium thiosulphate. This results in the production of a Bunte salt polymer with mono-, bi- and trifunctionality (Fig. 7.4). This polymer can be applied by a pad–dry–heat cure procedure. The reaction between Bunte salt groups and wool cysteine residues occurs at the curing stage to form polydisulphide

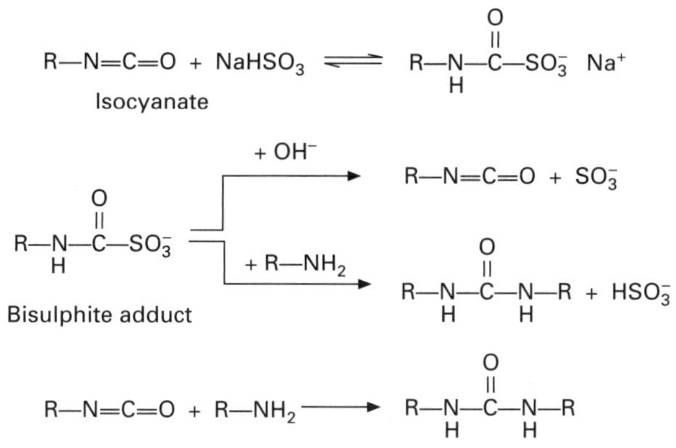

7.3 Chemical reactions of isocyanate and bisulphite adduct (Guise and Jackson, 1973; Guise, 1977).

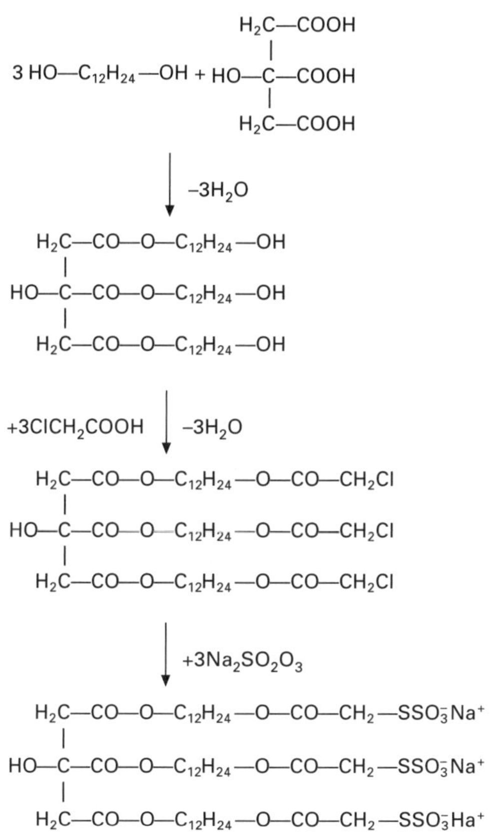

7.4 Preparation of a Bunte salt polymer with trifunctionality (reproduced with permission from Lewis, 1999).

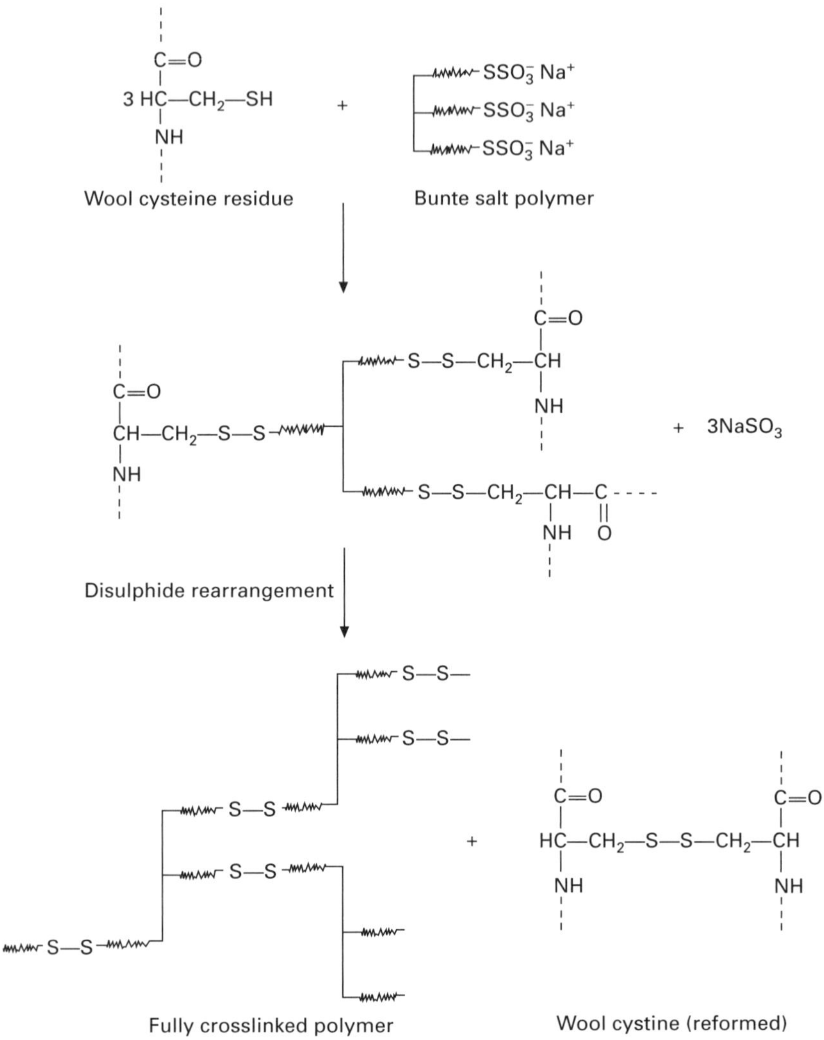

7.5 Curing reaction of Bunte salt polymer with wool (reproduced with permission from Lewis, 1982).

crosslinked polymer films at the fibre surface as well as 'spot welding' fibre–fibre bonding and polymer–fibre bonding, resulting in the shrink-resist effect (Fig. 7.5). The pad–batch application method can be achieved by adding sulphites or bisulphites as catalysts. The preparation of Bunte salt-terminated surface-active agents and their curing mechanism are discussed in detail by Lewis (1982, 1999). Recently Cognis has launched a water-soluble Bunte salt polymer, Securlana, for shrink-resist finishing of wool garments by an 'exhaust' process. This system uses magnesium chloride to promote polymer exhaustion at 40–50 °C; when the exhaust is nearly complete, ammonium hydroxide is added to bring about crosslinking (Lewis, 2005; Cognis technical data sheet).

Aqueous silicone-based polymer systems are also produced, but silicone polymer does not adhere strongly to wool. A previous study (Cook, 1984) claimed that the level of shrink-resistance conferred on wool by an aqueous silicone–polymer emulsion could be dramatically improved by adding a small amount of Synthappret BAP. The treatment also increased the smoothness and softness of the treated fabrics. More recently Wang *et al.* (2005) have introduced a novel bisulphite adduct of siloxane-modified aqueous polyurethane. The introduction of polysilicone segments into a polyurethane aqueous system can significantly improve the handle of the treated fabrics. Fabrics treated using this process are reported to possess good shrink resistance and handle, with the additional benefit of permanency of these effects.

Ciba Speciality Chemicals (now Huntsman Textile Effects) have introduced an aqueous solution of modified polyurethane and modified polydimethyl siloxane (Dicrylan WSR) for polymer shrink-resist finishing and soft handle for wool fabrics. Further development of the process led to a novel product, Dicrylan 7702, a combination of crosslinkable silicone rubber with silicone-modified polyurethane which can react on the wool fibre when a special metal-free catalyst (Phobotone Catalyst 7639) is used. This finish allows tumble-drying after washing, but this performance is not achievable using silicone rubber alone (Holme, 2007a; Ciba technical data sheet).

It is widely believed that the use of plasma treatment for the production of shrink-resistant wool will become an alternative to the existing processes. This technology has been attracting worldwide attention because it is eco-friendly, uses a lower volume of water and the effluent is AOX-free (*Wool Record*, 2004). Modification of the wool surface by plasma treatment has been the subject of a number of publications (Gregorski and Pavlath, 1980; Bradley *et al.*, 1992; Byrne and Godau, 1995; Rakowski, 1995; Zuchairah *et al.*, 1997; Erra *et al.*, 1999; Dai and Kviz, 2001; Höcker, 2002; Kan *et al.*, 2003; Sun and Stylios, 2005; Kan and Yuen, 2006; Holme, 2007b; Shishoo, 2007; Thomas, 2007).

Low-temperature plasma (LTP), generated by a strong electrical discharge at atmospheric pressure (corona discharge), or at low pressure (glow discharge) has been used for the surface modification of wool fabric for printing and for conferring shrink-resistance. During the plasma treatment, wool is exposed to a highly reactive gas containing free radicals, electrons and ions, among other reactive intermediates. Recent work (Hesse *et al.*, 1995a; Mori and Inagaki, 2006) has suggested that the plasma treatment not only disrupts the outermost lipid layer of the epicuticle, but also attacks one-third of the highly crosslinked disulphide bonds in the exocuticle A-layer. The outer 30–50 nm of the fibre surface appears to disintegrate as a result of plasma etching. The extent of etching is in the order of gas used: N_2 < air < O_2. New hydrophilic groups such as sulphonate and carboxylic acid groups are created and it is suggested that the disulphide groups are oxidised to intermediate cystine oxides, i.e. —SO—S— and —SO_2—S— groups, on the surface of wool fibres. The shrink-resist effect of the plasma treatment is due to oxidation and etching reactions on the fibre surface which enhance the fibre wettability by the formation of polar groups and partial removal of the outermost lipid layer (Hesse *et al.*, 1995a; Mori and Inagaki, 2006).

Recently the German Wool Research Institute (DWI, Aachen) examined a continuous atmospheric plasma treatment for wool fabrics over a width of 150 cm at 5–7 m/min. Homogeneous treatment of fabric was achieved by passing fabric through the grounded barrier layer on both sides for a double-sided exposure of fabric with energy output up to 3.5 kW (see Fig. 7.6). This treatment demonstrated improved wettability and shrink resistance of the wool (Thomas *et al.*, 2005).

Plasma modification of the fibre surface is known to cause a significant increase in the friction coefficients of wool, in both with- and against-scale directions, resulting in a rather harsh handle. Subsequent polymer application is necessary to restore fabric softness and fabric tear strength, and to improve shrinkproofing efficiency. A low level of Synthappret polymer finishing has been used effectively to coat the scale edges on plasma-treated wool fibres and achieve a high level of shrink resistance. It was recently reported that the application of softener Arristan 64 (a reactive polysiloxane-based softener) improved the handle properties of plasma-treated wool as well as enhancing wool shrink resistance and colour fastness performance (Thomas *et al.*, 2005).

Other approaches using solvent media for modifications of the cuticle scales of wool fibres have previously been investigated. Leeder and Rippon (1985) studied the reaction of wool with an alkaline reagent (potassium *tert*-butoxide) or a reducing agent (sodium sulphite) under anhydrous conditions using a non-swelling organic solvent. It was reported that the treatment of wool fabric in 0.1 M potassium tert-butoxide in *tert*-butanol at 40 °C can achieve a high level of shrink resistance with no measurable weight loss. It appears that alkaline degradation was restricted to the fibre surface, and lipid materials from the outmost layer of the fibre surface were removed. Julia *et al.* (1985) investigated the treatment of wool with sodium sulphite in the presence of a cationic surfactant (*N*-cetyl-*N*,*N*,*N*-trimethylammonium bromide) in a 50:50 isopropanol/water medium. This treatment was found to confer good shrink resistance on wool fabrics. However, the use of expensive solvent in place of water is questionable. It is necessary to reduce costs and environmental impact by recycling the solvents.

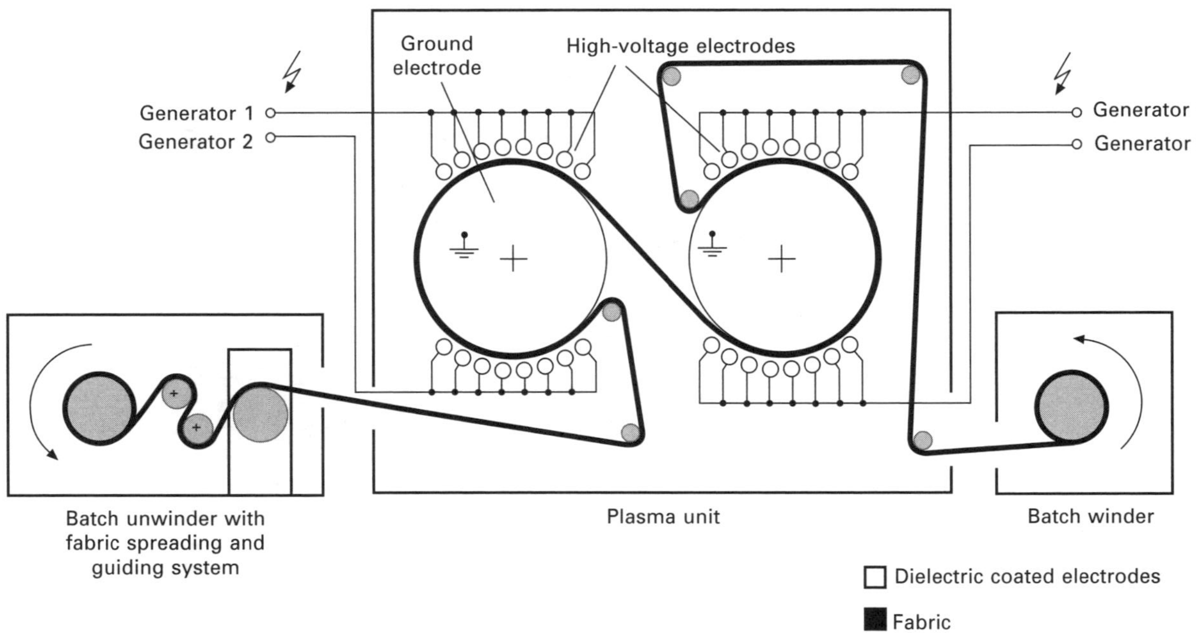

7.6 Schematic setup of the plasma unit with a treatment width of 160–170 cm (Thomas *et al.*, 2005).

7.3 Stain-repellent and stain-release finishing

In recent years, stain-repellent and stain-release treatments have become popular for apparel, upholstery and floor coverings. These markets are continuing to grow. Industrial uniforms and military uniforms are other market areas where stain-repellent/releases are desirable. The terms 'soil' and 'stain' 'repellence', 'block' or 'protection' and 'release' have always been used in textile industries but sometimes their meanings are ambiguous, and they may be interchangeable. Thiry (2005) describes a soil as the medium that goes on a textile substrate. It may or may not create a stain. A stain is the deposit from the soil on the fabric. Therefore, successful stain-repellent and stain-release finishes should (a) cause the fabric to repel the soiling substances and thus prevent staining, (b) prevent stains from forming weak bonding to fibres and (c) allow easier removal of stains during domestic washing.

There are a variety of treatments available to protect fabrics from stains and maintain a good appearance during use. Three mechanisms are involved:

1. *Stain repellence:*
   - reducing the surface energy of the fibre to such an extent that both aqueous and oily substances (polar and non-polar liquids) are repelled, also making the fabric surface difficult to wet, thus preventing the spread of contaminants;
   - chemical modification of the fibre surface to confer a negative charge which will repel most of the anionic colorant acids used in drinks. Chemical modification may also be used to block sites for the adherence of contaminants, while retaining the hydrophilic characteristics of fibres.
2. *Stain release:* treatments to improve fibre wetting, lower the adherence of contaminants and allow them to be transferred to detergents used in washing.
3. *Dual stain repellence and stain release:* surface treatment to provide dual stain repellent and release properties as a flip-flop mechanism for fabric from dry to wet washing.

Stain-repellent and stain-release finishing have been discussed and reviewed by Thiry (2005), Holme (2003), Schindler and Hauser (2004a) and Baldwinson (1990).

Wool carpets are required to be protected against accidental staining due to spillages of drinks and foods. The predominant source of stains from food and drinks are anionic colorant acids. The stain-repellent treatment can block the charged protonated amine groups in the fibre and thus achieve repulsion of anionic stains. One major group of 'stain blockers' is the sulphonated acid dyes used to block the dye sites in wool. Holme (2003) described a fibre-reactive dichlorotriazine compound containing a sulphonate group as stain-repellent agent covalently bound to the wool, and the mechanism of action for further improvement in stain-blocking performance by applying with a Thiotan-type resist compound.

Recently Trichromatic Carpet Inc. of Canada and Canesis Network Ltd developed a new stain-protection product, Lanashield, for use on wool and wool blends. Lanashield, an anionic stainblocker, provides wool carpets with durable resistance to a wide range of coloured anionic stains from soft drinks and cordials. It is reported that Lanashield stain protection does not cause yellowing of wool and is durable to shampooing.

Wool fibres have a naturally low surface energy, because the very thin layer of covalently bonded lipids make the surface hydrophobic. This means that wool fibres possess a natural water repellence, allowing time to remove liquid spills before they can cause permanent staining. However, most oil-based soils are difficult to remove; these tend to stick to the fibres mainly by non-polar bonding, a predominantly hydrophobic interaction. Therefore both water and oil-based soil-repellent treatment is essential for wool. Application of a fluorochemical finish can reduce the surface energy of wool fibre and further enhance water and oil-based stain resistance. Fluorochemical repellent finishes for textiles are now available that provide improved durability to washing and dry cleaning. The chemical structure of a fluorochemical is typically based upon a polyacrylate. The perfluorinated side-chains of the polyacrylate are oriented to point directly away from the fibre surface and hence the —$(CF_2)_n$—CF_3 terminal groups form a low-energy repellent surface (Fig. 7.7). However, in commercial practice there are various monomers that can be incorporated to produce a synergistic improvement for special end use performance that may be required (Holme, 2003). There are a wide variety of components that can be combined in the final active product formulation, such as dimethyl siloxane for softness of handle or ethylene oxide for hydrophilicity and stain-release properties. The durability of fluorochemical finishes can be improved by incorporating a crosslinkable monomer or other functional groups into the fluorochemical polymeric 'backbone'.

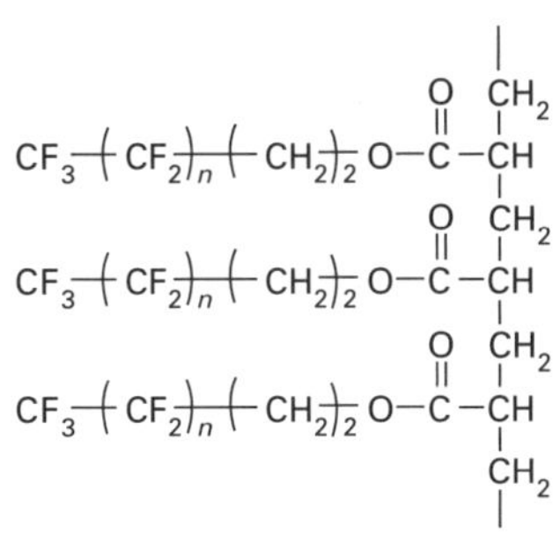

7.7 A fluorochemical polymer molecule.

In order to achieve dual stain-repellent and stain-release properties, unique polymers for textile finishing have been developed by incorporating both highly hydrophilic polymer segments and highly fluorinated polymer segments within a single chain molecule, so-called 'hybrid copolymers'. An example of a successful hybrid copolymer based on a conventional fluorochemical-based block and a hydrophilic block is shown in Fig. 7.8. This functions effectively as a hydrophobic stain repellent in air under dry conditions and also as an effective stain release during the laundering process. A 'flip-flop' mechanism of the dual stain-repellent and release properties has been suggested. The hydrophilic blocks are shielded by the fluorocarbon segments when dry, presenting a repellent surface. After immersion in the wash bath, the hydrophilic blocks swell and reverse the interfacial characteristics of the surface, yielding the hydrophilic surface necessary for oily soil release (Schindler and Hauser, 2004b). During drying, a re-orientation of the copolymer occurs and the stain-repellent fluorinated polymer parts re-assert their dominance at the fabric/air interface.

Perfluorooctane sulphonate (PFOS), used as a polymerisation aid in the production of fluorinated resins and occurring in trace amounts as a by-product of the manufacturing process, has, however, raised environmental concerns. PFOS has been found to be a persistent organic pollutant and bioaccumulates in wildlife and humans. The problem with different perfluorochemicals, including a large number of PFOS-related compounds and PFOA (perfluorooctanoid acid) compounds, is that the chemicals may degrade in the environment to PFOS and PFOA respectively, but no further degradation of PFOS or PFOA will occur. With increasing attention focused on environmental issues relating to the sources and pathways of human

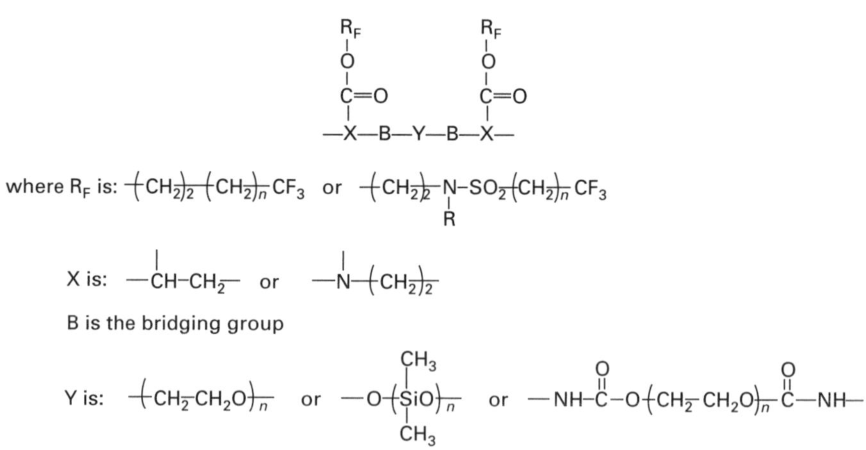

7.8 Hybrid copolymer with fluorine-based hydrophobic and hydrophilic blocks.

exposure to PFOA, alternatives for PFOA-free stain-repellent polymers or reformulated fluorochemicals have been developed in recent years.

Nanotechnology increasingly attracts worldwide attention and has potential for application on wool textiles by provision of a range of novel properties, such as water repellence, stain resistance and self-cleaning. The 'lotus effect' has been applied to textiles to enable self-cleaning by creation of a superhydrophobic surface on fabrics. Wool has a natural hydrophobic surface with a micro-scale of roughness due to the cuticle scales. Application of nanoparticles onto the fibre surface can bring multi-level roughness to enhance the hydrophobicity of wool fabrics. Nanotechnology approaches to create superhydrophobic surfaces on wool have recently been reported (Zhang *et al.*, 2004; Zhang and Lamb, 2005). A sol–gel process can be used for the application of nanometre sized particles onto wool (McNeil *et al.*, 2005). The potential for self-cleaning textiles has also been explored with titanium dioxide (TiO_2) nanoparticles which can act as a photocatalyst to break down organic dirt and kill micro-organisms (Xin *et al.*, 2004; Xin and Daoud, 2005; Holme, 2006a; Daoud *et al.*, 2008). These nanotechnologies have the potential for application to wool textile materials while causing minimum change in other properties such as handle.

7.4 Insect-resist and insect-repellent treatments

Woollen goods are susceptible to attack from the larvae of certain species of moth and beetle. This can be a particular problem for goods which lie undisturbed for long periods of time, for example carpets, blankets, upholstery fabrics, insulation in buildings and museum specimens. There are some 30 species of moths (Lepidoptera), 15 species of beetles (Coleoptera) and numerous chewing lice (Mallophaga) which have developed the ability to derive nutrition from keratin. In some cases this ability and a preference for a keratin diet have resulted in particular moths and beetles becoming major textile pests. Lewis and Shaw (1987) reviewed the important wool-damaging moths and beetles, as shown in Table 7.1.

Table 7.1 Significant wool-damaging insect species

Common name	Scientific name
Common or webbing clothes moth	*Tineola bisselliella* (Hummel)
Case making clothes moth	*Tinea pellionella* (Linnaeus)
	Tinea translucens (Meyrick)
Brown house moth	*Hofmannophila pseudospretella* (Stainton)
Carpet beetles	*Anthrenus flavipes* (Le Conte)
Fur beetles	*Attagenus pellio* (Linnaeus)
	Attagenus piceus (Oliver)

The generally accepted explanation of the mechanism of wool digestion by keratophagous insects is that larval guts contain an alkaline-reducing environment together with highly active proteolytic enzymes. The highly reducing conditions provided from the larval midgut are capable of breaking down the disulphide crosslinks of the ingested wool, and the proteolytic enzyme systems then digest the reduced wool almost completely (McPhee, 1971).

Early developments in insect-resist finishes have been reviewed by Lewis and Shaw (1987), McPhee (1971), Lewis (1992) and Barton (2000a). Researchers have approached insect resistance in different ways, including chemical modification of wool, chemical control of insect behaviour, antimetabolites (disruption of the insect's metabolic cycle), insect growth regulators, microbial pathogens and biological control.

Chemical modification of wool to reduce or eliminate insect attack involves inhibiting enzyme attack and protein digestion by the insect. Insects are typically attracted by the leftover smell of food stains and body oils on wool material. Their survival relies on the protein nutrition which the insects digest from the interior of fibre. Reduction of the cystine linkages in the fibre and replacement by non-reducible crosslinks has been attempted in the past (Moncrieff, 1950) to resist the enzymatic digestion by insects. The technique has not been developed and used commercially since the chemical changes degrade the physical and chemical properties of the treated wool.

Biological control of many insects can be achieved by use of microbial pathogens that attack the insect population. Certain types of microbial pathogens such as bacteria and fungi can be devastating to insect populations; however, little research has been undertaken into the use of this method to control insect attack on textiles. Initial studies have indicated that the powerful proteolytic enzyme systems in keratophagous insects render these systems ineffective.

Chemical methods of insect control have involved the use of chemicals to control insect behaviour, 'antimetabolites' which disrupt the insect's metabolic cycle, and insect growth regulators. Treatments such as juvenile hormones that prolong the insect's life in the harmful larval stages are counterproductive. In order to protect wool from insects during storage or use as an insulation material, manufacturers have explored the use of diatomaceous earth and boron compounds to eliminate insects by attacking the protective waxy outer layer of the insect, resulting in dehydration and death. It was also found that treatment of wool with cationic surface active agents may cause the death of the larvae, resulting in insect resistance (Gibb *et al.*, 2005). Unfortunately cationic agents are incompatible with the usual anionic dyeing systems used on wool.

To date, industrial control of keratophagous insects has been through use of pesticides that kill or control the pests. There are two types of pesticides (digestion-affecting poisons and nerve poisons) used in insect-resistant finishing. Digestion-affecting poisons interfere with the keratin-digesting process of the larvae and include chlorinated triphenylmethane, chlorphenylids, sulcofuron and flucofuron. Nerve poisons are to kill insects by preventing nerves from functioning, thereby stopping muscle function, such as permethrin, cyfluthrin and hexahydropyrimidine (Fig. 7.9). Both types of chemical methods function by entering the larvae's digestive tract with ingested wool. These commercial insect resist agents have typically been applied in the dyebath when the wool is dyed where they penetrate the wool fibre. They are released in the insect gut when the fibre is totally digested, thereby providing a degree of specificity as all other animals (including humans) are incapable of digesting and degrading wool if they ingest it.

Unfortunately application of insect resist agents in the dyebath cannot be 100% effective and some release into the environment is inevitable. All of the current treatment chemicals are toxic to aquatic life and are distinctly 'un-green'. Permethrin has been the main insect-resist agent used commercially since its development in the mid-1970s because of its cost, its safety to humans, its versatility of application, its suitability for wool and wool/synthetic blends and its effectiveness against a wide range of insects. Current permethrin-based commercial insect-resist agents are listed in Table 7.2. An alternative

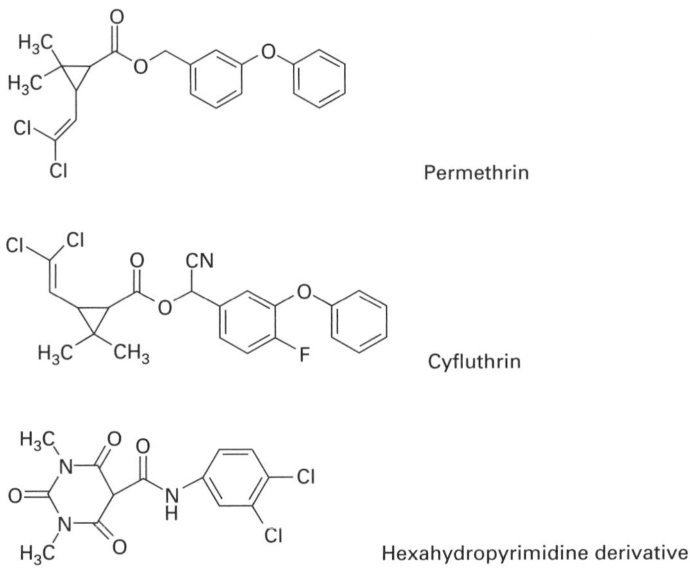

7.9 Chemical structures of insecticide: permethrin, cyfluthrin and hexahydropyrimidine derivative.

Table 7.2 Commercially available insect resist agents for wool

Mothproofing agents	Commercial products
Permethrin (synthetic pyrethroids)	Perigen New (Stephenson Thompson) Eulan SPA (Lanxess) Mystox MP (Catomance) Crosproof PEM (Eurodye-CTC)

insect-resistant agent is based on bifenthrin, another synthetic pyrethroid insecticide. Formulated by Melbourne Aniline & Lye Pty Ltd, bifenthrin is claimed to be more effective than permethrin because of better exhaustion onto the wool and this decreases the amount of active ingredient remaining in the waste effluent.

Currently, the wool textile industry applies insect-resistant agents mainly to carpet wools. The UK Environmental Agency has set increasingly stringent restrictions on the level of permethrin and other insect-resist agents permitted in effluents. Therefore achieving effective insect-resistance in wool carpets without compromising the environment has been an ongoing priority. Different approaches have attempted to improve permethrin fixation during application and to limit release of permethrin into wastewater. It was reported by Barton (2000a) that a low-temperature dyeing auxiliary, Valsol LTA-N (Asia Pacific Specialty Chemical Ltd) can improve the pyrethroid exhaustion onto the wool and thus lower the concentration of the insect-resistant agent required in the dyebath.

Foam and dry applications have been developed for effluent-free application of insect-resistant agents. The Wool Research Organisation of New Zealand (now AgResearch) developed the Lanaguard technology for insect-resistance treatment using a powder coating method. The dry powder-containing pyrethroid and polymer carrier is sprinkled onto the carpet pile. During the curing stage, the polymer carrier melts and adheres to the wool fibre surfaces to give insect resistance. There is no water involved (Barton, 2000a).

The major suppliers of insect-resistant agents are urgently searching for green technology or alternative methods that can result in less environmental impact. In the absence of an alternative 'safe' product, pesticide-based mothproofing finishes have been significantly reduced.

A new insect-resistant agent based on chlorfenapyr (Fig. 7.10), Mystox MP, has recently been developed by Catomance Technologies, UK, in conjunction with AgResearch. Mystox MP is claimed to have the dual function of contact insecticide and stomach poison. It can be used for protecting wool, wool blends and carpets. It is claimed to have a low environmental impact on waterways and environmental risk assessments based on toxicity to aquatic organisms are reported to show that Mystox MP is up to 30 times

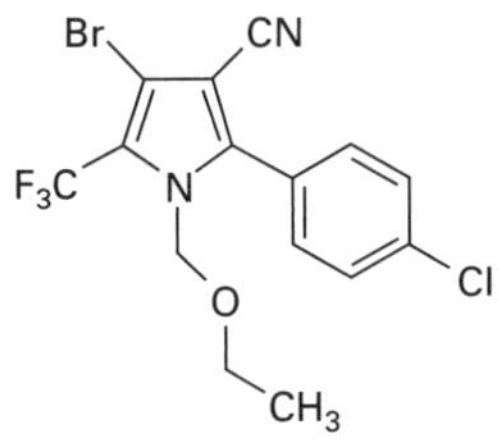

7.10 Chlorfenapyr.

less toxic than currently used permethrin-based insect resist agents (*Wool Record*, 2007). Detailed data are not yet available.

Microencapsulation technology has recently been utilised in insect-resist finishing for wool carpets in order to reduce the level of insecticide agents in wastewater. Insect-resist agents were encapsulated in yeast and applied to wool carpets. It was reported that moth larvae are attracted to yeast as a source of nutrients. Using these targeted microcapsules, the levels of insecticide on carpets could be reduced (Nelson, 1991). Thor SARL (Salaise sur Sanne, France) has developed a method of microencapsulating the synthetic pyrethroid permethrin (Holme, 2007a).

With increasingly stringent effluent consent levels for toxic pesticides, it is certain that time is fast running out for the permethrin-based products that have served the industry for almost half a century. Research and development into alternative insect-resist technologies is urgently required to ensure that alternative mothproofing agents are effective on moth and beetle larvae but have no toxic effects on other species and have minimum impact on the environment.

7.5 Flame-retardant finishing

Textiles including upholstery fabrics such as carpeting, curtains, wall panels and even bedding used in public places must meet flammability regulations throughout the world. Demand for flame-resistant fibres and fabrics has been growing rapidly in the market in recent years, especially for transportation textiles used in automobiles to airplanes, and for multi-functional protective wear including military combat apparel and uniforms.

Among natural textile fibres and many polymers, wool is regarded as the most flame-resistant fibre due to its unique protein structure and chemical composition. Compared with most textile fibres and polymers which contain mainly carbon and hydrogen that can burn easily, wool contains high levels of moisture, nitrogen and sulphur. In fact many fire-retardant additives used for other materials are high in nitrogen. Therefore it is difficult to ignite wool. Wool requires higher levels of oxygen in the surrounding atmosphere to accelerate combustion. The limiting concentration of oxygen (LOI) required to support combustion of wool in standard tests is about 25% which is higher than the ambient oxygen concentration in air (21%) (Table 7.3). When wool is heated to the point of combustion, it forms an intumescent char which provides an insulating layer of pyrolysed material separating heat and oxygen from the fuel. The fact that wool does not melt or drip prevents further flame spread, and releases only low heat, therefore the flame is easy to extinguish.

Horrocks and Davies (2000) describe the pyrolytic reactions that occur during the heating of wool. Wool pyrolyses by a complex series of reactions to yield a number of products at increasing temperatures. Initially at 230–240 °C, rupture of the helical structure occurs and the major ordered part of the wool protein undergoes a solid to liquid phase change. At 250–295 °C an endothermic reaction occurs associated with the release of sulphur compounds due to the breaking of the cystine disulphide bonds and simultaneous release of hydrogen sulphide. Above 250 °C general pyrolytic decomposition occurs, including char-forming reactions with dehydration and loss of other volatiles. In the presence of air, formation of sulphur dioxide occurs between 270 and 320 °C. Cleavage of the cystine disulphide bond is seen to play a very important role in the thermal degradation and combustion of keratin. It has been suggested that the oxidation of cystine may be the initial exothermic reaction in the burning of wool.

The sensitivity to oxidation of the cystine disulphide bond between adjacent protein polymeric chains supports any burning mechanisms and so pre-oxidation of cystine to cysteic acid residues can improve flame retardancy, especially if 60% or more of the disulphide bonds are oxidised. However such a high degree of crosslink rupture significantly reduces the wet strength retention of the fibre.

Despite the fact that wool fibres are inherently less flammable than most other fibres, the inherent flame-retardant properties of wool need to be enhanced in order to meet specific flammability tests or specific end uses such as

Table 7.3 Flammability properties of wool fibre

LOI	25.3 %
Ignition temperature	570–600 °C
Heat of combustion	4.9 kcal/g
Flame temperature	680 °C
Melting point	No melting
Burning behaviour	Difficult to ignite, burns slowly with formation of char, supports combustion with difficulty

aircraft carpets and seat covers. Flame-retardant finishes for wool fibres are mainly focused on enhancement of char-formation in the condensed phase, although bromine-containing, vapour phase-active surface treatments are effective for most textile materials. Currently there is interest in the use of intumescents, and research and development in this area has been undertaken by Horrocks and Davies (2000).

Ammonium phosphates and organophosphorus species with Lewis acidic properties are effective flame retardants for wool, and each enhances char formation. One well-known process, Zirpro (developed by the former International Wool Secretariat), based on the reaction of zirconium and titanium salts with wool, also enhances char formation. Zirpro treatments are based on the exhaustion of negative charged zirconium or titanium salts, under acid conditions, onto positively charged wool. This results in the deposition of only about 3% of flame retardant inside the fibre with negligible effect on properties such as handle. These treatments stabilise and further crosslink the protein structure. K_2ZrF_6 and K_2TiF_6 are the current commercial flame retardants used. Their application along with zirconium acetate solution can achieve low smoke emission for wool. Another flame-retardant treatment for wool is the use of tetrabromophthalic anhydride (TBPA). TBPA can be incorporated with the Zirpro treatment to reduce after-flaming times and minimise heat release. The current commercial flame retardant finishes for wool are summarised in Table 7.4.

Flame-retardant finishes for textiles, including wool-based materials, have been comprehensively reviewed by Horrocks (1986, 2003, 2005), Schindler and Hauser (2004b) and Pure Strategies Inc. (2005). Environmental issues associated with flame-retardant finishes, especially brominated flame retardants, have been extensively discussed during the past decade (Barton, 2000b; Holme, 2001, 2006b; *International Dyer*, 2003; Dombrowski, 2006). A comprehensive assessment of the impact of chemicals on the environment and human health has been conducted (Barton, 2000b; Holme, 2001; *International Dyer*, 2002; Dombrowski, 2006). The requirements of flammability legislation and regulation have changed significantly. With increasing environmental demands as well as demands for higher technical performance, much effort is being expended by manufacturers and scientists to improve existing products and develop new eco-friendly products for the flame-retardant finishing of textiles.

7.6 Enzyme applications in wool finishing

Biotechnology has recently found application within the textile industry in order to lessen the environmental impact of textile wet processes. The restrictions on AOX in effluent have become part of the legislation in EU countries. Even in other countries where there is no such legislation, there is

Table 7.4 Flame-retardant finishes for wool

Trade name	Chemical constitution	Process and durability
Huntsman		
Flovan® CGN	Phosphorus and nitrogen-based compound, halogen free	Non-durable
Pyrovatim® PBS		Semi-durable
Thor Specialities Ltd.		
Aflammit® ZR	Potassium hexafluorozirconate	Applied by an exhaust process, durable to washing and dry cleaning
Aflammit® TI	Potassium hexafluorotitanate	
Aflammit® ZAL	Zirconium acetate solution	Used in combination with Aflammit® ZR for low smoke emission
Aflammit® WPA	Sodium salt of tetrabromophthalic acid	Suitable for wool/nylon blend, durable to several washes at 40 °C, used in combination with Aflammit® ZR/TI for better performance
Firestop Chemicals Ltd		
Noflan	Organophosporus flame retardant based on complex alkyl phosphonate	Applied using pad–dry–cure method, durable to dry cleaning up to 25 cycles
Schill & Seilacher		
Flacavon® R neu	An organic phosphorus and nitrogen-containing compound	Applied by padding, dipping and spraying methods and drying at 100 °C, durable to dry cleaning.
CHT R Beitlich GmbH		
Apyrol ZFK	Potassium hexafluorozirconate	Applied by exhaustion method, durable to washing and dry cleaning
Apyrol ZAC	Zirconium acetate solution	Used in combination with Apyrol ZFK for low smoke finishing
Avocet Dye & Chemical Co. Ltd		
Cetaflam® PHFZ	Potassium hexafluorozirconate	Applied by exhaustion process, fast to washing and dry cleaning
Cetaflam® PHTI	Potassium hexafluorotitanate	
Cetaflam® ZAS	Zirconium acetate solution	Applied by exhaustion in combination with Cetaflam® PHFZ to give low-smoke FR finishes, durable to washing and dry cleaning

Cetaflam® DTB Liquid	Tetrabromophthalate derivative	Used in combination with the Zirpro® process to reduce after-flaming times and minimise heat release, applied by exhaustion
Cetaflam DB WN 240	Blend of flame retardant components	Durable to repeated dry cleaning and standard neutral launderings
Rhodia Consumer		
Amgard RD	Phosphorus and nitrogen-based compound	Applied by pad–dry, brushing, dipping or spraying technique, non-durable

pressure on textile manufacturers and retailers to limit or even eliminate this kind of contamination in the effluent. Because of such environmental concerns, much research effort has gone into searching for environmentally friendly processes for wool processing. The environmental benefits gained by using enzymes as bio-catalysts in wool processes to replace harsh chemicals, especially for wool shrink resistance, and simultaneously to improve dyeability, handle and whiteness of wool are well recognised.

Enzymes are biological catalysts for specific chemical reactions and require comparatively mild conditions. All enzymes are proteins and biodegradable. The precise reaction specificity of an enzyme can be used for specific or targeted textile finishing without causing undesirable effects. There are thousands of enzymes available. Classification of enzymes is very important for enzymes users as well as the scientific community. The Enzyme Commission of the International Union of Biochemistry devised a rational system for classification of enzymes. The classification is made by code numbers, consisting of four digits separated by dots, on the basis of the total reaction catalysed. The first digit shows the main class to which the enzyme belongs. In this system, enzymes are divided into six groups, which are presented in Table 7.5.

Most of enzymes used in the textile industry belong to Group 3, hydrolases. This Group includes the amylases, cellulases, pectinases, catalyses and proteases which are used for various textile applications such as desizing, bioscouring, bio-polishing, bleach cleaning-up and wool shrink resistance. Proteases, proteolytic enzymes or peptidases are general terms for enzymes which catalyse the hydrolysis of certain peptide bonds in protein molecules (forming the group EC 3.4.XX of hydrolases). These have been suggested for incorporation in wool processing for improving scouring efficiency, handle properties, imparting shrink resistance and low temperature dyeability.

Table 7.5 The Enzyme Commission's system of classification of enzymes and assigning code numbers (Palmer, 2001)

First digit	Enzyme class	Type of reaction catalysed
1	Oxidoreductases	Oxidation/reduction reactions
2	Transferases	Transfer of an atom or group between two molecules
3	Hydrolases	Hydrolysis reactions
4	Lyases	Removal of a group from substrate (not by hydrolysis)
5	Isomerases	Isomerisation reactions
6	Ligases	The synthetic joining of two molecules, coupled with the breakdown of pyrophosphate bond in a nucleoside triphosphate

Proteolytic enzymes can be divided into exopeptidases (which hydrolyse terminal peptide bonds) and endopeptidases (which hydrolyse peptide bonds inside the substrate molecule). Proteolytic enzymes can be further grouped according to the chemical nature of the catalytic site. They are divided among 13 sub-subclasses (Table 7.6). Four distinct families are the serine endopeptidases (such as chymotrypsin, trypsin and subtilisin), the cysteine endopeptidases (such as papain), the aspartic endopeptidases (such as pepsin) and the metalloendopeptidases (such as thermolysin). In particular, serine endopeptidases have been extensively studied. The application of enzymes in wool processing to replace harmful chemicals has been extensively studied and developed in last decade, and has been recently reviewed by Heine and Höcker (1995, 2001).

Early studies and more recent work (Moncrieff, 1953; Bishop *et al.*, 1998; Shen *et al.*, 1999) have confirmed that the action of proteases with undamaged wool is slow. This is due to the protective nature of the hydrophobic epicuticle surface containing fatty acid molecules and highly crosslinked cuticle cell components. However, once some of the cystine disulphide crosslinks in the cuticle cells are broken, the rate of enzyme reaction is greatly increased. Early claims suggested that proteases were unable to penetrate into wool fibre even when the fibre is wet and swollen owing to the large size of the enzyme molecules (Moncrieff, 1953). However, it was later found that during treatment with proteolytic enzymes, enzyme attack occurred preferentially at the highly swellable cell membrane complex by penetrating between cuticle cells and then between cortical cells. Once the enzyme has diffused into the membranes between the cells, the enzyme can rapidly disrupt the cell membrane complex, and eventually damage the fibre if the reaction is prolonged. A

Table 7.6 The Enzyme Commission's system of classification of peptidases (NC-IUBMB, 2008)

Sub-subclass	Type of peptidase	Number of entries
3.4.11	Aminopeptidases	20
3.4.13	Dipeptidases	12
3.4.14	Dipeptidyl-peptidases and tripeptidyl-peptidases	9
3.4.15	Peptidyl-dipeptidases	4
3.4.16	Serine-type carboxypeptidases	4
3.4.17	Metallocarboxypeptidases	20
3.4.18	Cysteine-type carboxypeptidase	1
3.4.19	Omega peptidases	9
3.4.21	Serine endopeptidases	98
3.4.22	Cysteine endopeptidases	54
3.4.23	Aspartic endopeptidases	38
3.4.24	Metalloendopeptidases	80
3.4.25	Threonine endopeptidase	1

fluorescence microscopy study by Heine (1991) demonstrated the diffusion of fluorescently labelled enzymes between the cuticle scales, through the cell-membrane complex and into the cortical cells. Therefore it is difficult to limit enzymatic degradation to the cuticle scales and to achieve machine-washable wool without significant fibre damage (Heine *et al.*, 2000).

Heine (Heine, 1991; Nolte *et al.*, 1996) investigated the removal of lipids from the wool fibre surface outer layer by treating grease-free wool with a lipoprotein-lipase. It was suggested that the aliphatic-hydrocarbon content of the cuticle surface was reduced by 20%, and this led to an improvement in wettability. The lipoprotein-lipase-treated wool top also showed a reduced felting capacity compared with the untreated top, but this effect did not constitute shrink resistance according to the IWS standard.

The enzyme kinetics do not only rely on the concentration of the reaction substrate (fibre), reaction temperature and pH, but also on the diffusion of enzyme to and into the solid phase of the fibre and the diffusion of the reaction products out of the solid phase into the solution. The complex structure of the wool fibre is important in the reaction. In addition, buffer and surfactants could be other important factors affecting the activities of enzymes. It is well known that proteolytic enzymes are compatible with non-ionic surfactants, which are widely recommended to be used in the enzymatic treatments. A recent study (Zhang *et al.*, 2006) has shown that the activity of the protease Esperase towards wool can be promoted by an ethoxylated alkyl phosphate anionic surfactant. Therefore, this anionic surfactant can provide additional benefits as an alternative choice to the widely used non-ionic surfactants. This could lead to the development of a more efficient enzymatic scouring process. Use of a buffer can maintain the optimum pH for protease activity during the enzymatic treatment, but it is found that different buffer systems and their ionic strengths have different effects on the activity of proteolytic enzymes. This is due to the interference of buffer cations and anions with the conformation of enzymes and their biological reactions.

An early study (Moncrieff, 1953) showed that the attack on wool fibres by proteolytic enzymes is variable. Enzymes appear to attack weathered fibres more rapidly and therefore are inclined to attack the tips of fibres more than the root ends. Irregularity of damage of fibres by enzymes can be clearly shown by scanning electron microscopy.

Bishop *et al.* (1998) reported that carefully controlled treatments with proteolytic enzymes can reduce the buckling load and collapse energy of wool yarns. These treatments were shown to improve the softness and reduce the subjectively perceived prickle of wool fabric knitted from the treated yarns. Combination of chlorination and treatments with proteolytic enzyme was also reported to improve handle properties, especially for coarse wool and mohair fibres, as well as improving fibre whiteness without causing any real damage to the wool (Holme, 2006c).

In the past decade, scientists have made a great effort to use different wool pre-treatments prior to enzyme treatment in order to limit enzymatic degradation to the cuticle scales and thus to achieve machine-washable wool without causing significant fibre damage. Different combinations of enzymes, oxidants/reductants and polymers have been proposed to improve felting and shrink resistance of wool (Levene and Shakkour, 1995; Levene *et al.*, 1996; Jovancic *et al.*, 1998; Cardamone *et al.*, 2006). In some of the early enzyme finishing processes, wool was pre-treated by chlorine or hydrogen peroxide prior to incubating the fibres with proteases. Oxidative treatments of wool can disrupt disulphide bonds and open up the wool fibre surface assisting enzymatic attack on the cuticle. On the other hand, oxidative pre-treatment probably induces a partial removal of the fatty acid barrier from the epicuticle, which confers hydrophilicity to wool (Cardamone *et al.*, 2005). Consequently the enzymatic attack on the cuticle can be selectively activated. It has been claimed that a combination of chlorination with chlorine gas or dichloroisocyanuric acid and subsequent enzyme treatments with a protease such as papain showed that pre-oxidation will limit enzymatic attack to the cuticular layer, resulting in the enzymatic descaling of wool fibres to enhance not only lustre but also shrink resistance (Moncrieff, 1953; Levene and Shakkour, 1995). However, the process might cause fabric yellowing.

Recently a two-step process which combines bleaching, shrinkage prevention and biopolishing was suggested as a way to make wool feel silky smooth. This involved a pre-treatment using hydrogen peroxide enhanced by dicyandiamide and stabilised by gluconic acid for powerful oxidation, and followed by enzyme treatment with proteases in the presence of sodium sulphite in triethanolamine buffer. Benefits claimed were a high level of whiteness, the removal of protruding fibre ends for fabric smoothness and shrink-resistance (Cardamone *et al.*, 2004b, 2005).

Other oxidising agents used as a pre-treatment for wool are peroxymonosulphuric acid, peracetic acid and potassium permanganate. Alkaline and reducing agents such as bisulphite can alternatively be used to open disulphide crosslink bonds in the cuticle scales to make the fibre more susceptible to enzymatic attack. These pre-treatments have been reported to enhance the activities of proteases and improve the efficiency of proteases in conferring anti-felting and shrink resistance of wool (Levene and Shakkour, 1995; Levene *et al.*, 1996; El-Sayed *et al.*, 2001). However, these processes caused severe fibre damage when shrink resistance was reaching adequate levels for machine-washable wool. Recently, Lenting *et al.* (2005, 2007) used the addition of a high concentration of sodium salt in the peroxide pre-treatment for restricting the oxidative reaction to the surface scale of wool fibres. This resulted in an improvement in the susceptibility of the outer surface protein layer to subsequent proteolytic hydrolysis. This agreed with an earlier study which suggested that the presence of high concentrations of salt reduced the extent of swelling of the fibre and reduced the rate of diffusion of the oxidant through the cuticle to the cortex. This led to preferential attack on the cuticle (Maclaren and Milligan, 1981) and wool shrink resistance was claimed without substantial loss of fibre tensile properties.

Jovancic *et al.* (2001) demonstrated that incorporating an enzyme in the alkaline peroxide treatment bath enhanced wool wettability and improved the effectiveness of subsequently applied chitosan biopolymer in achieving a degree of shrink resistance.

A two-step treatment consisting of a low-temperature plasma treatment (LPT) and subsequent enzymatic treatment by proteases has also been investigated to achieve wool shrink resistance (Dybdal *et al.*, 2001; Jovancic *et al.*, 2003). X-ray photoelectron spectroscopy (XPS) analysis revealed that the LPT /enzymatic treatment completely removed the outermost lipid layer (the F-layer) of the epicuticle. However, in order to avoid the excessive damage of wool fibres, the subsequent enzymatic treatment was mostly used to remove fibres protruding from the surface of the fabric and thus increased its softness.

Although there is currently considerable interest in the use of enzymes to achieve a shrink-resist finishing effect on wool, it is apparent that the results of enzymatic treatments, especially with proteases, can be unpredictable and may sometimes lead to unacceptable degradation of the fibre (see Fig. 7.11).

7.11 Scanning electron micrograph of wool fibres treated with proteases.

A new approach (Cavaco-Paulo and Silva, 2003; Silva *et al.*, 2004, 2005, 2006; Smith *et al.*, 2008) has increased the size of proteases to try to limit the enzymatic degradation of wool fibre to their cuticle scales. The increase in the size of the proteases was achieved by covalently attaching them to soluble Eudragit polymer (Fig. 7.12). Chemical modification of proteases with Eudragit improved the thermal stability of the enzymes as well as enzyme recycleability due to soluble–insoluble reversibility of the Eudragit polymer attached to the enzyme.

Shen *et al.* (2007) have published the results from the bulk trials on wool fabrics carried out using the modified proteases. It was demonstrated the modification of the protease enabled the reaction of the enzyme with wool to be controlled, so that less degradation of the wool occurred than in similar treatments with the unmodified protease. An anti-felting effect has been achieved without any significant weight loss being caused by the modified protease during the treatment. This novel enzymatic process leads to environmentally friendly production of machine-washable wool.

There is increasing interest in use of the protein-crosslinking enzymes transglutaminases (EC2.3.1.13) for surface modification of wool fibres. Transglutaminases belong to the class of transferases and act by a different mechanism than hydrolases. These enzymes can catalyse the post-translational modification of proteins by the formation of isopeptide bonds. This occurs either through protein crosslinking via epsilon-(gamma-glutamyl)lysine bonds

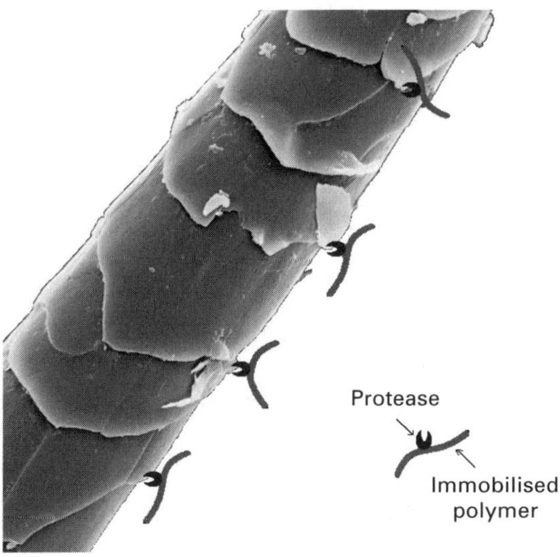

7.12 Illustration of protease immobilisation on soluble polymer and their enzymatic degradation of wool fibres limited to the cuticle scales.

or through incorporation of primary amines at selected peptide-bound glutamine residues (Griffin *et al.*, 2002a). This crosslinking leads to increased protein stability and increased resistance to chemical and proteolytic degradation. In the enzymatic treatment of wool, transglutaminases can be used for protein crosslinking within the fibres to compensate for the reduction of tensile strength and degradation of wool after treatment with an oxidative agent (e.g. chlorination for shrinkage prevention), a reducing agent (sodium sulphite) and/or a protease.

It is also reported that treatment of wool with transglutaminases can decrease the tendency to felting of wool without causing any negative effect of stiff and harsh handle (McDevitt and Winkler, 1998; Griffin *et al.*, 2002b; Cortez *et al.*, 2004; Du *et al.*, 2007). Transglutaminase has been shown to graft protein polypeptides, or fluorescein cadaverine onto the wool fibre surface for special effects. It is reported that the crosslinking within wool fibres by transglutaminase can improve the resistance to enzymatic cleavage, and increase the resistance of wool to chemical and mechanical disruption. This was claimed to be beneficial for protecting wool garments from detergent damage (Cortez *et al.*, 2005, 2007). However, the amount of lysine and glutamine residues available may be limited, especially on the surface of wool fibre, restricting the extent of enzyme reaction. Therefore this may affect the wide application of transglutaminases on wool.

7.7 Future trends

Increasing use of the domestic washing machine means that consumers continue to seek comfort, greater softness and easy-care properties in wool products, especially machine-washability and tumble-dryability. Wool manufacturers have extended their product ranges to include wool garments for babies and children, and bedding products including blankets and duvets, due to wool's natural comfort, warmth, moisture retention and wicking properties. In order to accommodate consumer requirements, finishing processes are required to ensure that greater softness and performance properties such as easy care and antipilling are maintained throughout the lifetime of the product.

It is anticipated that lightweight and soft wool fabrics will be increasingly included in the summer wear market for garments to be worn next to the skin. Such a scenario means that characteristics as durability, antistatic behaviour, deodorising performance and easy-care finishing will be increasingly important.

Environmental issues have been an ongoing concern for the textile industry and will necessarily become of increasing importance. Within the wool industry, green technology has been rapidly developed in order to replace the harmful chemicals traditionally used in wool processing. Plasma treatment is an innovative approach for the replacement of the chlorination process for machine washable wool. Indeed, the use of plasma has a promising future due to significant benefits such as its requirement for a lower volume of water and the absence of AOX in the waste effluent.

Another option when considering environmentally preferable methods of wool processing is the use of enzyme treatments. The reaction of conventional protease formulations on wool is difficult to control and may cause unacceptable damage. However the modification of enzymes or alteration of the absorption and penetration pattern of enzymes during treatment can confer specific finishing to wool including soft handle, shrink-resistance and low-temperature dyeing.

The development of different functional finishes to produce high-performance characteristics without causing environmental impacts is urgently required. Reduction of the levels of AOX and FPOS in processing effluents must be addressed by the textile industry as a whole, while insecticide discharges from insect-resist treatments are specific to wool. The use of wool in carpets has significantly expanded in the recent marketplace. As a consequence it has become increasingly important to develop self-cleaning and stain-repellence technology finishing for wool carpets.

Innovations in chemical finishes have recently been expanded and notably include nanotechnology, microencapsulation, plasma treatments and polymer finishing for textiles. Adoption and adaptation of these advanced technologies by the wool industry will facilitate the addition of more desirable properties to natural wool fibres. With increasing demands for sophisticated functionality and protective fabrics for military apparel and transportation textiles, a diverse functionality for wool is demanded and dictated by a host of special end uses. Advanced technologies enable attachment of a range of micro- and nanoparticles to the surface of wool fibres to develop smart and intelligent wool materials with multifunctionality. Plasma finishing, polymerisation and grafting technology with multifunctional and multi-reactive polymers are fast developing technologies and represent the way forward for smart wool performance and fashion in what promises to be an exciting future.

7.8 Acknowledgements

The author would like to thank Dr Ian Russell, Dr Jane Harwood, Dr Peter Speakman and Dr Edward Smith for their valuable comments and preparation of the figures.

7.9 References

BALDWINSON T M (1990), 'Soil-release and soil-repellent agents', in Shore J, *Colorants and Auxiliaries: organic chemistry and application properties – Vol. 2: Auxiliaries*, Bradford, Society of Dyers and Colourists, 463–467.

BARTON J (2000a), 'It's a bugs life – or is it?', *International Dyer*, **185**(9), 14–16.

BARTON J (2000b), 'New research calms toxicity fears', *International Dyer*, **185**(7), 16–20.

BENISEK L (2006), 'Scientists developing new ways to make softer textiles', *Wool Record*, **165**(3736), 38.

BELL V A and LEWIS D M (1975), 'The use of Bunte salt terminated polymers in wool finishing, with special reference to their use in dyeing processes', *Proc. 5th Int. Wool Textile Res Conf*, Aachen, Schriftenreihe, III, 595–606.

BISHOP D P, SHEN J, HEINE E and HOLLFELDER B (1998), 'The use of proteolytic enzymes to reduce wool-fibre stiffness and prickle', *J Text Inst*, **89**, Part 1, No. 3, 546–553.

BRADLEY R H, CLACKSON I L and SUTHERLAND I (1992), 'Oxygen plasma modification of wool fibre surfaces', *J Chem Tech Biotechnol*, **63**, 221–226.

BYRNE K M (1996), 'Easy-care wool woven products', *J Soc Dyers Color*, **112**, 350–353.

BYRNE K M and GODAU J, (1995), 'Commercial plasma treatment for woven fabrics', *Proc. 9th Int Wool Text Res Conf*, Vol IV, 415–421.

CARDAMONE J M and YAO J (2004), 'DCCA shrinkproofing of wool, part II: improving whiteness and surface properties', *Textile Res J*, **74**(7), 565–570.

CARDAMONE J M, YAO J and NUNEZ A (2004a), 'DCCA shrinkproofing of wool, part I: importance of antichlorination', *Textile Res J*, **74**(6), 555–560.

CARDAMONE J M, YAO J and NUNEZ A (2004b), 'Controlling shrinkage in wool fabrics: effective hydrogen peroxide system', *Textile Res J*, **74**(10), 887–898.

CARDAMONE J M, YAO J and PHILLIPS J G (2005), 'Combined bleaching, shrinkage prevention, and Biopolishing of wool fabrics', *Textile Res J*, **75**(2), 169–174.

CARDAMONE J M, NUNEZ A, ASHBY R and DUDLEY R (2006), 'Activated peroxide for enzymatic control of wool shrinkage, part I: elucidation', *Textile Res J*, **76**(2), 99–108.

CAVACO-PAULO A and SILVA C J S M (2003), 'Treatment of animal hair fibres with modified proteases', World Patent WO03097927.

COOK J R (1984), 'Shrink-resisting wool with aqueous silicone polymers: the effect of added Synthappret BAP', *J Text Inst*, **75**(3), 191–195.

COOK J R and FLEISCHFRESSER B E (1985), 'Shrink resisting wool with Synthappret BAP: the effect of drying conditions', *Textile Res J*, **55**, 607–614.

CORTEZ J, BONNER P L R and GRIFFIN M (2004), 'Application of transglutaminases in the modification of wool textiles', *Enzyme Microb Techol*, **34**, 64–72.

CORTEZ J, BONNER P L R and GRIFFIN M (2005), 'Transglutaminase treatment of wool fabrics leads to resistance to detergent damage', *J Biotechnol*, **116**, 379–86.

CORTEZ J, ANGHIERI A, BONNER P L R, GRIFFIN M and FREDDI G (2007), 'Transglutaminase mediated grafting of silk proteins onto wool fabrics leading to improved physical and mechanical properties', *Enzyme Microb Techol*, **40**(7), 1698–1704.

DAI X J and KVIZ L (2001), 'Study of atmospheric and low pressure plasma modification on the surface properties of synthetic and natural fibres', *Textile Institute 81st World Conference: An odyssey in fibres and space*, Melbourne, Australia, April.

DAOUD W A, LEUNG S K, TUNG W S, XIN J H, CHEUK K and QI K (2008), 'Self-cleaning keratins', *Chem Mater*, **20**(4), 1242–1244.

DOMBROWSKI R (2006), 'Beyond perception: brominated flame retardants', *AATCC Review*, **6**(10), 33–35.

DU G, CUI L, ZHU Y and CHEN J (2007), 'Improvement of shrink-resistance and tensile strength of wool fabric treated a novel microbial transglutaminase from *Streptomyces hygroscopicus*', *Enzyme Microb Techol*, **40**, 1753–1757.

DYBDAL L, HEINE E and HÖCKER H (2001), 'Method for enzymatic treatment of wool', US Patent 6258129.

EL-SAYED H, KANTOUCH A, HEINE E and HÖCKER H (2001), 'Developing a zero-AOX shrink-resist process for wool. Part I: preliminary results', *Color Technol*, **177**, 234–238.

ERRA P, MOLINA R, JOCIC D and JULIA M R (1999), 'Shrinkage properties of wool treated with low temperature plasma and chitosan biopolymer', *Textile Res J*, **69**(11), 811–815.

FELDTMAN H D, LEEDER J D and RIPPON J A (1983), 'The role of fibre structure in wool fibre and fabric performance' in Postle R, Kawabata S and Niwa M, *Objective Evaluation of Apparel Fabrics*, Japan, The Textile Machinery Society of Japan, 125–134.

GIBB R, SHEN J and SPEAKMAN P T (2005), 'Insect resistance brought about by chemical modification of wool', *The 11th Int Wool Text Res Conf*, 4–9 September, Leeds, UK.

GRIFFIN M, CASADIO R and BERGAMINI C M (2002a), 'Transglutaminases: nature's biological glues', *Biochem J*, **368**, 377–396.

GRIFFIN M, CORTEZ J M and BONNER P (2002b), 'A method for enzymatic treatment of textiles such as wool', *World Patent* WO0204739.

GUISE G B (1977), 'Poly(carbamoyl sulfonates) (bisulfite adducts of polyisocyanates)', *J Appl Polym Sci*, **21**, 3427–3443.

GUISE G B and JACKSON M B (1973), 'The shrinkproofing of wool with the bisulfite adducts of polyisocyanates', *J Text Inst*, **64**, 665–667.

GREGORSKI K S and PAVLATH A E (1980), 'Fabric modification using the plasmod, the effect of extensive treatment in nitrogen and oxygen plasmas at low pressure', *Textile Res J*, **50**, 42–46.

HEINE E (1991), Grundlegende untersuchungen zur einwirkung von enzymen auf wolle, PhD Thesis, Technischen Hochschule Aachen.

HEINE E and HÖCKER H (1995), 'Enzyme treatments for wool and cotton', *Rev Prog Coloration*, **25**, 57–63.

HEINE E and HÖCKER H (2001), 'Bioprocessing for smart textiles and clothing' in Tao X, *Smart Fibres, Fabrics and Clothing*, Cambridge, Woodhead Publishing Ltd, 254–263.

HEINE E, HOLLFELDER B, SHEN J and BISHOP D (2000), 'Parameters affecting the morphology of wool during enzyme treatment', *The 10th Int Wool Text Res Conf*, 26 Nov–1 Dec, Aachen Germany.

HESSE A, THOMAS H and HÖCKER H (1995a), 'Zero-AOX shrinkproofing treatment for wool top and fabric, part I: glow discharge treatment', *Textile Res J*, **65**(6), 355–361.

HESSE A, THOMAS H and HÖCKER H (1995b), 'Zero-AOX shrinkproofing treatment for wool top and fabric, part II: collagen resin application', *Textile Res J*, **65**(7), 371–378.

HÖCKER H (2002), 'Plasma treatment of textile fibers', *Pure Appl Chem*, **74**(3), 423–427.

HOLME I (2000), 'Challenge and change in wool dyeing and finishing', *Rev Prog Color*, **33**, 88–92.

HOLME I (2001), 'FR and the environment', *International Dyer*, **181**(1), 34–37.

HOLME I (2003), 'Water-repellency and waterproofing' in Heywood D, *Textile Finishing*, Bradford, Society of Dyers and Colourists, 135–213.

HOLME I (2006a), 'Nanotechnology: the future of finishing', *International Dyer*, **191**(6), 8–10.

HOLME I (2006b), 'Flame retardancy: driven by legislation', *International Dyer*, **191**(8), 7–9.

HOLME I (2006c), 'New frontiers in wool dyeing and finishing', *International Dyer*, **191**(10), 9–11.

HOLME I (2007a), 'Innovative technologies for high performance textiles', *Color Technol*, **123**, 59–73.

HOLME I (2007b), 'Plasma technologies for innovative treatments', *International Dyer*, **192**(5), 7–9.

HORROCKS A R (1986), 'Flame-retardant finishing of textiles', *Rev Prog Coloration*, **16**, 62–101.

HORROCKS A R (2003), 'Flame-retardant finishes and finishing', in Heywood D, *Textile Finishing*, Bradford, Society of Dyers and Colourists, 214–250.

HORROCKS A R (2005), 'Thermal (heat and fire) protection', in Scott R A, *Textiles for Protection*, Cambridge, Woodhead Publishing Ltd, 398–440.

HORROCKS A R and DAVIES P (2000), 'Char formation in flame-retarded wool fibres. Part 1. effect of intumescent on thermogravimetric behaviour', *Fire and Materials*, **24**, 151–157.

International Dyer (2002), 'Brominated FRs – new study results', **187**(11), 6–9.

International Dyer (2003), EU bans 'Penta' and 'Octa', **188**(11), 17.

JOVANCIC P, JOCIC D and DUMIC J (1998), 'The efficiency of an enzyme treatment in reducing wool shrinkage', *J Text Inst*, **89** Part 1, No. 2, 390–400.

JOVANCIC P, JOCIC D, MOLINA R, JULIA M R and ERRA P (2001), 'Shrinkage properties of peroxide-enzyme-biopolymer treated wool', *Textile Res J*, **71**(11), 948–953.

JOVANCIC P, JOCIC D, MOLINA R, JULIA M R and ERRA P (2003), 'The combined low-temperature plasma/enzyme wool shrink-resist treatment', *AATCC Review*, **3**(2), 25–28.

JULIA M R, ERRA P, DOMINGUEZ J G and INFANTE M R (1985), 'The influence of a cationic surfactant on the shrink-resist treatment of wool with sodium sulphite in aqueous solution', *J Soc Dyers Color*, **101**, 66–71.

JULIA M R, COT M and ERRA P (1998), 'The use of chitosan on hydrogen peroxide pretreated wool', *Textile Chemist and Colorist*, **30**(8), 78–83.

JULIA M R, PASCUAL E and ERRA P (2000), 'The influence of the molecular mass of chitosan on shrink-resistance and dyeing properties of chitosan-treated wool', *J Soc Dyers Color*, **116**(2), 62–67.

KAN C W and YUEN C W M (2006), 'Low temperature plasma treatment for wool fabric', *Textile Res J*, **76**(4), 309–314.

KAN C W, CHAN K, CHUN-WAH M Y (2003), 'Surface characterisation of low-temperature plasma treated wool fibre', *AUTEX Res J*, **3**(4) 194–205.

LEEDER J D and RIPPON J A (1985), Changes induced in the properties of wool by specific epicuticle modification, *J Soc Dyers Color,* **101**(1), 11–16.

LENTING H B M, SCHROEDER M, GÜBITZ G M, CAVACO-PAULO A and SHEN J (2005), 'A new enzyme technology based method for obtaining machine washable wool', *The 11th Int Wool Text Res Conf*, 4–9 September, Leeds, UK.

LENTING H B M, SCHROEDER M, GÜBITZ G M, CAVACO-PAULO A and SHEN J (2007), 'New enzyme-based process direction to prevent wool shrinking without substantial tensile strength loss', *Biotechnol Lett*, **28**, 711–716.

LEVENE R and COHEN Y (1996), 'An oxidative batchwise shrink-resist treatment for wool using monoperoxyphthalic acid', *J Soc Dyers Color*, **112**, 44–49.

LEVENE R and SHAKKOUR G (1995), 'Wool fibres of enhanced lustre obtained by enzymatic descaling', *J Soc Dyers Color*, **111**, 352–359.

LEVENE R, COHEN Y and BARKAI D (1996), 'Applying proteases to confer improving shrink-resistance to wool', *J Soc Dyers Color*, **112**, 6–10.

LEWIS D M (1977), 'Continuous dyeing and polymer shrink-resist processes for wool', *J Soc Dyers Color*, **93**, 105–113.

LEWIS D M (1982), 'Applying shrink resist polymers to wool fabrics at room temperature using pad-store methods', *Text Res J*, **52**, 580–586.

LEWIS D M (1992), *Wool Dyeing*, Bradford, Society of Dyers and Colourists, 126–136.

LEWIS D M (1999), 'The preparation of Bunte salt-terminated surface-active agents and their use in wool fabric shrinkproofing', *J Soc Dyers Color*, **115**, 59–61.

LEWIS D M (2005), 'Recent trends in the chemistry of wool finishing and wool dyeing processes', *The 11th Int Wool Text Res Conf*, 4–9 September, Leeds, UK.

LEWIS D M and HAWKES J A (2007), 'Improvements in and relating to wool treatment', World Patent WO2007/017668.

LEWIS D M and SHAW T (1987), 'Insectproofing of wool', *Rev Prog Coloration*, **17**, 86–94.

MACLAREN J and MILLIGAN B (1981), *Wool Science; the Chemical Reactivity of the Wool Fibre*, Australia, Science Press, 299–300.

MCDEVITT J P and WINKLER J (1998), 'Method for enzyme treatment of wool', United States Patent 6051033.

MCNEIL S J, STANDARD O C and PAILTHORPE M T (2005), 'Hybrid organic-inorganic coatings for wool textiles', *The 11th Int Wool Text Res Conf*, 4–9 September, Leeds, UK.

MCPHEE J R (1971), *The Mothproofing of Wool*, Watford, Merrow Publishing Co. Ltd, 28.

MCPHEE J R and SHAW T (1984), 'The chemical technology of wool processing', *Rev Prog Coloration*, **14**, 58–68.

MONCRIEFF R W (1950), *Mothproofing*, London, Leonard Hill Ltd.

MONCRIEFF R W (1953), *Wool Shrinkage and its Prevention*, London, The National Trade Press Ltd, 322–332.

MORI M and INAGAKI N (2006), 'Relationship between anti-felting properties and physicochemical properties of wool fibers treated with Ar-plasma', *Textile Res J*, **76**(9), 687–694.

NC-IUBMB (Nomenclature Committee of the International Union of Biochemistry and Molecular Biology) (2008), Enzyme Nomenclature: Recommendations of Peptidase Nomenclature, http://www.chem.qmul.ac.uk/iubmb/enzyme/EC34/

NELSON G (1991), 'Microencapsulates in textile coloration and finishing', *Rev Prog Coloration*, **21**, 72–85.

NOLTE H, BISHOP D P and HÖCKER (1996), 'Effects of proteolytic and lipolytic enzymes on untreated and shrink-resist-treated wool', *J Text Inst*, **87**, Part 1, No 1, 212–226.

PALMER T (2001), *Enzymes: Biochemistry, biotechnology and clinical chemistry*, England, Horwood Publishing Ltd., 3–14.

PASCUAL E and JULIA M R (2001), 'The role of chitosan in wool finishing', *J Biotechnol*, **89**, 289–296.

PURE STRATEGIES INC. (2005), *Decabromodiphenylether: An investigation of non-halogen substitutes in electronic enclosures and textile applications,* The Lowell Center for Sustainable Production, University of Massachusetts Lowell.

RAKOWSKI W (1995), 'Plasma treatment of wool – from lab curiosity to industrial reality', *Fibres & Textiles in Eastern Europe*, September/December, 45–49.

ROBERTS G A F and WOOD F A (2001), 'A study of the influence of structure on the effectiveness of chitosan as an anti-felting treatment for wool', *J Biotechnol*, **89**, 297–304.

SAVILLE B P (2000), *Physical Testing of Textiles*, Cambridge, Woodhead Publishing Ltd, 168–183.

SCHINDLER W D and HAUSER P J (2004a), *Chemical Finishing of Textiles*, Cambridge, Woodhead Publishing Ltd, 87–97.

SCHINDLER W D and HAUSER P J (2004b), *Chemical Finishing of Textiles*, Cambridge, Woodhead Publishing Ltd, 98–116.

SHAW T (1990), 'Environmental issues in the wool textile industry', *Proc 8th Int Wool Text Res Conf*, Christchurch, New Zealand, vol. 4, 533–547.

SHEN J, BISHOP D, HEINE E and HOLLFELDER B (1999), 'Some factors affecting the control of proteolytic enzyme reactions on wool', *J Text Inst*, **90**, Part 1, No. 3, 404–411.

SHEN J, RUSHFORTH M, CAVACO-PAULO A, GUEBITZ G and LENTING H (2007), 'Development and industrialisation of enzymatic shrink-resist process based on modified proteases for wool machine washability', *Enzyme Microb Techol*, **40**, 1656–1661.

SHISHOO R (2007), *Plasma Technologies for Textiles*, Cambridge, Woodhead Publishing Ltd.

SILVA C J S M, SOUSA F, GÜBITZ G and CAVACO-PAULO A (2004), 'Chemical modifications on proteins using glutaraldehyde', *Food Technol Biotechnol*, **42**(1), 51–56.

SILVA C J S M, GÜBITZ G and CAVACO-PAULO A (2005), 'Optimization of a serine protease coupling to Eudragit S-100 by experimental design techniques', *J Chem Technol Biotechnol*, **81**, 8–16.

SILVA C J S M, ZHANG Q, SHEN J, GUBITZ G and CAVACO-PAULO A (2006), 'Immobilization of proteases with a water soluble-insoluble reversible polymer for treatment of wool and enzyme recycling', *Enzyme Microb Techol*, **39**, 634–640.

SMITH E, ZHANG Q, SHEN J, SCHROEDER M and SILVA C (2008), 'Modification of Esperase by covalent bonding to Eudragit polymers L 100 and S 100 for wool fibre surface treatment', *Biocatalysis and Biotransformation*, **26**(5), 391–398.

SUN D and STYLIOS G K (2005), 'Investigating the plasma modification of natural fiber fabrics - the effect on fabric surface and mechanical properties', *Textile Res J*, **75**(9), 639–644.

THIRY M C (2005), 'A stainless virtue', *AATCC Review*, **5**(11), 20–24.

THOMAS H (2007), 'Plasma modification of wool' in Shishoo R, *Plasma Technology for Textiles*, Cambridge, Woodhead Publishing Ltd., 228–246.

THOMAS H, LEHMANN K-H, HÖCKER H, MOLLER M, THODE C and LINDMAYER M (2005), 'Plasma pre-treatment at atmospheric conditions for improved processing and performance characteristics of wool fabrics', *The 11th Int Wool Text Res Conf*, 4–9 September, Leeds, UK.

VELDSMAN D P and SWANEPOL O A (1971), 'The influence of the rate of chlorination on the shrinkproofing of wool with DCCA', *Appl Polym Symp*, **18**, 691–700,

WANG W, XU P, YAN K and CHEN S (2005), 'A siloxane modified polyurethane aqueous system for shrink-resist treatment of wool', *The 11th Int Wool Text Res Conf*, 4–9 September, Leeds, UK.

Wool Record (2004), 'Interest in plasma treatment', **163**(3717), 39.

Wool Record (2007), 'Breakthrough in wool insect-resist treatment', **166**(3749), 40.

XIN J H and DAOUD W A (2005), 'Method of providing a coating of titanium dioxide to an article and the article with this coating', World Patent WO2005/113443.

XIN J H, DAOUD W A and TAO X M (2004), 'Multifunctional surface treatment for textiles', US Patent US2004/0117915.

ZHANG H and LAMB R (2005), 'Superhydrophobicity – a nanotechnology application to wool textiles', *The 11th Int Wool Text Res Conf*, 4–9 September, Leeds, UK.

ZHANG H, LAMB R N and ASHLEY J W (2004), 'Durable superhydrophobic coating', World Patent WO2004090065.

ZHANG Q, SMITH E, SHEN J and BISHOP D (2006), 'An ethoxylated alkyl phosphate (anionic surfactant) for the promotion of activities of proteases and its potential use in the enzymatic processing of wool', *Biotechnol Lett*, **28**, 717–723.

ZUCHAIRAH I M, PAILTHORPE M T and DAVID S K (1997), 'Effect of glow discharge-polymer treatments on the shrinkage behaviour and physical properties of wool fabric', *Textile Res J*, **67**(1), 69–74.

8
The coloration of wool

D. M. LEWIS, The University of Leeds, UK

Abstract: Wool coloration has to respond to real and perceived environmental threats and this aspect tends to drive the selection of dyes and dyeing processes. Reactive dyes bring benefits such as excellent wet-fastness and some classes of reactive groups give excellent fibre protection, especially when dyeing medium to deep shades. The fundamental chemical principles behind the above important effects are described. The role of anti-setting agents such as hydrogen peroxide or maleic acid in the dyebath is described and explanations are given of their influence on the disruption of cysteine thiol breakdown to hydrosulphide anion, which is the trigger for the permanent setting observed following wool dyeing in boiling dye-baths. The chapter concludes with a description of the great opportunities provided by printing wool fabrics using ink-jet printing.

Key words: acid dyes, chrome dyes, reactive dyes, dyeing machine-washable wool, anti-setting agents, ink-jet printing of wool.

8.1 Introduction

Wool fabrics and wool-blend fabrics are highly prized by consumers because of their excellent comfort properties during wear, their drape characteristics and, not least, the ability of the fibre to be dyed and finished to give a range of high-fashion colouristic effects. Leeder[1] points out that wool has been bio-engineered over millions of years to be worn next to an animal's skin, and is thus better than other types of fibres in terms of comfort in wear; it possesses the properties of absorbing up to 30% of its own weight of water without feeling wet, and even giving out heat when it absorbs water. Wool scientists have met the challenges of possible performance deficiencies such as felting in household laundering procedures, and attack by moths and beetles during garment storage, by applying innovative chemistry.

Wool thus enjoys a position in the marketplace as a fine natural product whose reputation from the environmental impact point of view must be secure. This simple statement actually has profound implications for the finisher, dyer and printer of wool materials. The environmental implications result in the necessity for both the finisher and colourist to understand the whole chemical finishing chain as well as to be an expert in the science of wool dyeing.

Heavy metals in dyes or the dyeing processes are important issues; although reactive dyes are being used more and more as replacements to produce dyed materials having high wet-fastness properties they do not give the same level of light-fastness in pale depths as do dyeings produced with pre-metallised or after-chrome dyes.

The lack of brilliance in whites and pastel shades has been identified as a major market issue; solutions must be based on research into the photo-stability of the base wool fibre and on optimising the application of novel bright chromophores, even those not normally suited for dyeing wool. Millington has recently reviewed the factors influencing the photo-yellowing of wool and possible solutions.[2,3]

Increasingly shrink-resist processes based on chlorination as a pre-treatment are under pressure due to high absorbable organo-halogen (AOX) residues – the colourist has to be aware that up to 1000 mg/L AOX can be discharged from the dyehouse contracted to dye these substrates; it is vital that such pre-treatments are replaced with simple oxidation processes without adversely affecting dyeing properties. Combined dyeing and shrink-resist processing should be on the agenda, especially if one notes that wool is the only fibre requiring such anti-shrink processing.

Some of the remarkable findings in wool setting chemistry and their implications in dyeing will be reviewed, emphasising the value of using so-called anti-setting agents in the dyeing process to improve final wool quality. Novel technologies for ink-jet printing textile fabrics are gradually gaining a foothold; in particular this procedure makes printing more cost effective in short runs, a scenario ideally suited to wool printing. The special factors limiting the rapid adoption of such digital technologies for wool fabric printing will be considered.

8.2 The theoretical basis of wool dyeing

8.2.1 Anionic or so-called acid dyes in wool dyeing

Wool is a heterogeneous polymer, mainly made up of many different amphoteric proteins; it may be coloured with a variety of water-soluble dyes but in practice it is invariably dyed and printed with sulphonated dyes. Wool has been, and still is to some extent, dyed with simple, low molecular weight (circa 400–800 dalton), sulphonated chromophores, termed 'acid dyes'; as the dye-manufacturing and the dye-using industries grew more sophisticated these acid dyes were modified, usually by increasing their molecular size, to give better washing and light fastness properties. Thus acid-milling dyes, chrome dyes, pre-metallised dyes and reactive dyes all find use in various application sectors. The simplest way to write an acid dye structure is as $D(—SO_3^-Na^+)_n$ where D is a chromophoric entity which may be substituted azo, anthraquinone, phthalocyanine or others, and n is an integer, usually 1–3.

Since it is made up of many different amphoteric protein macromolecules, according to the pH and temperature of the surrounding solution, wool can show clear differences in charge; these differences are brought about by protonation or deprotonation of carboxylic and amino side-chain functionalities. Figure 8.1 schematic describes this situation. It is of course an approximation and is governed by the acid dissociation constants of the —NH_3^+ and —COOH residues; for example at 25 °C lysine ε-amino has a pK_a value of about 10.5 and the glutamic acid terminal carboxylate is about 4.2.

Acid levelling dyes are generally monosulphonated dyes with little substantivity for wool and nylon under neutral or weakly acidic conditions but which exhaust well from boiling dyebaths under acidic conditions (pH 3-4). The structure shown in Fig. 8.2 is typical. The small molecular size of acid dyes means that there is very significant dye migration during their boil application, allowing good coverage of tippy wools and also for their application in dyeing systems where there is limited interchange between liquor and goods, e.g. piece dyeing in winches and hank dyeing of yarns.

Acid milling dyes are of greater molecular size and exhibit high 'neutral' (pH 5–7) substantivity for the wool fibre but can give unlevel dyeings due to their poorer levelling properties; the dyeings produced with these dyes have good fastness to water and mild washing treatments coupled with good light-fastness. A typical example of this type of dye structure is shown in Fig. 8.3. Acid milling dyes, owing to their lower migration properties, find use in dyeing machines where there is good liquor/fibre interchange and these include package dyeing of wool yarn and soft-flow jet dyeing of wool piece goods.

Pre-metallised dyes fall into two categories; the older 1:1 dye : metal complex and the later 2:1 dye:metal complexes. The complexing metal cations are

NH_3^+ COOH	NH_3^+ COO^-	NH_2 COO^-
Wool fibre +ve charge	Wool fibre zero charge	Wool fibre –ve charge
pH < 4	pH 4–8	pH > 8

8.1 Wool surface charge vs pH.

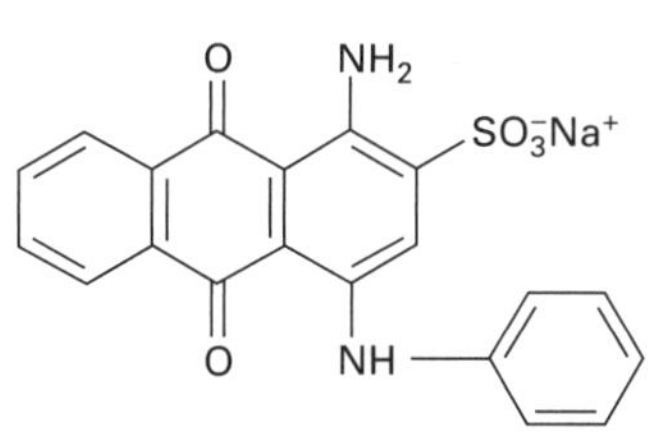

8.2 CI Acid Blue 25.

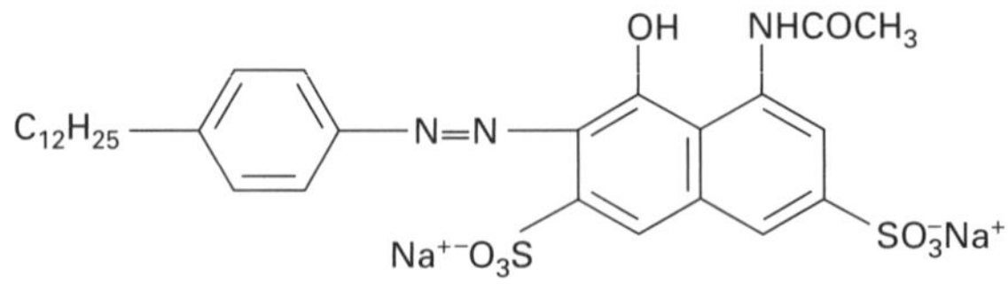

8.3 CI Acid Red 138.

either chromium$^{(3+)}$ or cobalt$^{(3+)}$. The application conditions for these two classes of dye are quite different – the former are dyed at pH 1–2 to achieve maximum levelling, as required for piece dyeing, whereas the latter are dyed at pH 6–7 since they have very high substantivity due to their large molecular size and hydrophobicity. Comprehensive reviews are available.[4,5]

8.2.2 Physical chemistry of wool dyeing

Classical wool dyeing theory attempts to model the physical chemistry of wool dyeing according to electrostatic principles, the Gilbert/Rideal approach or according to the Donnan membrane theory; it is not the intention to cover these approaches in detail and therefore readers are referred to texts which deal very fully with these analyses.[6,7] The assumptions made in applying the above physical chemical equations to the wool dyeing system are too general and do not take into account the bewildering complexity and heterogeneity of the wool fibre proteins – let alone the fact that the fibre structure changes physically and chemically during the boiling process itself.

The following is therefore an attempt to explain some of the more puzzling aspects of acid dye absorption by wool. As mentioned above, the initial driving force for dyeing to occur with a simple acid dye/polyamide fibre combination is undoubtedly Coulombic, but, depending on dye structure, non-polar interactions are also capable of playing an important role. In terms of Coulombic interactions, an important factor which affects the rate of acid dye uptake and its final saturation value on the fibre is the total number of protonated amino sites in the fibre. In this context it is valuable to make a comparison of wool, silk and nylon (Table 8.1).

Table 8.1 Amino groups in wool, silk and nylon

Wool	(lysine primary amino plus histidine secondary/tertiary amino plus arginine guanidino plus α-terminal amino)	0.820 moles per kilogram
Silk	(lysine, plus histidine plus α-terminal amino)	0.150 moles per kilogram
Nylon 6,6	(α-terminal amino)	0.036 moles per kilogram

Thus one may expect the saturation values of acid dyes on these three polyamide fibres to be in the same ratio as the above basic group contents, i.e. wool : silk : nylon = 23:4:1. This effect was verified experimentally by Skinner and Vickerstaff[8] who studied the equilibrium uptake of CI Acid Blue 45 (1,6-diamino-4,8-dihydroxy-3,7-disulpho-anthraquinone) on the above three polyamide fibres at pH 1.6 and 85 °C; the corresponding ratio obtained was 14:4:1.

The effect of varying the number of sulphonate groups in the dye molecules on the substantivity of acid dyes for nylon, wool and human hair, has been examined by numerous workers.[6–12] These workers clearly demonstrated that dye substantivity for the polyamide substrates decreases in the following order:

monosulphonate > disulphonate > trisulphonate > tetrasulphonate

In the light of the above observations it is thus important to discuss the various types of molecular interactions which may occur in dye–fibre systems. In terms of non-covalent interactions, the different forces between two molecules, like or unlike, are usually divided into five categories:[13] van der Waals forces, electrostatic interactions, induction forces, charge transfer stabilisation effects and solvophobic interactions.

The three categories large enough to be of importance in terms of dyeing processes are as follows:

1. *Van der Waals (VDW) interactions*: these non-covalent molecular interactions, are the sum of dispersion and repulsive energies. The thermodynamic strength of VDW interactions is smaller than 8 kJ mol^{-1}.[14]
2. *Electrostatic interactions between static molecular charge distributions*: in dyeing processes these include not only, for example, obvious attractions such as those between protonated amino groups in polyamide fibres and sulphonated anionic dyes, but also, according to classical definitions, hydrogen bonding.[13] A recent review points out that this may be an inadequate definition[15] but for the purposes of this analysis it is suitable. Hydrogen bond strengths vary considerably, but for neutral molecules they lie in the range 10–65 kJ mol^{-1} and when one of the components is ionic this range rises to 40-190 kJ mol^{-1}.
3. *Solvophobic or hydrophobic interactions*: these arise from the effect of non-polar parts of water-soluble solutes on the structure of water. When such molecules are brought into aqueous solution through a relevant solubilising group (e.g. sulphonate) then the water structure must change to accommodate the non-polar or hydrophobic residues. This change represents a gain in entropy for the whole system. Since most dyeing processes are restricted to aqueous systems, hydrophobic interactions

are likely to play an important role in determining both dye uptake by fibres and subsequent wet-fastness. Zollinger,[11] in his George Douglas lecture, highlighted hydrophobic interactions as being responsible for over-dyeing effects on nylon as well as the unusually high substantivity of dyes containing bulky aryl or alkyl residues for wool.

Since dyes and fibres such as wool, silk and polyester both contain aromatic residues it is worthwhile to consider current thinking regarding π–π interactions. Hunter[14] points out that these are commonly used to explain interaction between two or more aromatic molecules, but current evidence shows that they are negligible compared with electrostatics.[16]

On first sight it might be expected that, since aromatic groups are planar, maximum VDW interactions occur in a perfectly flat stacked arrangement. In water, hydrophobic interactions will also favour stacking, since the flat π-electron surfaces of the dye molecules are non-polar.[17] Thus authors of papers, e.g. Giles[18] on dye-aggregation in aqueous solution tend to draw the aggregates as perfectly stacked molecules with maximum overlap of free aromatic rings. Hunter and Sanders[19] have studied the geometry and energy contour plots of two stacked porphyrins and noted that the aromatic residues were in an offset or staggered arrangement. To explain these observations the above authors concluded that electrostatics provide a large repulsive force which pushes the π-systems away from the usually accepted maximum overlap position; the best model involved a positively charged σ framework sandwiched between two idealised 'π-atoms'. Continuing this analysis led Hunter[14] to the conclusion that certain face-to-face arrangements of aromatic systems lead to repulsion and other arrangements such as face-to-edge and offset lead to attraction. He was thus able to construct a very useful diagram, which is reproduced in Fig. 8.4.

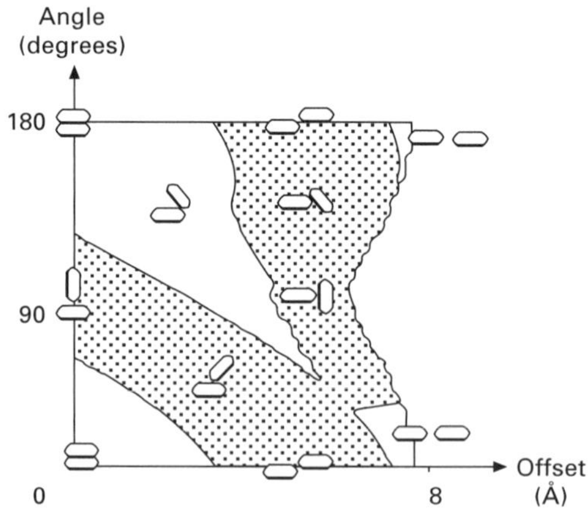

8.4 Electrostatic interaction between two benzene rings as a function of orientation (from Hunter[14]).

This new way of thinking about π–π interactions[14] allowed a ready explanation of the characteristic herringbone packing of aromatic hydrocarbons in the crystalline state[20] and also explained the phenyalanine–phenylalanine geometries found in X-ray crystal structure analysis results from certain proteins.[21] The maximum electrostatic interaction for two π electron systems was subsequently calculated as 6 kJ mol^{-1}, which is relatively weak. However, dyes are poly-aromatic systems and in fibres such as wool the aromatic side-chain residues in phenylalanine, tyrosine and tryptophan often occur in the same region; thus the overall small contributions, when summed, become significant. The author[22] contends that the previous explanations of non-polar interactions in dye–fibre systems have neglected the special case of aromatic π–π interactions. Attention should be drawn to the anomaly that in aqueous systems hydrophobic interactions increasingly break down above about 60 °C and many of the synthetic polyamide fibre overdyeing studies were carried out in boiling aqueous solution. It is thus proposed that, rather than hydrophobic interactions, aromatic or π–π interactions are largely responsible for the over-dyeing effects observed when dyeing synthetic polyamides with mono-sulphonated dyes.

8.2.3 Pathways to dye diffusion

The most widely accepted view regarding the pathway to dye diffusion, developed by CSIRO workers,[23–25] is that the dye molecules pass rapidly through the heavily water swollen cell membrane complex (CMC) proteins, initially between the cuticle scales, and thence into the cortex. In one approach[23] the above workers prepared anionic metal-complex dyes as heavy metal (Pt, Pd and U) chelates which, owing to their nuclear denseness, have high electron scattering power, and hence show up in transmission electron microscopy (TEM); they were able to determine the location of dye according to dyeing time and dyebath temperature. The results showed that these anionic dye molecules enter the wool fibre between cuticle cells and then diffuse into the non-keratinous endocuticle and cell membrane complex material, into the inter-macrofibrillar material and into the nuclear remnants; in the case of the dyes studied significant amounts were finally associated with the more hydrophobic proteins of the A-layer in the exocuticle and with the high sulphur matrix proteins in the cuticle. Brady[26,27] used fluorescence spectroscopy to following the diffusion pathways of a fluorescent rhodamine dye (C I Acid Red 52) and came to very similar conclusions – reinforcing the thesis for dye diffusion via the inter-cuticular, non-keratinous route into the fibre. It is clear that fibre reactive dyes may show a different diffusion profile since they would form covalent bonds and become immobilised earlier than non-reactive dyes. This latter proposal was proven by Lewis and Smith[28] using a vinylsulphone derivative (VS), a sulphatoethylsulphone derivative (SES) and a non-reactive sulphanilic acid/hydroxytriazine derivative of a sulphonated di-amino-stilbene as fluorescent brightener models for reactive and hydrolysed dyes. The uptake pathways of such colourless models when dyed at the boil for different time periods, could thus be followed using fluorescence microscopy – UV excitation at circa 340 nm resulted in the emission of intense blue visible light. It was thus shown that the highly reactive free VS form did not penetrate the fibre but reacted with the endocuticule and the endocuticular intercellular regions; the less reactive SES form partially penetrated the fibre but was also mainly associated with the above regions; the non-reactive model gave full penetration of the fibres after dyeing for 1 h at the boil. Figure 8.5, taken from the original thesis,[29] shows these fluorescence microscope results after dyeing 15 min and 60 min at the boil. These are very important findings and demonstrate that the distribution of fibre-reactive dyes has to be very different than acid dyes without fibre-reactive residues (acid, acid milling and metal-complex dyes). It is particularly significant that reactive dyes tend to selectively covalently bond to and hence modify proteins in the endocuticular regions.

The important question as to why the wool fibre cuticle surface does not allow trans-cellular diffusion of dyes was actually addressed prior to the above studies by such workers as Hall;[30] this author proposed that dyes gain entry to the wool fibre interior via the junctions between the scales rather than directly through the scale surface. The presence of a barrier to dyeing

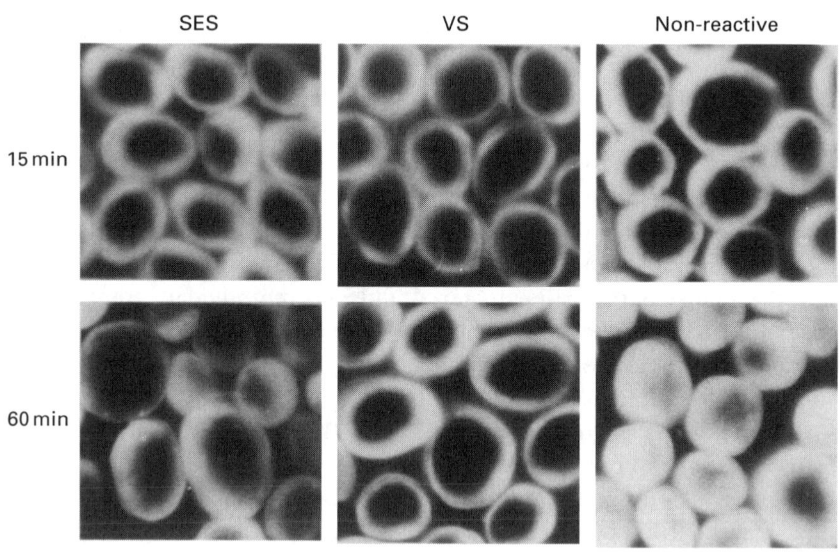

8.5 Fluorescence-microscope images of fibre cross-sections from wool 'dyed' with reactive/non-reactive dye models showing diffusion pathways.

at the scale surface is widely accepted [e.g. Rippon[25]]; the nature of this barrier has become the subject of intensive research since it is expected that a simple process to remove it would render wool more readily dyeable and printable – and possibly even more importantly readily shrink-resisted. Current evidence is that the surface barrier is a proteo-lipid with the lipid being covalently bonded to the protein via a thioester linkage to cysteine.[31] A simple process to remove the lipid would produce a modified fibre[32] which would be more readily dyed and printed, and even shrink-resisted compared with the untreated fibre; chlorination does partially remove this surface lipid but its use may give environmental problems related to the production and discharge of organo-halogen pollutants. Rippon[25] has developed an amphoteric surfactant, Valsol LT or Sirolan LTD, for treating wool in a simple pre-scour prior to dyeing – since this treatment brings about some lipid removal the wool becomes more readily dyeable at 80–90 °C. Perachem Ltd have patented a method for rapid removal of covalently bonded surface lipid which is currently being trialled.[33]

8.3 Environmental impact of wool coloration processes

8.3.1 Chrome dyeing processes

In order to achieve very black shades on wool, chrome dyes have found widespread usage since they give the desired shade of black, coupled with outstanding wash- and light-fastness. It is interesting to note that the blackest blacks produced for the Japanese tuxedo market are invariably produced by dyeing chrome dyes on chlorinated wool. It has been estimated that chrome dyes still share some 20% of the total wool dye market, predominantly for dyeing black and navy blue shades. Invariably the dyeing is after-treated with sodium dichromate to produce the metal complex dye in the fibre. It is the toxicity of Cr(VI) which should give rise for concern both from the point of view of its effect on worker's health and its effect in the environment following discharge of exhausted dyebath. Chromium(III) is relatively safe, it is naturally present in soils and is even considered an essential human dietary component. If Cr(III) could be used then chrome dyeing would be a much more environmentally friendly procedure.

The Cr(III) cation cannot be applied directly as the chroming agent as under the acidic conditions used in dyeing it has low substantivity for the positively charged, protonated wool fibre. However, the trivalent chromium cation can be readily converted to an anionic complex by reaction with suitable ligands; such complexes have substantivity for wool under hot acidic conditions. The complexes must not be too stable since the carrier ligands have to readily exchange the chromium with the ligand groups in the dye.

Suitable complexing compounds include α-hydroxycarboxylic acids such as lactic acid and salicylic acids, and polycarboxylic acids such as maleic acid.[34–41] Much of the published work has focused on lactic acid[34–38] and 5-sulpho-salicylic acid.[39,40]

An important factor to be taken into consideration when replacing Cr(VI) with anionic Cr(III) complexes is the oxidising effect of the dichromate anion. During after-chroming dyed wool with dichromate, absorbed Cr(VI) is reduced by the wool fibre to a Cr(III) species which complexes with the dyestuff to give full rich shades of excellent wet-fastness. The exact nature of the oxidising effect of dichromate is not fully understood, but it has been shown that in order to produce equivalent dyeings with the Cr(III) complex system a substantial amount of an oxidising agent (usually hydrogen peroxide) must be added during the chroming stage to effect full shade development and to produce dyeings of the expected wet-fastness. The amount of hydrogen peroxide required is influenced by the dye used; in some cases the dye molecule itself should be oxidised to give the expected chrome-dyed shade as is the case when dyeing with CI Mordant Black 9 (Diamond Black PV – DyStar).[38] This oxidation process is shown in Fig. 8.6.

Xing and Pailthorpe have described the use of the anionic complex prepared from 5-sulpho-salicylic acid and chromic sulphate[39] for after-chrome dyeing of wool and have recently further developed the method for the industrial processing of cashmere fibres.[40] To achieve the correct shade and wet-fastness properties these authors proposed the addition of a small amount (0.6% on mass of fibre, o.m.f.) of Cr(VI) to the after-chrome process 5 minutes after the addition of the Cr(III) complex. The structure of this latter complex was not revealed, simply being coded SCA-Cr; this addition of Cr(VI) can be interpreted as an oxidation step. Lewis, *et al.*[41] have described an after-chrome dyeing system using sodium maleate as the preferred complexing agent for Cr(III) cations and oxidation at the end of the after-chroming process with hydrogen peroxide.

The maleate complex was readily prepared by dissolving maleic anhydride (18.7 g) in water (100 mL) and the solution boiled for 5 minutes, then cooled; this solution was neutralised using solid sodium carbonate (20 g); chrome alum (48.4 g) was dissolved in water (200 mL), this solution was added to the sodium maleate solution and the volume was made up to 500 mL with water; boiling for 5 minutes ensured complex formation (ligand : chromium = 2.1). When dyeing 1 kg of wool, 500 mL of this solution would be added to the dyebath at the after-chrome stage giving a Cr factor of 0.35. The nature of the complex is shown in Fig. 8.7.

Lewis *et al.*[41] developed the following chrome dye application process: dyeing was carried out in a 1 kg Obermaier sample package dyeing machine. The bath, containing 1 kg of wool yarn wound on a package, was set at 50 °C at a liquor to goods ratio of 16:1, and at pH 4.0; chrome dye was added, the

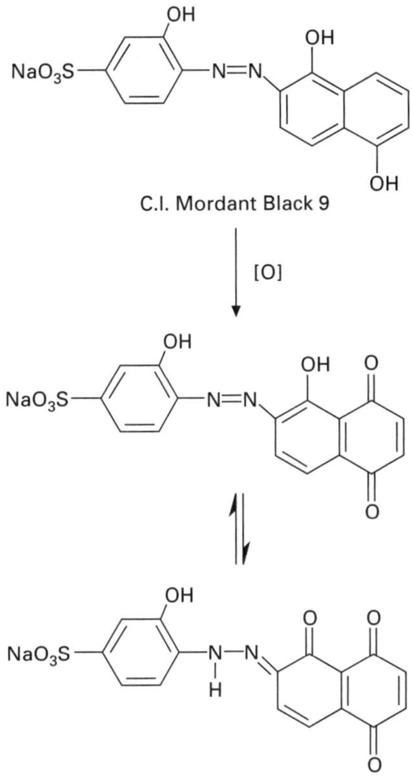

8.6 Oxidation of Diamond Black PV (CI Mordant Black 9).

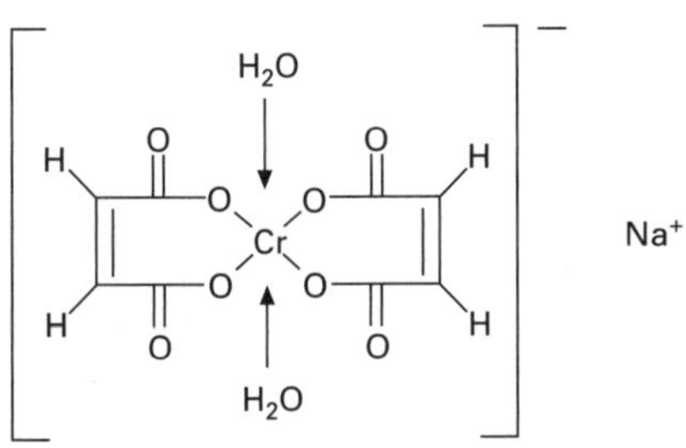

8.7 Maleic acid/chromium (III) complex with Na^+ counter-ion.

bath raised to the boil over 40 minutes and maintained at the boil for 1 hour. The temperature was then lowered to 80 °C and the appropriate amount of Cr(III)/maleate complex added; the pH was adjusted to 3.5–4.0 with formic acid and the bath returned to the boil. After 15 minutes hydrogen peroxide (65 mL of 35% w/w material) was added and after another 15 minutes a further 65 ml H_2O_2 (35%) added; boiling was then continued for a further 25 minutes. The dyed package was washed-off for 20 minutes at 50 °C (pH 8.0

with ammonium hydroxide, 1 g/L 0.880 NH_3); final rinsing with dilute acetic acid (pH 5) at 40 °C completed the process.

For comparative purposes the following standard chrome dyeing process was carried out. In this case the first dyeing stage was as described but the after-chrome process was performed by cooling the dyebath to 80 °C and adding the requisite amount of dichromate (for 4% o.m.f. Diamond Black PV (200%) a Bayer G Cr factor of 0.35 was used); the bath was returned to the boil and sodium thiosulphate ($Na_2S_2O_3{\cdot}5H_2O$) added (1.7 × mass of $K_2Cr_2O_7$ used). Boiling was continued for a further 30 minutes.

In the case of those dyes that do not change structure on oxidation, for example, Diamond Fast Bark Blue RRN (CI Mordant Blue 9) and Diamond Fast Bordeaux BL (CI Mordant Red 30), the total amount of peroxide required for a 1 kg package dyeing must be reduced to 120 ml of 35% H_2O_2. The requirement to add peroxide is necessary to produce dyeings having a wet-fastness level comparable to those dyeings produced with the usual dichromate process. In particular, potting fastness tests critically distinguish the non-peroxide after-treated samples.

It was proposed that the necessity to use oxidising agents to achieve the maximum wet-fastness properties from the Cr(III)/maleate system is due to the strong binding of Cr(III) to thiol ligands. During dyeing of wool at the boil significant setting occurs due to thiol–disulphide interchange reactions;[42] undoubtedly the small amount of cysteine-free thiol (circa 39 μmoles per kg of wool) undergoes β-elimination to produce dehydroalanine and H_2S. The latter attacks intact disulphides to produce further free thiols, initiating a chain reaction which produces significant permanent set. There is thus a significant concentration of free thiols in the wool at the end of the dyeing cycle which, being highly nucleophilic, selectively binds the chromium when the Cr(III) maleate complex is introduced. An oxidant is thus necessary to free the chromium from this thiol ligand binding in order to make it available for complex formation with the dye. Figure 8.8 summarises this mechanism.

If oxidation in the after-chrome cycle re-forms toxic Cr(VI) then these research results would not be too useful practically. Thus the sensitive diphenyl-1-carbazide colour test for Cr(VI) formation was carried out on residual dyebaths from dyeings produced by the new process; no positive results were seen. However when peroxide oxidation of the Cr(III) complexes was carried out under alkaline conditions, clear evidence for Cr(VI) formation was seen.

Chromium content of residual dyebaths

Analysis by atomic absorption spectroscopy indicated that the effluent obtained from the chrome(III) complex dyeings contained considerably higher concentrations of residual chromium than the dichromate standards. However,

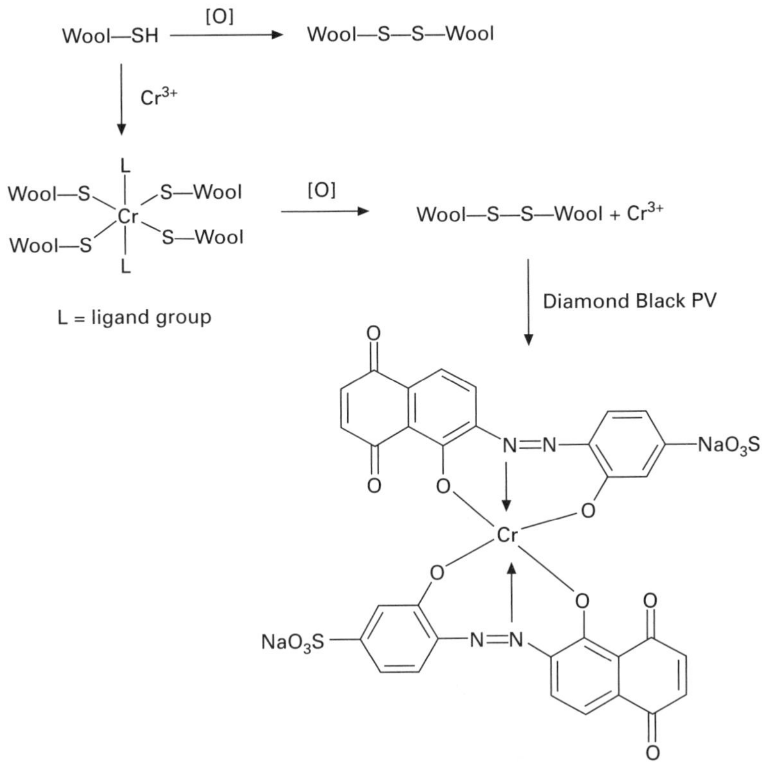

8.8 Proposed mechanism for the action of hydrogen peroxide in the Cr (III)/maleate complex chroming system.

Table 8.2 Residual chromium levels in effluent from 4% o.m.f. Diamond Black PV (200%) dyeings with varying Cr factors

Cr factor and method	Chromium level detected in effluent (ppm)
0.35 Dichromate-thiosulphate standard	12.0
0.35 Cr(III) complex	104.0
0.25 Cr(III) complex	53.6
0.20 Cr(III) complex	32.8
0.175 Cr(III) complex	24.6
0.15 Cr(III) complex	12.3
0.12 Cr(III) complex	11.5

a significant reduction in the concentration of residual chromium may be achieved by a reduction of the Cr factor when using the chrome(III)/maleate system (Table 8.2). Trials have shown that the Cr factor may be reduced to 0.20 without affecting the shade or fastness to potting. Reduction of the Cr factor to lower than 0.175 leads to poor chroming and inadequate shade development.

Further reduction in the chrome content of the effluent was achieved by fresh bath chroming. However, King and Brady[43] have already noted this effect when dyeing with dichromate and attributed it to soluble protein reacting with dye and chromium in the after-chrome bath. Rinsing prior to after-chroming will remove residues of dye and soluble protein which could compete in the bath for the Cr(III) complex. Amino acid residues such as aspartic acid, glutamic acid, cysteine and histidine will form strong complexes with Cr(III).

The reflectance spectra of 4% o.m.f. dyeings of Diamond Black PV chromed by the standard and the new Cr^{3+} maleate method were taken and *K/S* values calculated, where *K* is the coefficient of absorption and *S* is the coefficient of scalter. Figure 8.9 reproduces these spectra and it is clear that the colour yields of the two processes are very similar.

8.3.2 Absorbable organo halogens in wool dyeing and finishing effluents

Special attention is required when discharging compounds containing AOX. The concept of AOX has arisen from a German drinking water directive (DIN 38409414 – 1985). Consent limits as low as 2 mg/L have been applied.

Products used by the dyer and printer which fall into this category are the insect-resist agents applied to wool in dyeing, some types of carriers for disperse dyeing polyester, certain chromophores themselves and some halogen-containing classes of reactive dyes. This type of foreseeable legislation will therefore direct the colour chemist to develop new systems which will reduce

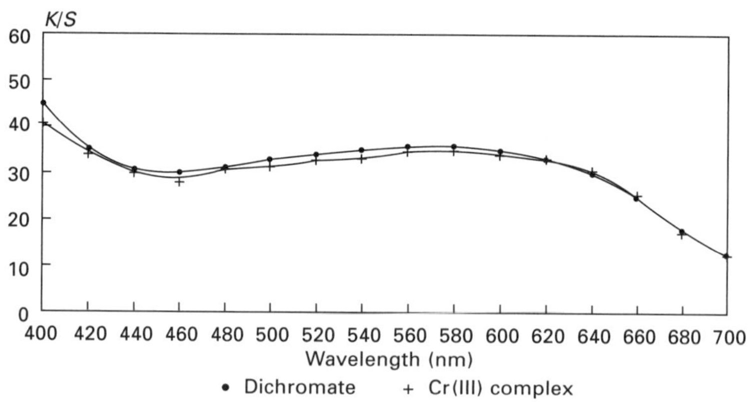

8.9 K/S spectra for Diamond Black PV, using standard dichromate and the chrome maleate complex (fresh bath chroming methods).

these discharges to the allowable limits. These implications are indeed far-reaching. For example, in the case of reactive dyeing cellulosic fibres, the demand for achieving virtually 100% dye uptake and covalent bonding is urgent. Research is therefore required whether to develop innovative reactive systems or to develop fibre pre-treatments which solve the problem of existing reactive dyes.

It is worthy of note that dyeings of chlorinated wool leads to effluents containing on average more than 1000 mg/L AOX. So far special dispensation, to discharge this concentration of AOX, has been given to dyers of machine-washable wool produced by the chlorine-Hercosett (TM Hercules Powder Corporation) continuous top process. An important research activity is the production of machine washable wool using non-chlorine procedures. In fact there is renewed urgency in this area as expected EU directives are likely to force processors to abandon the chlorination route.

Absorbable organo-halogen (AOX) free shrink-resist finishing and its relation to dyeing

Currently the most popular method to produce machine-washable wool tops is the so-called chlorine-Hercosett process. The chlorination pre-treatment removes much of the covalently bound lipid material from the wool surface and also oxidises surface cystine disulphides to cysteic acid; these effects cause the fibre surface to become both hydrophilic and anionic and highly receptive to the cationic polymer, Hercosett, which 'exhausts' and spreads onto the surface of each individual fibre.

Alternative systems for preparation of the fibre surface for shrink-resist treatments based on peroxide chemistries have been explored but to date have not realised the same commercial success as chlorine-Hercosett. Most studies have in fact explored the potential of Caro's salt ($KHSO_4 \cdot KHSO_5$) as a direct chlorine replacement[44] but in this case it is vital to include a sodium sulphite treatment not only to remove excess oxidant but to generate sufficient shrink resistance; undoubtedly this system produces anionic polypeptides at the fibre surface by a process of oxidative sulphitolysis[45] to give cysteine-*S*-sulphonate (wool Bunte salt) as the hydrophilic anionic residues. In continuous processing this technique has not allowed the application of Hercosett polymer, presumably because removal of covalently bonded lipid is inefficient compared to chlorination processes. Other workers[46] have trialled permonosulphuric acid/sulphite followed by silicon polymer application in an attempt to re-dress the above problem; unfortunately silicon polymers lead to subsequent problems in spinning and dyeing.

It is something of a hindrance, and an extra cost to be borne by the processor, that a separate process has to be carried out to shrink-proof wool, usually prior to dyeing. The author has reported on a possible system[47] to achieve these objectives; this system involved the co-application, in a pad–batch fabric dyeing process, of a tri-functional Bunte salt-acetoxy-polyol, with reactive halo-*s*-triazine reactive dyes. Providing that urea and sodium bisulphite were also present in the pad-liquor, excellent dye fixation and level dyeings were obtained – the fabrics were also machine washable and possessed a good soft handle. The Bunte salt 'capped' polyol is water soluble and surface active; reaction with wool thiol groups converts it to a crosslinked poly-disulphide-polyol which is water insoluble and imparts the shrink-proofing effect via fibre–fibre spot welding.

8.4 Reactive dyes for wool

The chemistry of reactive dyes for wool has been fully reviewed[48,49] but the most important features will be detailed. Studies on wool were of great significance in the early development of reactive dyes. In particular the ω-chloroacetylamino reactive group appeared as early as 1938 in the IG dye, Supramino Orange R, as shown in Fig. 8.10.

Ciba put together a small range of bright wool dyes containing this grouping in 1954 (Cibalan Brilliant). In 1952, following the patented work of Heyna and Schumacher,[50] Hoechst marketed Remalan and Remalan Fast, 2:1 premetallised reactive dyes for wool which contained pendant sulphatoethylsulphone residues capable of elimination to vinylsulphone reactive groups when applied from boiling dyebaths above pH 5. Hoechst even sold chrome dyes with the sulphatoethylsulphone residue being incorporated in the dye molecule, these being Metachrome Orange H3R, Metachrome Gey RL, Salicinchrome Orange H-3R and Salicinchrome Grey RL. However, it seemed that the marketing push for selling these dyes as fibre reactive systems was somehow lacking, probably because wool had only a small share of the textile fibre market. It was only when developments took place which allowed covalent fixation of dyes on the major textile fibre, cotton, that a major marketing effort commenced to sell the reactive dye concept for wool dyeing.

Also it should be noted that the above early reactive dyes for wool gave considerable problems of unlevel dyeing, especially in piece dyeing and yarn hank dyeing; thus their usage was mainly in the fields of loose-wool stock and wool top dyeing. A major advance in reactive dyes for wool

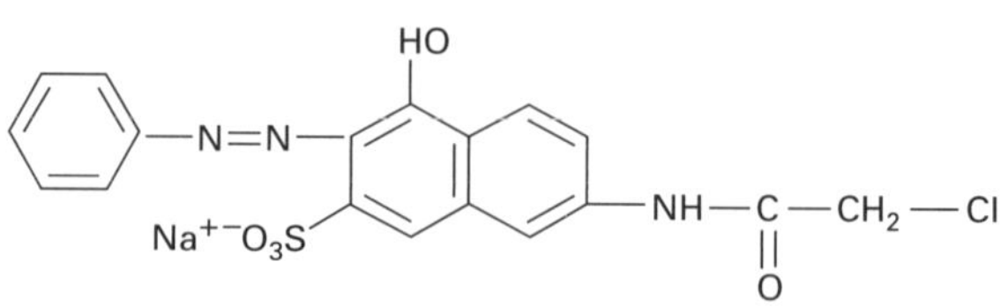

8.10 Supramino Orange R.

occurred in 1966 with the launch of the Lanasol dyes (Ciba-Geigy); these dyes were based on the α-bromoacrylamido group[51] and were sold with the very important level dyeing auxiliary, Albegal B. Albegal B is an amphoteric product which possibly has the general structure shown in Fig. 8.11.[52]

It is noteworthy that dyeing in the presence of this novel product alleviated the problems of unlevel dyeing, associated with reactive dyeing of wool. By happy coincidence the launch of the Lanasol dyes coincided with the marketing of truly machine-washable wool, produced by the continuous chlorine–Hercosett treatment of wool tops, a process developed by scientists at CSIRO and IWS.[53] Other companies followed suit, Bayer and Sandoz launching difluoro-monochloro-pyrimidine dyes (Verofix/Drimalan F)[54] and Hoechst launching a reassembled range of 'blocked' vinylsulphone dyes (Hostalan).[55] Dyeings produced on the chlorine-Hercosett treated substrate with acid dyes, acid milling dyes and premetallised dyes did not show adequate wash-fastness properties but dyeings produced with reactive dyes gave outstanding washing performance.[56]

Modern reactive dye systems for wool fibres

In the light of increasing environmental concerns with heavy metals it is desirable to use reactive dyes to match deep shades of black and navy blue in order to offer the dyer a real alternative to chrome dyes. In this context dye manufacturers have increased their efforts to offer wool dyers ranges of attractively priced reactive dyes; examples include Lanasol CE dyes from Ciba, Realan dyes from DyStar and Drimalan dyes from Clariant. Black and navy blue shades are often based on the popular and widely available dye, CI Reactive Black 5. This dye is a bifunctional reactive dye since it contains two sulphatoethylsulphone residues, which activate to vinylsulphone on boiling at pH 5.5.

Müller[57] has reviewed the nature and amount of AOX residues in textile effluents; various reactive dye systems such as halo-triazines and halo-pyrimidines can give problems and AOX discharges can be reduced by selecting dyes containing only sulphatoethylsulphone residues as the reactive group.

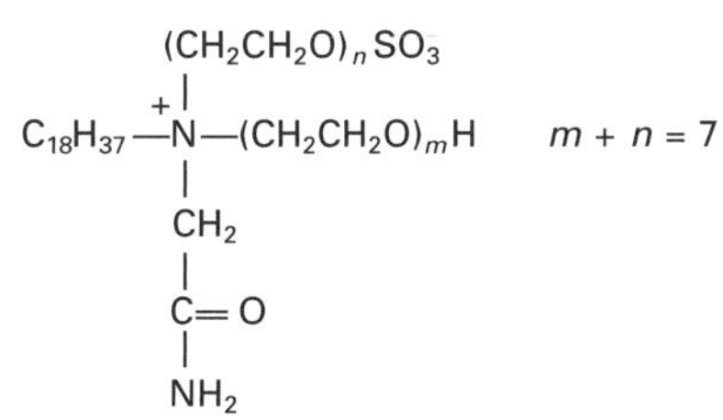

8.11 Amphoteric levelling agents for use in wool reactive dyeing.

8.5 Wool protection during dyeing by using reactive dyes and/or anti-setting agents

Permanent setting is a contributory factor leading to loss in wool fibre strength during dyeing. It is also the main cause of increased hygral expansion of wool fabrics following piece dyeing, the reason for surface marks such as 'crow's-feet' in piece dyeing and the source of reduced bulk or yarn leanness following package dyeing.[58–60] Apart from the above wholly negative effects, permanent setting can sometimes be seen as beneficial: such instances include the setting of wool yarn in hank form when using hank dyeing machines – this gives extra bulk and resilience to yarns and explains why this dyeing route remains popular for the production of wool carpets. It is necessary to summarise the various chemistries involved in the production of permanent set in wool dyeing.

Wool is a heterogeneous material made up of keratin proteins, a small amount of non-keratinous proteins and even smaller amounts of lipid and fatty acid materials. The keratinous or cystine disulphide crosslinked proteins are responsible for most of wool's physical properties and thus the reactivity of the cystine disulphide residue is of paramount importance. The cystine content of wool varies but an appropriate average value is 450 μmol g^{-1} and for its reduced form, cysteine, 30 μmol g^{-1}. The chemistry of these residues bound to the protein chains through the amide linkage is illustrated in Fig. 8.12.

Both these amino acid residues can readily undergo a *trans*-1,2-β-elimination reaction to form the dehydroalanine residue which contains an activated double bond capable of subsequent Michael addition with suitable nucleophiles. These reactions occur in water; the extent and nature of the reaction depends especially on the parameters pH, temperature and time.[61–63] The reactions are activated by the electron withdrawing character of the adjacent amide carbonyl residue. In general the thiol form, cysteine, more readily undergoes β-elimination than the disulphide form, cystine; even at pH 3, in boiling aqueous dyebaths, there is some elimination of hydrosulphide ion from the cysteine residue whereas the cystine disulphide only undergoes such reactions above pH 7. Figure 8.13 summarises these reactions.

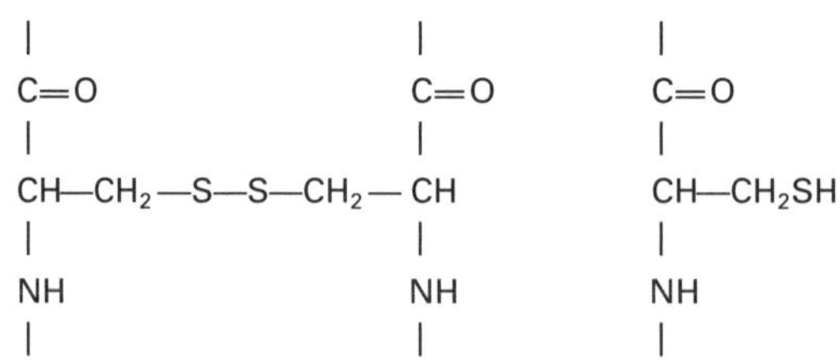

Cystine *Cysteine*

8.12 Representation of bound cystine/cysteine in wool keratin.

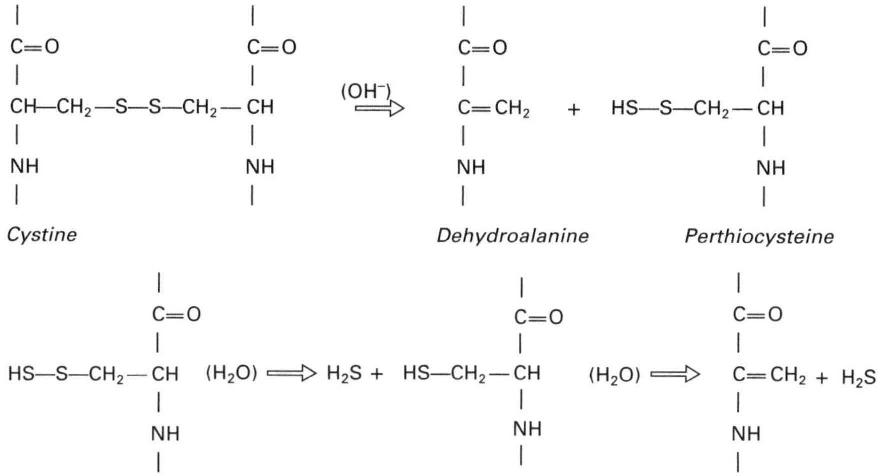

8.13 Elimination reactions of cystine and cysteine.

The hydrogen sulphide or hydrosulphide anion produced in the reaction is capable of ready reaction with cystine disulphide residues to produce further cysteine thiol residues which will undergo rapid β-elimination to dehydroalanine and hydrogen sulphide in boiling dyebaths; this is clearly the start of a runaway degradation reaction. The reactive entity, dehydroalanine, will undergo Michael addition with amino nucleophiles present in histidine and lysine residues and with thiol nucleophiles present in cysteine; in this way, more stable crosslinks such as histidino-alanine, lysino-alanine and lanthionine are formed.[64,65] The extent and exact composition of these new crosslinking amino acids vary greatly with pH and temperature of treatment, since the nucleophilicities of amino and thiol residues increase with pH and temperature increases. The chemistry of these crosslinking reactions is exemplified, for lysine and cysteine residues, in Fig. 8.14. The crosslinks will not undergo degradation or elimination reactions under conditions normally encountered in wool dyeing and are thus likely to be of great importance in explaining the phenomena of permanent setting during wool dyeing. Also of some importance in setting is the so-called thiol-disulphide interchange reaction.[66,67]

It is clear from the above discussion that control of setting in dyeing can be achieved by the addition of chemicals which scavenge hydrosulphide anions as they are liberated or which rapidly modify free cysteine thiol residues to prevent the elimination reaction; in practice this can be achieved in two ways:

1. Inclusion of oxidants in the dyebath.
2. Inclusion of fibre-substantive electrophiles in the dyebath.

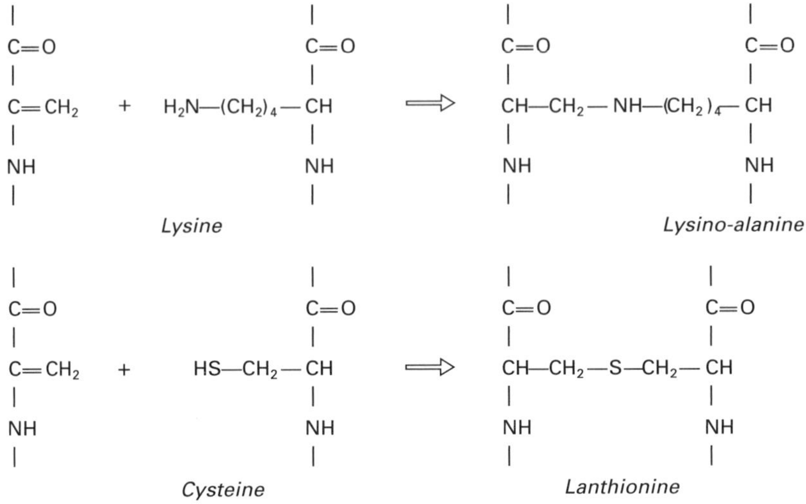

8.14 Formation of lysino-alanine and lanthionine crosslinks.

It is thus important to measure set following dyeing and most of the published research in this area uses Køpke's crease angle method to achieve this;[68] typically blank dyeings of wool fabric in pH 5 buffer for 1 hour at the boil, without anti-setting agent present, give set values of about 70%, whereas including an effective anti-setting agent gives a set value of about 30%.

Reactive dyes, dyed in moderate to heavy depths of shade, actively prevent damage in wool dyeing[69] especially those dyes which contain activated carbon–carbon double bonds and which thus react with fibre nucleophiles via a Michael addition mechanism (these dyes include acrylamido dyes and vinylsulphone dyes). The magnitude of this effect increases with increasing amounts of reactive dye applied being optimum at circa 3% dye o.m.f. The importance of this effect when dyeing wool fabric at pH 4 with the α-bromoacrylamido reactive dye, Lanasol Red 6G – 4% o.m.f., is demonstrated in Fig. 8.15.

It is interesting to reflect why reactive dyes based on reactive halogenated heterocycles, which react with wool fibre nucleophiles by a nucleophilic substitution reaction, are less effective in controlling wool damage in dyeing than are the activated carbon–carbon double bond type of reactive dye. From the above discussions it is clear that successful control of damage and set go hand-in-hand and it is thus necessary to look carefully at the reactivity/stability of the reactive dye-cysteinyl residue covalent bond. Thioether derivatives of triazine or pyrimidine heterocycles will react further with amines to form bonds of greater stability; the leaving group in this reaction being the substituted thiol.[70] The thioether formed from reaction with an activated carbon–carbon double bond is, however, resistant to nucleophilic

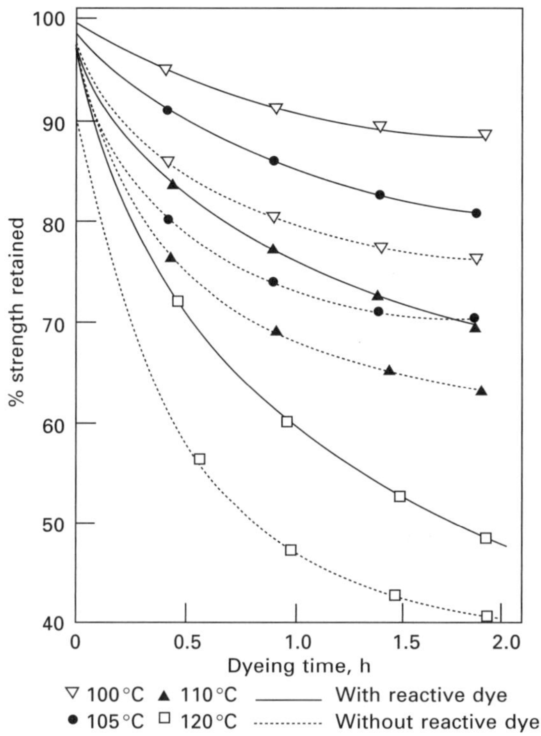

8.15 Effect of dyeing time and temperature on wool fabric strength.

attack or β-elimination under the mildly acidic conditions pertaining in wool dyeing.[71]

When set was measured from dyeings on wool fabric produced at pH 5 from the activated halogenated heterocycle type of reactive dye, Drimalan Red F-2G – Clariant (3% o.m.f.), a value of 74% was obtained; in contrast when the above dye was replaced with a reactive dye containing an activated carbon–carbon double bond, Lanasol Red 6G – CIBA (3%omf), a set value of 41% was obtained.[72] The reactions responsible for these differences are summarised in Fig. 8.16.

The importance of hydrogen sulphide as a catalyst to promote setting/wool damage under dyeing conditions was proven by Lewis and Smith[73] who demonstrated that a bis-(dye-sulphonylethyl)-thioether dye was present in the bath after dyeing with a model vinylsulphone dye. This dye arises from the reaction of the vinylsulphone with free hydrogen sulphide, according to the mechanism shown in Fig. 8.17.

Fluoro-chloropyrimidine dye:

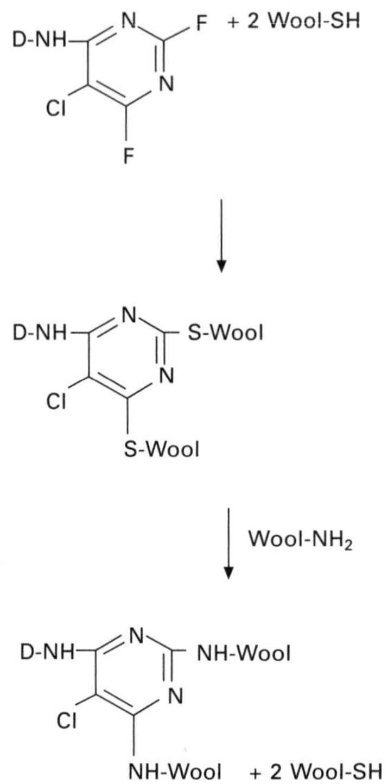

Activated carbon–carbon double bond dye:

$$\text{D-NH—CO—C(Br)=CH}_2 \quad + 2 \text{ Wool-SH}$$

$$\downarrow$$

$$\text{D-NH—CO—CH(S-Wool)—CH}_2\text{—S-Wool}$$

$$\downarrow \text{ Wool-NH}_2$$

No reaction

8.16 Cysteinyl reactions with active heterocyclic and double bond types of reactive dye.

$$\text{D—SO}_2\text{—CH=CH}_2 + \text{H}_2\text{S} \rightarrow \text{D—SO}_2\text{—CH}_2\text{—CH}_2\text{—SH}$$

$$(\text{D—SO}_2\text{—CH=CH}_2) \downarrow$$

$$\text{D—SO}_2\text{—CH}_2\text{—CH}_2\text{—S—CH}_2\text{—CH}_2\text{—SO}_2\text{—D}$$

Thioether dye

8.17 Thioether dye formation in wool dyebaths.

The role of oxidants to prevent setting in dyeing

One of the most useful anti-setting systems for use in wool dyeing was developed by workers at CSIRO, IWS and BASF.[74] This system, offering improved fibre physical properties, was based on a mixture of hydrogen peroxide and a special auxiliary Basolan AS; the latter auxiliary performed two functions:

- Inhibition of the degradative effect of hydrogen peroxide on some wool dyes.
- Enhancement of the stability of the oxidant in the boiling dyebath.

The BASF/IWS/CSIRO process[74] recommended dyeing wool in the presence of hydrogen peroxide (35%) at a level of 1 mL/L (minimum 2% on o.m.f.) and Basolan AS at 0.5 g/L (minimum 1% o.m.f.).

Kim and Lewis[75] studied the effect of hydrogen peroxide concentration on set after boiling wool fabric for 1 hour at pH values 3, 5 and 7 in the presence of citric acid/phosphate buffers (McIlvaine buffers); their results are reproduced in Fig. 8.18.

Hydrogen peroxide (30%) concentrations as low as $2\,cm^3\,L^{-1}$ gave set values of 21% at pH 3 (control without oxidant gave 56% set), set values of 27% at pH 5 (control 59%) and set values of 43% at pH 7 (control 67%). The effect of pH on set parallels the effect of pH on cystine degradation; if the

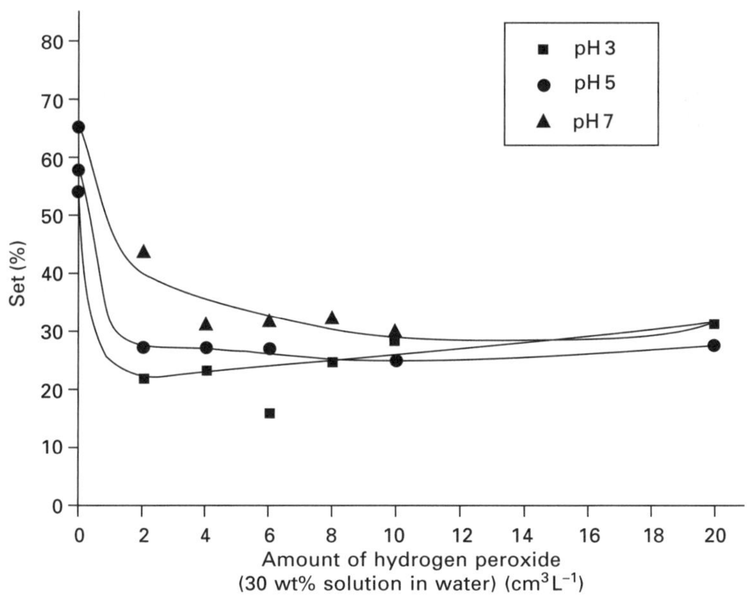

8.18 Percentage set after boiling wool, at pH values 3, 5 and 7, with H_2O_2.

mechanism is ß-elimination of cysteine as shown in Fig. 8.14, then a higher concentration of oxidant should be more effective at controlling set at pH 7 where elimination is more rapid – Fig. 8.18 confirms this hypothesis since 4 $cm^3\ dm^{-3}$ hydrogen peroxide gives the best control of set. The oxidant clearly functions by rapidly oxidising cysteine back to cystine or cysteic acid; of possible equal importance is the oxidation of hydrogen sulphide to bisulphate anion thus removing this reactive reducing nucleophile from the system.

Kim and Lewis[75] used Fourier transform infrared (FTIR) second derivative spectroscopy to follow the production of cysteic acid when treating wool at the boil with hydrogen peroxide (0-10 $cm^3\ dm^{-3}$); by measuring the sulphonate band intensity, attributed to cysteic acid, at 1040–1044 cm^{-1} it was shown that over-oxidation to cysteic acid was most significant in those treatments carried out at pH 7. Other oxidants shown to have anti-setting properties include sodium bromate and sodium tetrathionate.[76] It is not possible to use oxidants such as hydrogen peroxide when dyeing with reactive dyes since the perhydroxy anion is a potent nucleophile which reacts at the electrophilic site in the reactive dye to rapidly produce the hydrolysed, inactive dye.

The role of electrophilic reagents to control setting in dyeing

At the same time as the hydrogen peroxide-based Basolan AS-A system was launched, a system based on application of an electrophile Basolan AS-B was also marketed but the structure of the latter reagent remains undisclosed. Kim and Lewis[77] showed that sodium maleate (MAS) was a very effective anti-setting agent when included in boiling pH 3 wool dyebaths. Their results are reproduced in Fig. 8.19, which clearly demonstrates that the maleate anion is most effective at controlling set at pH 3; it shows only modest effect on set at pH 5 and has no effect on set at pH 7. The reason for this strong pH dependance lies in the modest substantivity of the maleate ion for wool under acidic conditions and its almost total lack of substantivity at pH 7 and above. It was estimated,[77] from capillary electrophoretic analysis of the treatment baths, that when using 2 g/L sodium maleate for the 1 hour boiling treatment, maleate anion uptake was 22% at pH 3, 15% at pH 5 and 4% at pH 7. The mechanism whereby maleate anions reduce the extent of wool setting in boiling aqueous treatments is described in Fig. 8.20.

Liao and Brady[78] synthesised *N*-naphthylmaleimide and showed it inhibited wool setting and also protected the wool component of a wool/polyester blend from serious damage during the high-temperature dyeing step required to adequately dye the polyester component with a disperse dye. These authors and others[74] attribute the main cause of setting, hence wool damage, in hot aqueous wool treatments to the thiol-disulphide interchange reaction without implicating cysteine residue ß-elimination reactions.

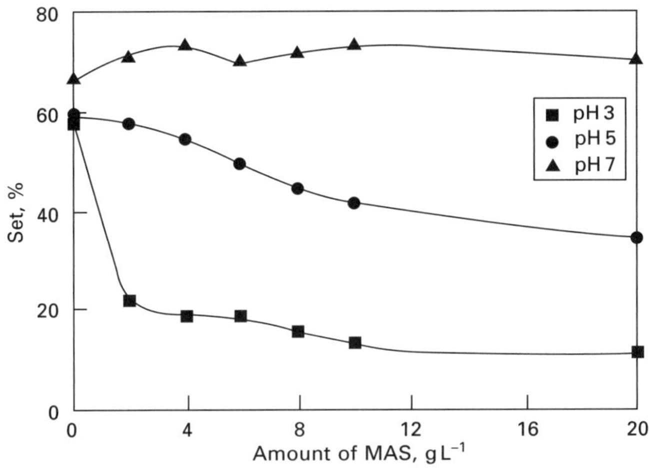

8.19 Percentage set after boiling wool, at pH values 3, 5 and 7, with sodium maleate.

Wool-SH + $Na^+ \ ^-OOC—CH{=}CH—COO^-Na^+$

Cysteine residue

⇓

Wool-S—$CH(COO^-Na^+)—CH_2—COO^-Na^+$

Reaction of non-absorbed maleate with H_2S in solution:

H_2S + $Na^+ \ ^-OOC—CH{=}CH—COO^-Na^+$

⇓

$Na^+ \ ^-OOC—CH_2—CH(—COO^-Na^+)—SH$

⇓

$Na^+ \ ^-OOC—CH_2—CH—CH(—COO^-Na^+)—S—CH(—COO^-Na^+)—CH—COO^-Na^+$

8.20 Addition reactions of maleate with cysteinyl residues and H_2S.

8.6 Ink-jet printing of wool fabrics

Printing of wool fabrics has always been a small-lot, high-fashion activity; this has meant that wool printing did not sit easily in print mills equipped with high productivity rotary screen printing machines capable of printing speeds of 50 metres per minute. However, even in the cotton fabric printing industry runs of 500 metres are now common and likely to become even smaller. Drawbacks of rotary screen printing methods include low machine utilisation efficiency or downtime; a pattern set-up can take 1 hour and screen washing and drying processes can take 1–2 hours. Also of concern is the time delay from print design to screen commissioning and screen delivery; this can be 3 or 4 weeks. Given the demands for individual designs it could be that the age of mass production printing to a particular colour-way and design is coming to an end; the only way to meet this challenge is to develop digital printing processes for textiles. Successful development of these digital systems will advantage wool and wool-blend fabric printers with their traditional small-run production. Digital printing eliminates the screen costs, allows designers to make pattern and colour changes immediately and eliminates registration problems when printing.

Two issues are important before digital printing can commence commercially on wool fabric; the availability of suitable machinery and the formulation of fabric pre-treatments and ink formulations designed to give maximum colour yield, eliminate background yellowing and allow production of prints of high wet- and light-fastness.

8.6.1 Ink-jet printing machinery

Drop on demand ink-jet heads use either thermal or piezoelectric technology to deliver droplets to the substrate as needed. Thermal ink-jet heads have been the most popular for paper printing; they operate by thermally producing a bubble of water-based ink which is jetted out of the nozzle. High-resolution prints are produced due to the small drop size and up to 1440 drops per inch (2.5 cm) covering can be obtained.

Piezoelectric ink-jet heads rely on an electric charge pulse to warp the interior of the ink chamber, thus forcing the drops out of the nozzle. Piezo heads are seen as more robust and more suited for high-volume printing; additionally it is envisaged that even pigment and binder compositions could be printed from such heads.

Continuous ink-jet printing machines are also being trialled; these machines use a continuous stream of charged ink droplets which passes through a deflection region where they either are directed onto the substrate or returned to the reservoir by deflection into a gutter. It seems likely that this system will be most capable of meeting the production speeds ultimately required to move ink-jet printing of textiles from a mainly sampling activity into mainstream production.

Examples of machines currently on the market are the CIBA (now Huntsman)- Reggiani Dream machine, the Mimaki TX 1600, Aristri (DuPont), Sophis Silk Express, Chromojet (Zimmer) and Encad TX150; these are all based on the piezzo system. The most successful full-scale production machine is the Dream system. In terms of the sampling market the Mimaki systems seem to dominate; these are based on piezoelectric heads. The Mimaki TX is mainly used for sampling, giving 720 dots per inch coverage, and the TX2 wide format machine (1.6 m) is used for small production and fast sampling. The Mimaki machines are sold with a 1 year warranty which requires sole use of Mimaki ink; replacement ink cartridges cost about £60 for 220 mL of ink. Each cartridge has a chip that monitors ink level and polices cartridge use. Clearly there is a totally new philosophy of colour purchase being developed by some printing machine manufacturers which could detract from rapid development of the textile ink-jet printing field.

8.6.2 Print pre-treatments and ink formulations

Fabric pre-treatment for ink-jet printing is seen by many as requiring a different approach from that required for screen printing. Thus there is already an established business activity in which fabric is prepared by simply padding and drying thickening agent; in the case of cotton ink-jet printing with reactive dyes an alkali such as sodium carbonate is also included in the preparation.

Wool preparation may be Kroy chlorination but on steaming potentially unacceptable levels of yellowing are produced. In the case of wool and cotton fabric ink-jet printing the author holds the strong opinion that reactive dyes should be the product of choice; they offer the potential of giving prints of excellent wash-fastness and minimal wash-off providing conditions are optimised to achieve high fixation. Dye manufacturers offer ranges of high-purity reactive dyes by desalting commercial products, thus maximising solubility; for deep shades inks may have to contain up to 200 g/L of dye in aqueous solution. The latter may be difficult to sustain over a period of time without aggregation, precipitation and subsequent jet clogging; this problem was virtually solved by using up to 50% (w/w) *N*-methylmorpholine oxide as an ink additive.[79]

Again chlorination of wool is the favoured method to pre-treat the fabric to render it readily printable to give strong vibrant prints, which fix well following steaming – the environmental downside of chlorination has been discussed already. There is thus substantial scope for fabric preparation and ink R&D to allow wool fabrics to take part in the digital printing revolution. Investment in this activity is vital. Since machinery costs are so low, digital printers do not have to produce at the same volume as rotary screen printers; in fact a production speed of 20 m per hour would make the machines, when networked like looms, competitive on lots of any size.

8.7 References

1. J D LEEDER (1984), *Wool – Nature's Wonder Fibre*, Australasian Textiles, Belmont.
2. MILLINGTON K R (2006), 'Photo-yellowing of wool. Part 1: Factors affecting photo-yellowing and experimental techniques', *Coloration Technology*, **122**, 169–186.
3. MILLINGTON K R (2006), 'Photo-yellowing of wool. Part 2: Photo-yellowing mechanisms and methods of prevention', *Coloration Technology*, **122**, 301–316.
4. LEWIS D M (1977), 'The dyeing of wool with pre-metallised dyes', *Wool Science Review*, **54**, 30–45.
5. BURKINSHAW S M (1992), 'Dyeing wool with metal-complex dyes', in *Wool Dyeing*, Society of Dyers and Colorists, Bradford, pp 196–220.

6. RATTEE I D (1974), *The Physical Chemistry of Dye Absorption*, Academic Press, London and New York.
7. VICKERSTAFF T (1954), *The Physical Chemistry of Dyeing*, Oliver and Boyd, London.
8. SKINNER B G and VICKERSTAFF T (1945), 'The absorption of acid dyes by wool, silk, casein fibre and nylon', *J. Soc. Dyers and Colourists*, **61**, 193–201.
9. ATHERTON E, DOWNEY D A and PETERS R H (1955), 'Dyeing of nylon with acid dyes. 1. Measurement of affinity and the mechanism of dyeing', *Textile Res. J.*, **25**, 977–993.
10. MEYBECK J and GALAFASSI P (1970), 'The effect of hydrocarbon substituents in azo dyes on wool dyeing', *Proceedings of the 4th International Wool Textile Res. Conf., Applied Polymer Symposia*, No. 18, Part 1, 463–472.
11. ZOLLINGER H (1965), 'The dye and the substrate: the role of hydrophobic bonding in dyeing processes', *J. Soc. Dyers and Colourists*, **81**, 345–350.
12. MEGGY A B (1956), 'Some recent developments in the theory of dyeing', *J. Soc. Dyers and Colourists*, **66**, 510–521.
13. PAULING L (1940), *The Nature of the Chemical Bond and the Structure of Molecules and Crystals – an Introduction to Modern Structural Chemistry*, 2nd Edition, Oxford University Press, London.
14. HUNTER C A (1994), 'The role of aromatic interactions in molecular recognition', *Chem. Soc. Rev.*, **23**, 101–109.
15. AAKERÖY C B and SEDDON C R (1993), 'The hydrogen bond and crystal engineering', *Chem. Soc. Rev.*, **22**, 397–407.
16. COZZI F M, CINQUINTI M, ANNUZIATA R and SIEGEL J S (1993), 'Dominance of polar/π forces over charge-transfer effects in stacked phenyl interactions', *J. Amer. Chem. Soc.*, **115**, 5330–5331.
17. FORGENSON W L and SEVERENCE D L (1990), 'Aromatic–aromatic interactions; free energy profiles for the benzene dimer in water, chloroform and liquid benzene', *J. Amer. Chem. Soc.*, **112**, 4768–4774.
18. GILES C H (1989), 'Dye–fibre bonds and their investigation', in *Theory of Coloration of Textiles*, 2nd Edition, Soc. Dyers and Colourists, Bradford, pp 97–168.
19. HUNTER C A and SANDERS J A (1990), 'The nature of π–π interactions', *J. Amer. Chem. Soc.*, **112**, 5525–5534.
20. DESIRAJU G R and GAVEZOTTI A (1989), 'From molecular to crystal structure; polynuclear aromatic hydrocarbons', *J. Chem. Soc., Chem. Commun*, 621–623.
21. SINGH J and THORNTON J M (1990), 'An automated method for the analysis of the preferred packing arrangements between protein groups', *J. Mol. Biol.*, **211**, 595–615.
22. LEWIS D M (1996), Dyestuff–fibre interactions', *Internat. J. of Soc. Cosmetic Chem.*, **18**, 123–135.
23. LEEDER J D, RIPPON J A, ROTHERY F E and STAPLETON I W (1985), 'Use of the transmission electron microscope to study dyeing and diffusion processes', *Proc. 7th Int. Wool Textile Res. Conf.*, Tokyo, **5**, 99–108.
24. LEEDER J D, RIPPON J A and STAPLETON I W (1990), 'Diffusion of dyes and other reagents into the wool fibre', *Proc. 8th Int. Wool Textile Res. Conf.*, Christchurch, New Zealand, **IV**, 227–238.
25. RIPPON J A (1992), 'The structure of wool', in *Wool Dyeing*, Society of Dyers and Colourists, Bradford, pp 1–51.
26. BRADY P R (1990), 'Penetration pathways of dyes into natural protein fibres, *Proc. 8th Int. Wool Textile Res. Conf.*, Christchurch, New Zealand, **IV**, 217–226.

27. BRADY P R (1992), 'Diffusion of dyes into natural fibres', *Rev. Prog. in Col.*, **22**, 58–78.
28. LEWIS D M and SMITH S M (1990), 'Reactive dyes and wool damage', *Proc. 8th Int. Wool Textile Res. Conf.*, Christchurch, New Zealand, **IV**, 177–186.
29. SMITH S M (1990), 'The effect of reactive dyes and other alkylating agents on the inhibition of wool damage during dyeing', *PhD Thesis*, University of Leeds.
30. HALL R O (1937), 'Fibre structure in relation to fur dyeing', *J. Soc. Dyers and Col.*, **53**, 341–344.
31. NEGRI A P, CORNELL H G and RIVETT D E (1993), 'A model for the surface of keratin fibres', *Textile Res. J.*, **63**, 109–115.
32. MEADE S J, DYER J M, CALDWELL J P and BRYSON W G (2005), 'Covalent attachment of novel surface modifications to wool fabric via removal of surface lipids', *Proc. 11th Int. Wool Textile Res. Conf.*, Leeds, CD, **FWSA**, 1–10.
33. Perachem Ltd (2006), Improvements in and Relating to Wool Treatment, PCT/GB 2006/002955.
34. LEWIS D M and YAN G (1993), 'The effect of various chromium species on wool keratin', *J. Soc. Dyers and Col.*, **109**, 193–198.
35. LEWIS D M and YAN G (1994), 'Chrome dyeing using chrome (III) salts', *J. Soc. Dyers and Col.*, **110**, 281–290.
36. LEWIS D M, YAN G, JULIA R M, CORDERCH L and ERRA P (2000), 'Chromium distribution in wool by electron microscopy and X-ray energy dispersive analysis', *Textile Res. J.*, **70**, 315–320.
37. LEWIS D M and YAN G (1995), 'Chrome dyeing using anionic trivalent chromium complexes as mordants', *J. Soc. Dyers and Col.*, **111**, 316–323.
38. LEWIS D M and YAN G (1995), 'Chrome dyeing of wool with Cr^{III} salts', *Proc. 9th Int. Wool Textile Res. Conf.*, Biella, **111**, 39–51.
39. XING J and PAILTHORPE M T (1995), 'Low chrome dyeing – eliminating Cr(VI) in the exhausted bath by using Cr(III) as the mordant', *Textile Res. J.*, **65**, 70–74.
40. XING J and PAILTHORPE M T (2000), 'Chrome dyeing of cashmere with SCA-Cr as the mordant', *J. Soc. Dyers and Col.*, **116**, 91–93.
41. LEWIS D M, HORNETT C, SMITH P and YAN G (2002), 'Chrome dyeing of wool using anionic chromium (III) complexes', *Adv. in Col. Sci. & Techn.*, **5**, 13–17.
42. BURLEY R W (1955), 'Some observations of the extension, contraction and supercontraction of wool', *Proc. 1st Int. Wool Textile Res. Conf.*, Melbourne, **D**, 88–117.
43. KING D G and BRADY P R (1995), 'Scientific and practical aspects of chrome dyeing', *Proc. 9th Int. Wool Textile Res. Conf.*, Biella, III, 52–59.
44. DENNING R J, FREELAND G N and GUISE G B (1995), 'Developing a zero AOX continuous shrink-resist process', *Proc. 9th Int. Wool Textile Res. Conf.*, Biella, **I**, 208–216.
45. DOUTHWAITE F J and LEWIS D M (1994), 'The formation of cysteine-S-sulphonate in wool and the effect on shrink-resistance', *J. Soc. Dyers and Col.*, **110**, 304–307.
46. SCHÜREK P, METELERKAMP R G M and BARKHUYSEN F A (1995), 'Continuous AOX-free shrink-resist treatment of wool', *Proc. 9th Intl. Wool Textile Res. Conf.*, Biella, **I**, 199–207.
47. LEWIS D M (1973), 'A combined cold dyeing and resin finishing process for the production of washable wool fabrics, *Textilveredlung*, **8**, 151–158.
48. LEWIS D M (1982), 'The dyeing of wool with reactive dyes', *J. Soc. Dyers and Col.*, **98**, 165–175.
49. CHURCH J S, DAVIE A S, SCAMMELLS P J and TUCKER D J (1999), 'Chemical interactions of α-bromoacrylamide reactive dyes with wool', *Rev. Prog. in Col.*, **29**, 85–93.

50. HEYNA J and SCHUMACHER W (1949), German Patent 965 902, Farbwerke Hoechst.
51. BÜHLER A, HURTER R, MÄUSEZAHL D and PETITPIERRE J C (1975), 'Nachbargruppenbeteiligung beim Farben Stickstoffgruppenhaltiger Substrate mit Halogenacrylamidfarbstoffen', *Proc. 5th Int. Wool Textile Res. Conf.*, Aachen, **5**, 263–272.
52. CHRISTOE J O and DATYNER A (1971), 'The effect of surfactants on rates of hydrolysis and covalent bonding of some reactive dyes with selected amino acids and wool', *Applied Polymer Symposium*, No. 18, 447–462.
53. FELDTMAN H D and MCPHEE J R (1964), 'The spreading and adhesion of polymers on wool', *Textile Res. J.*, **34**, 634–642.
54. HILDEBRAND D and MEIER G (1971), 'Kinetics of wool dyeing by means of Verofix dyestuffs', *Textil Praxis Intl*, **26**, 499–504.
55. OSTERLOH F (1968), 'Neue Möglichkeiten des Wollechtfarbens mit Reaktivfarbsoffen', *Melliand Textilberichte*, **49**, 1444–1448.
56. LEWIS D M and SELTZER I (1971), 'The dyeing of wool treated with Hercosett 57', *Proc. 4th Int. Wool Textile Res. Conf., Applied Polymer Symposia No. 18*, Berkeley, 437–446.
57. MÜLLER B M (1992), 'Absorbable organic halogens in textile effluents', *Rev. Prog. Color.,* **22**, 14–21.
58. FARNWORTH A J and DELMENICO J (1971), *Permanent Setting of Wool*, Merrow, Watford.
59. COOK J R and DELMENICO J (1971), 'Permanent press effects in wool. 5. Methods of stabilizing set', *J. Textile Inst.*, **62**, 27–46.
60. COOKSON P G, FINCHER K W and BRADY P R (1991), 'Minimising the impairment of the physical properties of wool during dyeing by restricting the level of permanent set', *J. Soc. Dyers and Col.*, **107**, 135–138.
61. STEENKEN I and ZAHN H (1986), 'Investigation of the alkaline modification of wool using new analytical methods for the determination of degradation', *J. Soc. Dyers and Col.*, **102**, 269–273.
62. MILLIGAN B and MACLAREN J A (1981), 'Alkali, alkylamines and cyanides', in *Wool Science – the Chemical Reactivity of Wool*, Science Press, NSW, Australia, 99–102.
63. ZAHN H (1995), 'Wool chemistry and processing', *Plenary Lecture – 9th Int. Wool Textile Res. Conf.*, Biella, **I**, 1–16.
64. MIZELL L R and HARRIS M (1943), 'Nature of the reaction of wool with alkali', *J. Res. National Bureau of Standards*, **30**, 47–53.
65. SPEAKMAN J B (1936), 'The reactivity of the sulphur linkage in animal fibres. 1. The chemical mechanism of permanent set', *J. Soc. Dyers and Col.*, **52**, 335–346.
66. SANGER F, RYLE A P, SMITH L F and KITAI R (1955), 'The distribution of cystine residues in insulin', *Proc. Int. Wool Textile Res. Conf.*, Melbourne, **C**, 49–61.
67. RYLE A P, SANGER F, SMITH L F and KITAI R (1955), 'Disulphide bonds of insulin', *Biochem. J.*, **60**, 541–556.
68. KØPKE V (1970), 'The role of water in setting of wool – a study of setting at temperatures above 100 degrees C. 1. Degree of set', *Textile Res. J.*, **61** 361–387.
69. LEWIS D M (1990), 'The effect of reactive dyes on damage in wool dyeing', *J. Soc. Dyers and Col.*, **106**, 270–274.
70. BROCK D, YOUSEFF T and LEWIS D M (1999), WO 99/51684, WO 99/51686 and WO 99/51687, to Procter and Gamble.
71. LEWIS D M and SMITH S M (1995), 'Substituted thiol derivatives of vinyl sulphone dyes', *Dyes and Pigments*, **29**, 275–294.
72. LEWIS D M (2004), 'Mechanisms of setting/antisetting in wool dyeing', *Adv. in Col. Sci. and Technol*, **7**, 1–5.

73. LEWIS D M and SMITH S M (1991), 'The role of vinylsulphonyl dyes in prevention of wool damage', *J. Soc. Dyers and Col.*, **107**, 357–362.
74. COOKSON P G, BRADY P R, FINCHER K W, DUFFIELD P A, SMITH S M, REINCKE K and SCHREIBER J, 'The Basolan AS Process: a new concept in wool dyeing', *J. Soc. Dyers and Col.*, 111, 228–236.
75. KIM J and LEWIS D M (2002), 'The effect of various anti-setting systems in wool dyeing. part 1: hydrogen peroxide based systems', *Coloration Technol*, **118**, 121–124.
76. KIM J and LEWIS D M (2003), 'The effect of various anti-setting systems in wool dyeing. Part 4: Sodium tetrathionate based systems, *Coloration Technol*, **119**, 112–115.
77. KIM J and LEWIS D M (2002), 'The effect of various anti-setting systems in wool dyeing. Part 2: Sodium maleate based systems', *Coloration Technol*, **118**, 181–184.
78. LIAO Q and BRADY P R (2000), 'A study of the reaction between wool and *N*-naphthylmaleimide using Raman spectroscopy and measurement of changes in fabric tensile properties', *Proc. 10th Int. Wool Textile Res. Conf.,* Aachen, CD paper code F1–15.
79. LYE J (2002), Kimberley Clark, WO 02/31067.

Part II

New wool products and applications

9

Improving the whiteness and photostability of wool

K. R. MILLINGTON, CSIRO Materials Science and Engineering, Australia

Abstract: The colour and photostability of fibres are important qualities, particularly for the production of brilliant white and bright pastel shade products that are essential for new and rapidly expanding casual apparel markets. This chapter describes the evolution of the natural cream colour of wool, current methods of wool colour measurement and commercial treatments for improving colour, such as bleaching and fluorescent whitening. The photostability of wool, which is poor compared with cotton and synthetic fibres, is discussed together with aspects of its complex photochemistry, and the potential for achieving brighter, whiter wools with improved stability is presented.

Key words: wool, colour, photostability, free radical oxidation, bleaching, fluorescent whitening.

9.1 Introduction

Wool has many excellent qualities that other fibres are unable to emulate, but its achievable whiteness and its photostability when exposed to sunlight are both areas where wool compares rather poorly. Current commercial bleaching processes for wool are far less efficient than those used on cotton and synthetics and can sensitise the wool to photoyellowing on exposure to sunlight. The use of fluorescent whitening agents (FWAs) on bleached wool improves whiteness significantly, but commercial use of FWAs is limited owing to the very poor photostability of FWA-treated wool, in particular when garments are wet after laundering and exposed to direct sunlight during line drying. A single laundering and drying cycle can cause an unacceptable level of yellowing, rendering the garment unwearable. This inability to produce photostable products in bright white and pastel shades affects wool's ability to compete effectively with other fibres in some major textile markets, especially the current growth areas of casual fashion knitwear, sports and leisurewear.

The chapter opens with a section on wool colour and discusses the evolution and genetics of the natural cream colour of wool, how it is measured and the effects of wool processing on colour, including commercial bleaching and whitening processes (Sections 9.2 and 9.3). These sections include a comparison of the achievable whiteness on wool with cotton and synthetic fibres and highlights the technical challenges that exist to allow wool to compete on a level playing field with these fibres with regard to colour attributes (Section 9.3). The chemistry of wool colour and photostability is discussed in Section 9.4, with particular emphasis on the current understanding of the mechanisms involved in photoyellowing. Section 9.5 describes the current methods available for improving whiteness and photostability, followed by a discussion of the prospects for further improvement (Section 9.6).

9.2 Wool colour

9.2.1 Evolution and genetics of natural white wool colour

Most ancient sheep were not white but had a highly pigmented brown fleece, probably similar to the wild Mouflon sheep that still inhabit parts of Sardinia and Corsica. Sheep having a grey fleece, consisting of a mixture of white and coloured fibres, are also known from Iron Age textile remains. Pigmentation in the wool, hair and skin of mammals is due to the presence of active melanin-producing cells (melanocytes). Within melanocytes, melanin granules are produced (melanosomes) and this melanin can be transferred into adjacent keratin-producing cells (keratinocytes) via finger-like projections. Melanin-containing keratinocytes within the follicle bulb of a wool or hair fibre are incorporated into the growing fibre (Forrest *et al.*, 1985).

Melanin pigments exist in two chemically distinct forms, eumelanin and pheomelanin. Eumelanin is generally black or dark brown and derived primarily from tyrosine (Tyr) residues. Pheomelanin is a yellow to reddish brown pigment formed from Tyr and cystine (Cys). Both pigments are polymeric and are derived via a series of redox reactions from a common precursor, dopaquinone, which is formed by the action of the enzyme tyrosinase on Tyr residues (Ozeki *et al.*, 1996). Melanocytes can form both eumelanin and pheomelanin, and production of eumelanin is controlled by α-melanocyte stimulating hormone (α-MSH) (Jackson, 1994). In mammals a defect in the gene that codes for tyrosinase synthesis results in a total lack of pigmentation (albinism). In sheep albinism is very rare, but has been observed in Icelandic sheep (Adalsteinsson, 1977, 1978) and in a flock of Suffolk sheep raised in South Australia (Rowett and Fleet, 1993).

The stimulus for the selective breeding of sheep with white (unpigmented) wool only arose after the development of dyeing in the Middle East around 1000BC. Textiles from the first millennium BC in the Middle East indicate an increase in the number of white sheep, since almost all the wool textiles found from this period are of white dyed wool (Ryder, 1987). Numerous wool textile remains also exist from Roman times, and those from Mediterranean sites are almost exclusively dyed white wools.

Selective breeding of unpigmented sheep in the first millennium BC would have required the presence of some white animals in the domestic flocks that had developed earlier. Breeding could have been carried out either by selecting white-spotted sheep having large areas of white wool, or by selection of all-white mutants (Maijala, 1997). White-spotting in mammals occurs in areas where skin or hair follicles lack melanocytes. The absence of pigmentation in modern-day Merino sheep is reported to be due to the deficiency of melaninocytes resulting from the action of a complex of white spotting genes (Fleet *et al.*, 1991). A second general mechanism for achieving whiteness is through dilution of pigment, which is accomplished by selection for decreased efficiency of melanin production by melanocytes (Sponenberg, 1997).

The loci *Agouti* (*A*) and *Extension* (*E*) control the production of pigments in the skin and hair of mammals. In sheep, wool whiteness is a multigenic trait dependent on alleles of white spotting genes and modifiers which result in low levels of melanocytes in the skin, and genes such as *Agouti* which produce an inhibitor of melanocyte activity (Sutton *et al.*, 1998). The *Agouti* locus is responsible for production of a protein that nullifies the effects of α-MSH on melanocytes, thus preventing the production of eumelanin. The *E* gene is expressed in melanocytes and controls the synthesis of α-MSH.

One question that arises from knowledge of the evolution and genetics of sheep is whether further improvements in wool whiteness are possible by selective breeding. For Merinos a number of studies suggest significant variance in wool colour within sheep populations. Several studies have measured greasy and clean (scoured) wool colour and calculated the phenotypic and genetic correlations and heritability. Phenotypic and genetic correlations predict the direction and rate of change in a given trait that would result from different selection practices. Heritability is the proportion of phenotypic variation in a population that is attributable to genetic variation among individuals. For wool colour the phenotypic correlation between greasy wool colour and clean wool colour is not strong (<0.3), showing that greasy wool colour is not a good indicator of scoured colour. This is due to greasy wool colour being affected by the presence of high quantities of wool grease and dust which are removed after scouring. However, a study on South Australian Collinsville Merinos, the largest family group within the Australian Merino population, showed that the genetic correlation between greasy and scoured yellowness was moderate to high (0.4–0.9), indicating that selection on the basis of greasy yellowness should lead to genetic improvements in scoured yellowness (James *et al.*, 1990b). Heritabilities of 0.42 for greasy colour score and 0.35–0.54 for objectively measured clean wool colour parameters were measured (James *et al.*, 1990a,b). A recent study on New South Wales (NSW) Merinos has confirmed that greasy wool colour is moderately heritable (0.31–0.35) (Brown, 2006) and this is supported by work in New Zealand on Corriedale sheep (Benavides and Maher, 2003).

These results suggest that scoured wool colour would respond fairly rapidly to selection. Few Merino sheep breeders in Australia currently select their rams with the assistance of objectively measured scoured colour data, although approved international test methods for clean wool colour exist and measurement facilities are readily available from wool testing houses (see Section 9.2.3). Selection for wool colour therefore appears to be an area where significant improvements in wool quality could be achieved.

9.2.2 Environmental and storage effects on wool colour

The colour of wool is adversely affected by weathering while exposed to sunlight on the sheep's back. Colour measurements on clean wool from different sites on Merino fleeces grown in Armidale, NSW show a deterioration of ~3 brightness (Y) units and ~1 yellowness (Y–Z) unit for wool grown on the back of the sheep compared with mid-flank samples (Holt *et al.*, 1994). In contrast, on rugged sheep the back wool is slightly lighter than flank wool and less yellow. Maintenance of good colour attributes is clearly one benefit of rugging and shedding sheep. Holt *et al.* showed that a single application of a UV absorber to the backs of sheep offered protection against weathering equivalent to 70% of that offered by rugging, and maintained the brightness and yellowness of back wool to within 0.7 and 0.1 units respectively of the mid-flank values (Holt *et al.*, 1994).

Wool stored in its greasy state undergoes colour changes that vary with sheep breed (Baxter, 2001). Merino fleece wools deteriorate in yellowness (Y–Z) by ~0.01–0.1 units per month and brightness (Y) values also decrease slightly (see Section 9.2.3 for definitions). New Zealand crossbred wools deteriorate more rapidly, with the yellowness of Romney wool deteriorating by up to ~1.5 units per month. Greasy crossbred wools may change colour in a matter of days, depending on the bloodline/environment combination and the moisture content at the time of shearing. Although most of the colour deterioration that occurs during storage is removed during scouring, it is not clear whether protracted storage times in the greasy state can affect the achievable whiteness on wool after processing. Differences in wool colour within individual Romney and Merino fleeces have been linked to changes in the pH of an aqueous extract of greasy wool (Sumner *et al.*, 2003). A significantly higher pH was measured for belly wool than mid-side wool, and the lowest pH was found on back samples. Presumably weathering and gravity cause migration of soluble suint components from the back and mid-side regions of the fleece towards the belly. Normal (bleach-free) scouring would not be able to remove any yellowing caused by exposure to alkaline conditions.

Scoured wool colour is generally stable for years (unless the wool was bleached during the scouring process), provided that the wool is protected from exposure to sunlight.

9.2.3 Measuring wool colour

The colour of greasy and clean unprocessed wools is most often expressed in terms of the tristimulus values *X*, *Y* and *Z*. These values represent the amount of the red/orange (*X*), yellow/green (*Y*) and blue/indigo/violet (*Z*) components of the spectrum of white light reflected from the sample. Using these parameters, wool yellowness is normally quoted as *Y–Z* and brightness as *Y*. The colour of fibrous materials can be objectively measured using a reflectance spectrophotometer fitted with a standard illuminant (D65) that is equivalent to outdoor daylight. Nowadays a 10° field of view is normally used to analyse reflected light from the sample, and the measured tristimulus values are quoted as D65/10° data. A more detailed review of the issues involved in wool colour measurement has been published by Wood (2002).

The colour of wool is an important characteristic when assessing its suitability for a particular end use, and hence colour affects the selling price. The initial colour is particularly important to wool dyers for the production of bright shades, and wool for this use needs to be both bright and white. Using the same dyeing conditions, yellow wools produce significantly duller shades and the colours obtained can often be quite different from those that are desired.

The International Wool Textile Organisation (IWTO) has approved international test methods for clean wool colour (IWTO, 2003b) and for wool sliver and top (IWTO, 2003a) that use a special cell fitted at either end with glass windows and which is packed with wool to a constant density or pressure. A cell is unnecessary for textile fabrics and their colour is normally measured by placing samples directly onto the spectrophotometer port. The colour of yarn can be similarly assessed by winding a suitable quantity around a white card. Table 9.1 shows the standard range of brightness and yellowness values for clean wools and sliver and the descriptors that are used to define clean wool colour.

Table 9.1 Brightness (*Y*) and yellowness (*Y–Z*) ranges and descriptors for clean unprocessed wools.

D65/10° Brightness (*Y*)		D65/10° Yellowness (*Y–Z*)	
Very bright	>70	Very white	<9
Bright	68–70	White	9–10.5
Average	64–68	Slightly creamy	10.5–9.5
Slightly dingy	59–64	Creamy	9.5–14.5
Dingy	<59	Quite yellow	14.5–16
		Heavily stained/yellow	>16

9.3 The effect of wool processing on colour

The long processing chain from greasy wool to wool garment, by either the worsted or woollen system, has many stages and several of these can affect wool colour. In addition to photoyellowing, yellowing of wool can also occur when it is treated under alkaline conditions or exposed to heat during drying. Also several chemical treatments used in wool processing, in particular shrink-resist treatments that use chlorine or the disodium salt of dichloroisocyanuric acid (DCCA) can lead to significant yellowing of the fibre. Wet bleaching treatments can also be used commercially to improve the whiteness of wool. The effect of the various stages of wool processing on its colour is summarised in the following subsections.

Scouring and mechanical processing

Efficient and effective scouring is important to obtain the brightest possible wool colour. A study has shown that by simply rewashing commercially scoured wool, in most cases, further improvements in both brightness (Y) and yellowness (Y–Z) can be obtained (Thompson and Teasdale, 1986). Effective scouring has been shown to depend on adequate refreshment of the scouring liquor and failure to do this impacts particularly on scoured wool brightness (Westmoreland *et al.*, 2006). The use of hydrogen peroxide in the final rinse of the scour before drying improves the colour of the scoured wool but this practice can lead to unnecessary fibre damage if further oxidative processes such as shrink-resist treatment or bleaching are performed in subsequent downstream processing at the top or yarn stage. The practice can also lead to colour instability.

A study on the processing of greasy Merino wools through to the top or noil stage has shown very similar brightness and yellowness values at the greasy and the top stage (Mahar and Osborne, 1996). In this study the scoured wool values were significantly inferior to those of the greasy wool and top, presumably due to somewhat inefficient scouring. The authors note that commercial wool cleaning can be viewed as a two-stage process, with the majority of contaminants removed during scouring and the remainder during the mechanical carding and combing operations. A more recent study on Australian, New Zealand and US wools has confirmed the improvements in scoured wool colour after carding (Cameron and Stobart, 2006).

Wet processing and finishing

Thermal yellowing of wool can be a serious problem during wool processing, in particular during extended dyeing at the boil, in setting with superheated steam (decatising), and in drying for extended times, particularly under alkaline conditions (Maclaren and Milligan, 1981). The thermal yellowing of wet wool is far more rapid than for dry wool, which is similar to the behaviour observed for photoyellowing. Thermal yellowing of wool during dyeing is influenced by pH, temperature and time, with chlorinated wools being especially sensitive. Yellowing during dyeing can be counteracted by adding a bleaching agent, based on sodium bisulfite or hydroxylamine sulphate, to the dyebath (Duffield, 1996). Addition of hydrogen peroxide to the dyebath after exhaustion of the dyestuffs can also be effective.

Commercial bleaching of wool is usually carried out either by an oxidative process using alkaline hydrogen peroxide or by a reductive process using stabilised sodium dithionite ('hydros') or thiourea dioxide (Duffield and Lewis, 1985). The whitest wool is obtained by first carrying out an oxidative hydrogen peroxide bleaching directly followed by a reductive treatment with sodium dithionite, the so-called double bleaching process. Nevertheless the ultimate whiteness achievable even after double bleaching is poor compared with bleached cotton and synthetics, since some of the natural cream chromophores of wool remain highly resistant to bleaching.

To further improve its whiteness, wool can be treated with a fluorescent whitening agent (FWA) either after oxidative bleaching or by including the FWA in the reductive bleaching bath. FWA-treated wool absorbs UV light and emits blue fluorescence, which makes it appear much whiter than bleached wool. Commercial FWAs for wool are usually based on a sulphonated stilbene, distyrylbiphenyl or pyrazoline derivative. Wool fabrics that have been treated with FWAs yellow rapidly when exposed to sunlight, especially when wet.

Chlorination of wool is carried out during commercial shrinkproofing at the top, fabric or garment stage, and to prepare wool fabrics for printing. The degree of wool yellowing that occurs during chlorination is dependent on pH, temperature and the concentration of the chlorinating agent. Using DCCA, a low pH (~3) at room temperature produces the least amount of yellowing (Veldsman and Swanepoel, 1971). The mechanism of wool yellowing by chlorine is unknown.

9.3.1 Wool colour compared with cotton and synthetic fibres

Reflectance spectrophotometers capable of measuring tristimulus values also provide a reflectance spectrum which represents the amount of light of any given wavelength reflected from a sample. A perfect white material (such as a ceramic MgO plate) exhibits close to 100% reflectance across the entire wavelength range and is used to calibrate the instrument. Figure 9.1 compares the reflectance spectra of wool, cotton and polyester (PET) woven fabrics of similar weight, and includes data after bleaching for the two natural fibres. The fabrics that appear bright and white to the eye (i.e. bleached cotton and

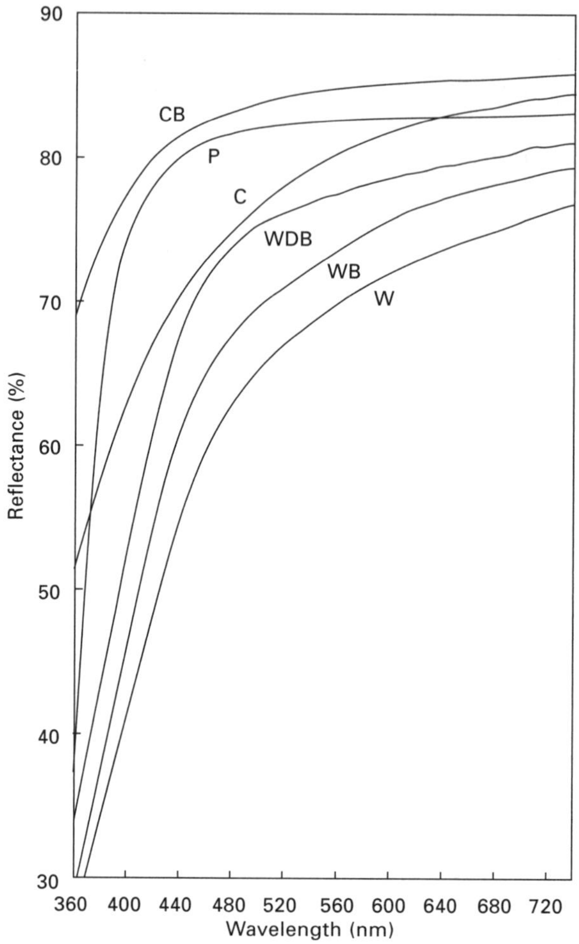

9.1 Reflectance spectra of wool (W), cotton (C) and polyester (P) fabrics, including H_2O_2-bleached wool (WB), double-bleached wool (WDB) and H_2O_2-bleached cotton (CB).

PET) have a very flat spectrum with reflectance >80% that begins to tail off only below ~460 nm. By contrast, the spectrum of natural wool is not flat, having a lower level of reflectance across the whole wavelength range that tails off further below ~500 nm. This low reflectance below 500 nm (equivalent to absorption of blue and UV wavelengths) gives natural wool its characteristic cream shade.

It is also clear from Fig. 9.1 that the improvement in whiteness for cotton fabric after bleaching is far superior to wool. Even after double bleaching (see above), wool's reflectance across the wavelength range remains inferior to unbleached cotton, and thus still appears cream-coloured rather than white.

A more efficient bleaching process for wool capable of radically reducing its yellowness would be of commercial value.

Treatment with a fluorescent whitening agent (FWA) improves the brightness and yellowness of all fibres, as shown in Table 9.2 for typical lightweight woven wool, cotton and polyester fabrics. It can be seen that the range of brightness (Y) values is lowest for wool (68.8) and highest for FWA-treated polyester (86.2) and also that this range is rather narrow since the difference in perceived whiteness between these two samples is very significant. In the textile industry it is more appropriate to use a whiteness index value (W) to compare bleached and FWA-treated fibres, as this has a much wider range than brightness values. There are a large number of different whiteness indices available, and the CIE Ganz 82 index provides a good correlation between the subjective visual whiteness and the measured value (Ganz, 1979).

The Ganz index is derived from the brightness value Y:

$$W_{\text{Ganz}} = Y + Px + Qy + C$$

where $x = X/(X + Y + Z)$, $y = Y/(X + Y + Z)$ and P, Q and C are constants that are specific to the spectrophotometer and its light source (particularly its UV content) and are set during calibration of the instrument.

Table 9.2 again highlights one of the major challenges for wool, the development of a more effective bleaching process capable of achieving whiteness values similar to cotton and synthetics. The other major challenge for wool is to maintain a comparable level of photostability after application of the FWA. This is discussed in the following section.

Table 9.2 Comparison of brightness (Y), yellowness (Y–Z) and CIE Ganz 82 whiteness index (W) for wool, cotton and polyester fabrics

Fibre	D65/10 Brightness (Y)	D65/10 Yellowness (Y–Z)	CIE Ganz 82 Whiteness (W)
Natural wool	68.8	8.2	17.0
H_2O_2-bleached wool	74.4	6.0	31.0
Double-bleached wool	76.9	2.7	48.7
Double-bleached/FWA-treated wool	81.7	−15.4	104.6
Greige (scoured) cotton	76.9	1.7	50.3
H_2O_2-bleached cotton	85.3	−3.3	75.5
Bleached/FWA-treated cotton	85.3	−20.0	122.2
PET	80.8	−3.7	104.4
FWA-treated PET	86.2	−27.2	138.5

9.4 The chemistry of wool colour and photostability

9.4.1 The natural chromophores in cream wools

Chromophores are molecules in a given material that absorb particular wavelengths of visible light, and in doing so confer colour on the material. In wool there are several ultraviolet (UV) chromophores present in the form of the aromatic amino acid residues tryptophan (Trp), tyrosine (Tyr) and phenylalanine (Phe), and also cystine (Cys) which can absorb light in the region between 250 and 320 nm. There are also visible chromophores present that absorb from the near-UV well into the visible region of the spectrum (350–550 nm), and the presence of these results in the cream colour of natural undyed wool. The reflectance spectrum of wool in Fig. 9.1 clearly shows that at wavelengths below ~500 nm its reflectance decreases rapidly, even after double bleaching. The presence of visible chromophores is also illustrated in the UV–visible absorption spectrum of a section from a single Merino wool fibre (Nicholls and Pailthorpe, 1976) shown in Fig. 9.2. Also shown is the calculated absorption spectrum based on the amino acid composition of

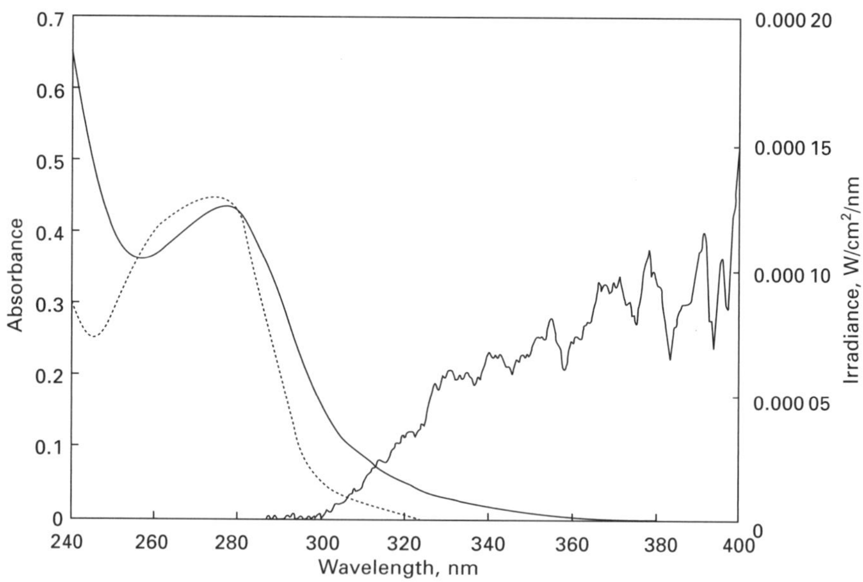

9.2 The UV/visible absorption spectrum of a 6 μm section of Merino wool keratin (solid line), and absorption spectrum calculated from its amino acid composition (dashed line) (redrawn from Nicholls and Pailthorpe, 1976). Also shown is the relative spectral intensity of Geelong summer sunlight.

wool. The visible chromophores are important even if their concentration in the wool fibre is low, because Fig. 9.2 also shows the solar spectrum contains significantly higher intensities of visible than UV wavelengths.

Surprisingly, very little is known regarding the identity of these natural yellow chromophores in wool. It is likely that they are a complex mixture of compounds that includes some wool protein oxidation products. Fundamental knowledge of the origin, chemical nature and location within the fibre of the natural yellow chromophores is well overdue, so that future research projects aimed at improving wool colour using both genetic and chemical approaches can be better targeted.

9.4.2 Influence of fibre morphology and physical properties on wool colour

Calculating the absorption spectrum of wool from its bulk amino acid composition assumes that the fibre is homogeneous. In reality the wool fibre has a highly complex morphology and consists of cells, with flattened overlapping cuticle cells forming a protective sheath around the inner cortical cells (Fig. 9.3). In coarser fibres, a central hollow medulla may be present. In fine wool, such as that obtained from Merino sheep, the cuticle is normally one cell thick (approximately $20 \times 30 \times 0.5\,\mu m^3$) and usually constitutes about 10% by mass of the total fibre. Sections of cuticle cells show an

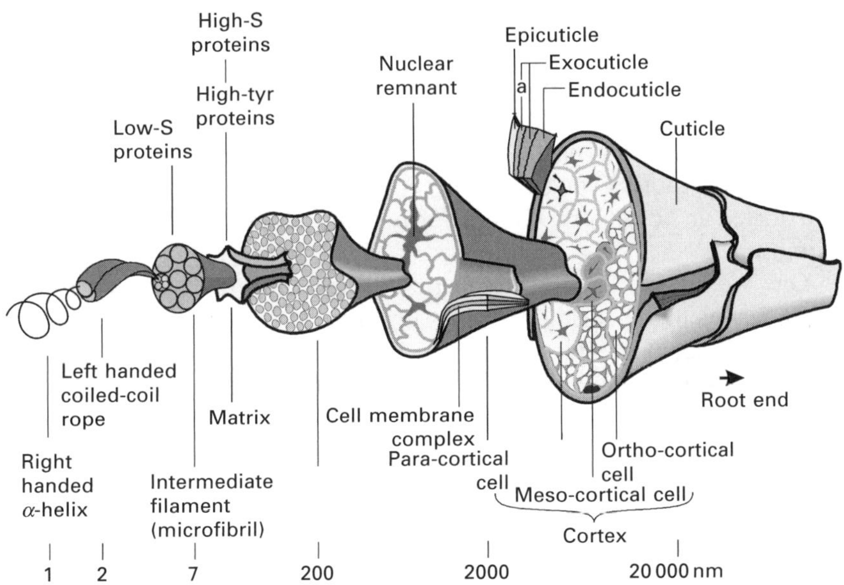

9.3 Schematic of the structure of a fine Merino wool fibre.

internal series of laminations, comprising outer sulphur-rich bands known as the exocuticle and inner regions of lower sulphur content called the endocuticle. On the exposed surface of cuticle cells, a membrane-like proteinaceous band (the epicuticle) and a covalently bound lipid component form a hydrophobic resistant barrier. Wool is composed of between 50 to 100 different proteins representing some 14 different protein families, making it by far the most complex fibre used for textile production.

While non-medullated wool fibres are substantially transparent to visible wavelengths, they appear opaque in air because some of the incident light is reflected or scattered at the interface between air and wool. When fibres are immersed in a colourless liquid with a refractive index very close to that of wool ($\eta = 1.553$), such as benzyl alcohol ($\eta = 1.540$), there is no reflection of light at the liquid–wool interface and the wool appears to be perfectly transparent. This property has recently been used to develop a test method for detecting the presence of dark fibre contamination in wool core samples, and also for detecting coarse medullated fibres in fine wools (Australian Wool Testing Authority, 2008). Unpigmented medullated fibres contain a central core of air-filled cells that are excellent light scatterers, making them appear whiter. This is particularly important for Arctic mammals to whiten their winter fur and act as camouflage, in addition to increasing the insulating properties of the fibres (Russell and Tumlinson, 1996).

Penetration of light into the wool fibre depends mainly on three factors, the distribution and concentration of chromophores in the different morphological regions of the wool fibre and the wavelength of the radiation. Higher-energy UV wavelengths (<320 nm) have very limited penetration into the fibre and are strongly absorbed mainly by the outer region of the cuticle cells, whereas most visible wavelengths above 400 nm that are not scattered or absorbed by the visible chromophores are able to pass through the fibre. Because wool has a very high absorption coefficient for high-energy UV, a fabric exposed specifically to UVB (280–320 nm) wavelengths is oxidised and discoloured specifically at the surface to a depth of ~1–2 μm (Simpson, 1999).

One would expect the sulphur-rich proteins in the exocuticle to absorb UVB wavelengths strongly. For an irradiated single fibre there will be a gradation of damage from the exposed surface to the core and this gradation may become more severe as new absorbing chromophores are formed at the fibre surface due to oxidation. This suggests that lower-energy UVA wavelengths (320–400 nm) that have greater penetration into the fibre are mainly responsible for the loss of tensile strength observed in fibres and fabrics exposed to sunlight, as fibre strength is determined by the cortical cells. This is consistent with research into the phototendering of wool which occurs following exposure to sunlight filtered by window glass. Phototendering (or simply tendering) refers to the effects of prolonged sunlight exposure on the mechanical properties of wool fabrics, such as loss of tensile strength, elasticity and abrasion resistance. Its effects are of commercial significance for wool upholstery, curtaining and carpets exposed to filtered sunlight through window glass, which absorbs light at wavelengths below 350 nm.

9.4.3 The photostability of wool: photoyellowing and photobleaching

The colour of natural wool is far less stable to sunlight than cotton and synthetics, and its photostability is further reduced if the fibre is oxidatively bleached with hydrogen peroxide. Application of an FWA to wool after bleaching results in a fabric that yellows very rapidly in sunlight.

The four principal factors which influence the observed colour change of wool exposed to radiation are the wavelength distribution of the incident light, the presence of oxygen, the presence of water and the processing history of the wool, in particular whether the wool has been bleached or whitened. There are two types of wavelength-dependent colour change that are observed for natural cream wool. UV radiation (280–380 nm) in the presence of atmospheric oxygen results in rapid wool photoyellowing that is accelerated in the presence of water, whereas exposure of natural cream wool to blue light (400–460 nm) causes photobleaching.

When natural wool is exposed to sunlight, which contains both UV and blue light wavelengths, photoyellowing and photobleaching occur concurrently, so that the overall effect is determined by the initial colour of the wool and the relative intensities of the component wavelengths of the light. For example, wool carpets photobleach because window glass attenuates the UV wavelengths of light below 350 nm that would cause photoyellowing, but transmits blue light wavelengths.

The intensity of UVB radiation in sunlight (290–320 nm) is variable and dependent on factors such as time of day, season, altitude and latitude. Seasonal variations in the levels of UVB radiation in sunlight explains why wool may be photoyellowed by summer sunlight but photobleached during winter (Maclaren and Milligan, 1981).

Wet photoyellowing of wool always occurs far more rapidly than under dry conditions. Figure 9.4 compares the wet photostability of natural, peroxide-bleached and bleached/FWA-treated wool fabric to simulated sunlight exposure. After three hours of exposure, the whiteness of bleached/FWA-treated wool is lower than natural wool, demonstrating that its photostability is extremely poor. The whiteness of bleached wool decreases most rapidly during the initial 30 minutes of exposure, whereas natural wool is slightly photobleached over the 3 hour irradiation period. A comparison of the relative rates of dry and wet photoyellowing is also shown in Fig. 9.5 (Leaver and Ramsay, 1969).

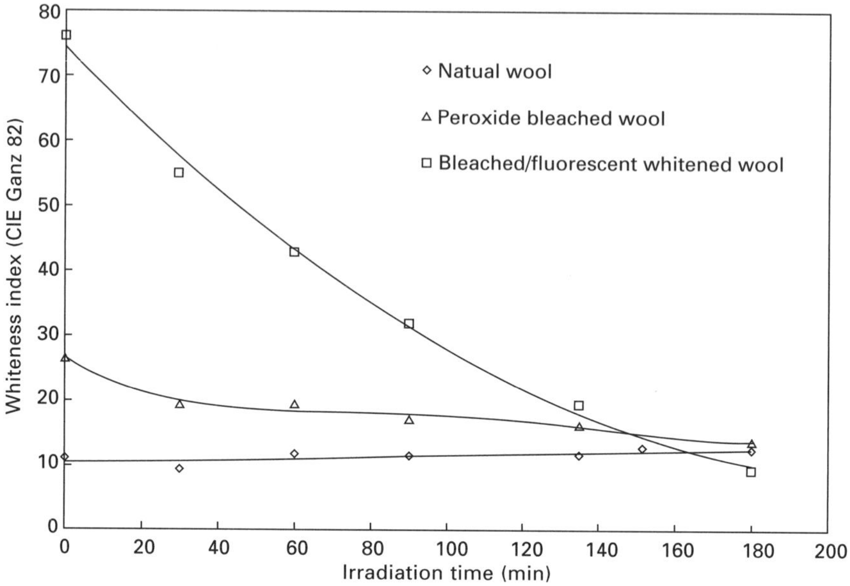

9.4 Comparison of the wet photostabilities of natural, peroxide-bleached and bleached/FWA-treated wool fabric to simulated sunlight exposure.

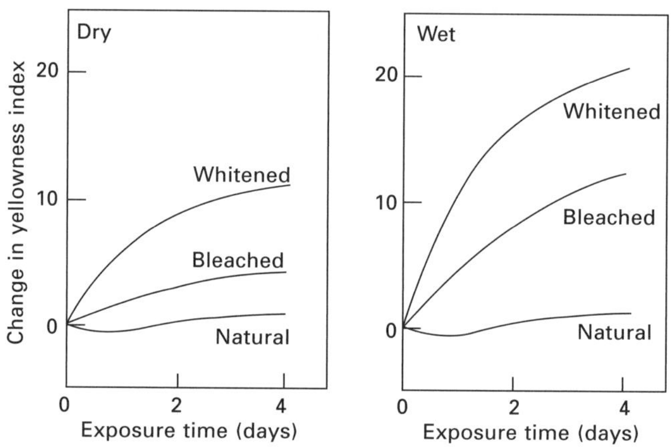

9.5 Comparison of the rates of yellowing of wet and dry treated wool fabric (redrawn from Leaver and Ramsay, 1969).

Most natural materials undergo photobleaching when exposed to blue light in the absence of UV radiation (Launer, 1968a,b). Such materials include several of commercial significance, such as paper, wood and leather, and other natural textile fibres such as cotton and silk. Blue light selectively destroys the yellow chromophores that absorb in this region, converting them to colourless products, resulting in increased reflection of blue light and hence improved whiteness. The most effective wavelengths for photobleaching of wool are 400–450 nm. The initial colour of wool affects the observed colour changes; deeply yellowed wool bleaches rapidly in sunlight, whereas the whitest wools may undergo little or no photobleaching (Lennox and King, 1968). This can be seen in Fig. 9.6, which shows the change in yellowness index for five wools covering a range of natural yellowness on exposure to sunlight and artificial light sources. Sources which include UV and visible radiation (Xenotest and sunlight) bleach the yellower wools during the early stages of exposure, whereas whiter wools are yellowed. When exposed to UV radiation only (Sunlamp) all wools undergo photoyellowing, whereas blue light produces only photobleaching.

The photobleaching of wool with blue light has been studied extensively although it is not used commercially. Launer used a small high-pressure mercury arc fitted with a UV filter to photobleach dry wool fabric in around 15 seconds (Launer, 1971). Other workers passed scoured wool through a conveyor system fitted with blue light fluorescent tubes overhead, but this process was not adopted by industry due to slow throughput and the need for a great many fluorescent tubes (Garrow *et al.*, 1971). Simpson (1992) found that photobleaching with similar low-intensity blue fluorescent tubes could be speeded up significantly by irradiating the wool in the presence of alkaline H_2O_2 at pH 10–11.5. Using a parallel array of low-power blue fluorescent tubes, irradiation times of 20 to 30 minutes were necessary to obtain a good bleaching effect using peroxide-padded wool, compared with over 24 hours for dry wool. However the irradiation times were still too long for a commercially viable continuous treatment. Recently a process for continuous photobleaching of wool fabric has been described that uses a modified UV-curing source (Millington, 2005). Doped medium-pressure mercury arcs developed for the rapid curing of polymer films offer very high intensities of blue light. Filtering the UV light at wavelengths below 350 nm from such a source and exposing wool fabric padded with either H_2O_2/oxalic acid or zinc formaldehyde sulphoxylate allowed continuous photobleaching at a rate of $2\,m\,min^{-1}$.

9.4.4 The mechanism of wool photoyellowing

Most natural and synthetic polymers undergo UV-induced discoloration, which results in an increase in yellowness on exposure. For synthetic polymers the photoyellowing mechanisms have been well studied, and in all cases the mechanism proceeds via a free radical route (Gugumus, 1993). Photodegradation of cellulosic materials such as cotton, paper and wood also proceeds via a free radical mechanism, with lignin playing the major role in both the absorption of UV light and the generation of yellow chromophores (Muller *et al.*, 2003).

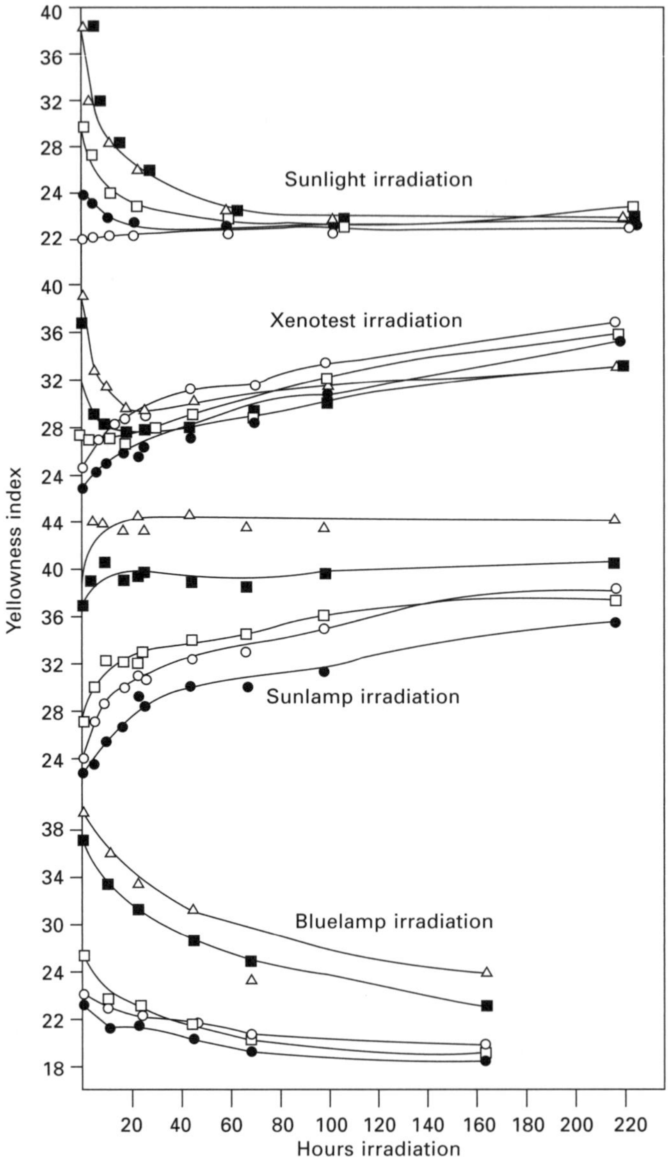

9.6 Exposure of scoured wools from five sheep, covering a range of initial yellowness indices, to sunlight and three artificial sources of radiation (redrawn from Lennox and King, 1968).

A number of different mechanisms have been described in the literature to account for the photoyellowing of wool, and these have recently been critically reviewed (Millington, 2006b). In common with other less complex polymers, most of the yellowing mechanisms proposed for wool involve free radical oxidation processes. Electron spin resonance (ESR) spectroscopy has shown that both carbon- and sulphur-centred free radicals are formed directly in wool after irradiation in the absence of atmospheric oxygen (Shatkay and Michaeli, 1970). In the presence of air or oxygen these signals are to some extent quenched, indicating that the radicals are reacting with oxygen.

It has been shown by studies where wet fibres have been irradiated with UVA or blue light in the presence of a highly specific fluorescent probe (the terephthalate anion), that wool, silk cotton, nylon and polyester all produce hydroxyl (•OH) radicals (Millington and Kirschenbaum, 2002). Wool produces higher concentrations of hydroxyl radicals than other fibres, and the attack of these highly oxidising species on the aromatic amino acid residues present in the protein leads to yellowing. Trace metal ions are involved in •OH generation since [•OH] was significantly reduced in the presence of the metal chelator deferoxamine. Analysis of hydrolysed Merino wool fabric has reported ~15 mg/kg iron, ~5 mg/kg copper and ~2 mg/kg manganese (Millington and Kirschenbaum, 2002).

In common with synthetic polymers and cellulosics, it has recently been shown that wool and other fibrous proteins emit light when heated above 40 °C due to chemiluminescence (CL) (Millington *et al.*, 2007). In a non-isothermal CL experiment in an inert nitrogen atmosphere, fibrous proteins show a CL peak near 130 °C similar to that observed in many synthetic polymers and characteristic of polymer hydroperoxides (POOH in Fig. 9.7)

Initiation	I	$\xrightarrow{h\nu}$	I*	(1)
	I*	$\longrightarrow$	A• + B•	(2)
	A• + PH	$\longrightarrow$	P• + AH	(3)
Propagation	P• + O_2	$\longrightarrow$	POO•	(4)
	POO• + PH	$\longrightarrow$	POOH + P•	(5)
Branching	POOH	$\xrightarrow{h\nu \text{ or } M^{n+}}$	PO• + •OH	(6)
	PO• + PH	$\longrightarrow$	POH + P•	(7)
	•OH + PH	$\longrightarrow$	H_2O + P•	(8)
	POOH + PH	$\xrightarrow{h\nu \text{ or } M^{n+}}$	PO• + P• + H_2O	(9)
	2 POOH	$\xrightarrow{h\nu \text{ or } M^{n+}}$	POO• + PO• + H_2O	(10)
Termination	P• + P•	$\longrightarrow$	P—P	(11)
	P• + POO•	$\longrightarrow$	POOP	(12)
	POO• + POO•	$\longrightarrow$	POOP + O_2	(13)

PH = polymer
I = primary photochemical absorber (radical initiator)
A• and B• = radical products from excited state of initiator
M^{n+} = catalytic metal ions

9.7 Free radical photo-oxidation mechanism in polymers.

(Kron *et al.* 1997). Hydroperoxides have previously been found in many oxidised proteins and amino acids (Gebicki and Gebicki, 1993; Dean *et al.*, 1997), but have not previously been reported in wool.

It seems likely that the photo-oxidation of wool proceeds via a free radical chain reaction similar to that which occurs in other polymers (Fig. 9.7). There may be a number of primary light-absorbing species (I in Fig. 9.7) in wool, including the Trp and Tyr residues in the UV region, and in the visible the natural yellow chromophores and yellow chromophores produced by photo-oxidation reactions.

Recent studies on the trypsin digests of heavily irradiated wool fabrics using high-performance liquid chromatography tandem mass spectrometry (HPLC/MS/MS) techniques have confirmed the presence of 13 different yellow chromophores in 25 photomodified peptide sequences (Fig. 9.8). All of these chromophores are derived from Trp and Tyr residues in the keratin intermediate filaments (KIFs) and high glycine tyrosine proteins present in the wool fibre cortex (Dyer *et al.*, 2005, 2006a,b).

Since high levels of absorption of the most energetic UVB wavelengths of sunlight will occur in the cystine-rich epicuticle and A-layer proteins, one might have expected to find a large number of photomodified residues in the cuticle proteins. Unfortunately far less is known about the structure of wool cuticle proteins than those in the cortex, despite the importance of the cuticle in commercial wool processing, and hence the amino acid sequencing data available for cuticle proteins is very limited.

Simpson is highly sceptical of the view that Trp photoproducts are the main contributors to wool photoyellowing (Simpson, 1999, 2002). He pointed out that most of the known products of Trp degradation, including the major product kynurenine, are colourless and that minor coloured products such as *N*-formylkynurenine (NFK) are very weak chromophores. Semi-quantitative calculations suggested that Trp degradation products could not account for more than 20% of the observed wool yellowness. Using a mild acid hydrolysis technique, Simpson found that almost all of the yellow chromophores in irradiated wool were released in the first 5–10% of wool to be solubilised (Simpson, 1999). The main yellow fraction was highly hydrophobic and was eluted from a reverse-phase chromatographic column as three components having similar amino acid composition, UV-visible absorption and weak fluorescence spectra. These fractions appeared to be polymeric peptides with different levels of polymerisation, having a very high proline content (~22 mol%) and low in aromatic amino acids. The nature of the chromophores present in these peptides is currently unknown. Simpson has speculated that the yellow material he extracted from irradiated wool consists mainly of polymeric proteins of high tinctorial strength derived from small proline-rich proteins (SPRPs) located in the outer regions of the wool surface (Simpson, 1999).

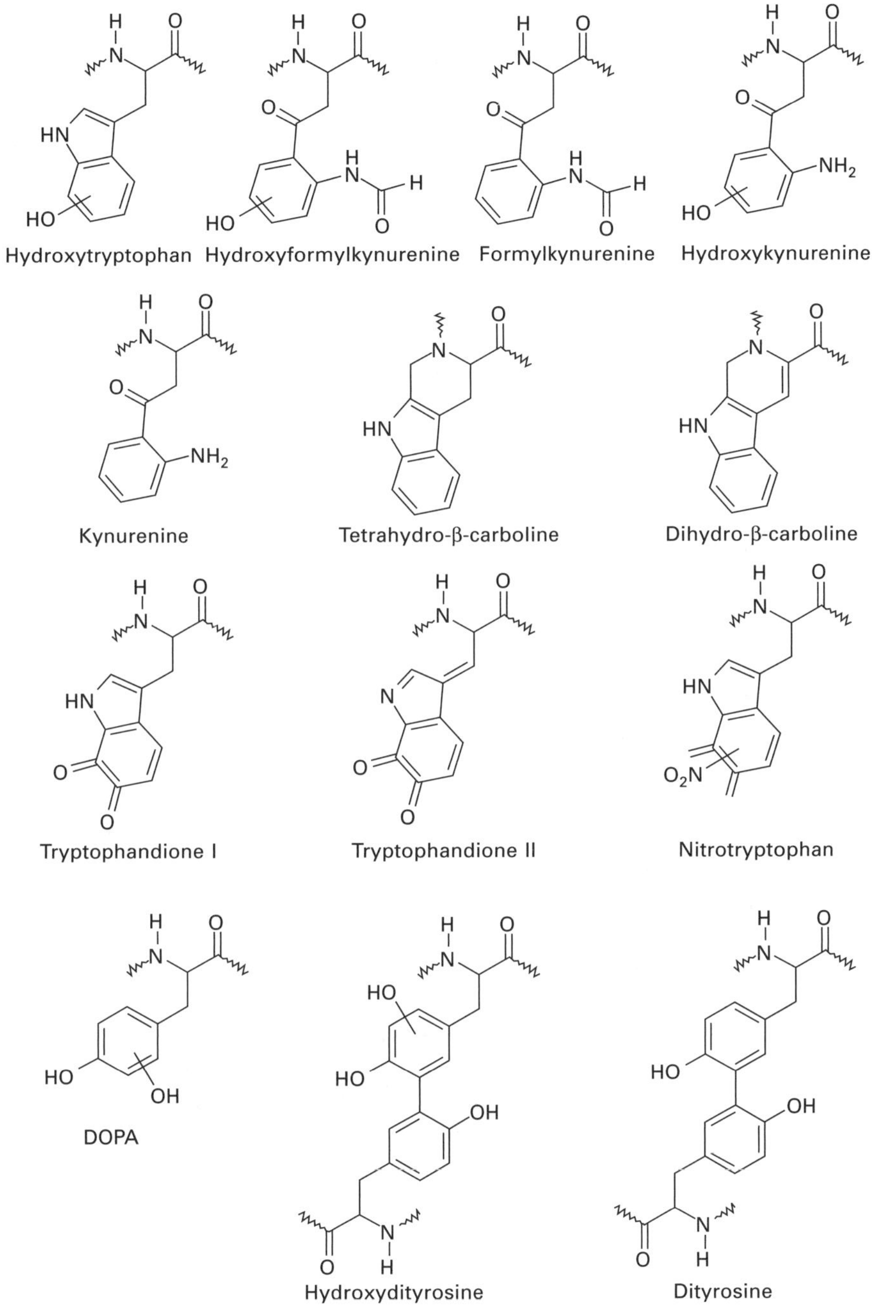

9.8 Thirteen yellow chromophores derived from Trp and Tyr residues identified in tryptic digests of photoirradiated wool fabric by HPLC/MS/MS.

9.4.5 The mechanism of photoyellowing of FWA-treated wool

Many previous studies have been carried out in order to understand the reasons why FWA-treated wool and silk yellow so rapidly, whereas other FWA-treated fibres, including cotton, nylon and polyester have high lightfastness. It has often been assumed that the presence of the FWA merely increases the rate of the photoyellowing reactions that occur in untreated wool, but this has recently been shown to be incorrect (Millington and Maurdev, 2005). Photoyellowing experiments were conducted using FWA-treated collagen I from bovine skin. Like wool keratin, collagen is a fibrous protein but it contains no Trp or Cys residues. Although untreated collagen was far more photostable than wool or silk, FWA-treated collagen photoyellowed rapidly and to a similar extent as FWA-treated wool. This clearly showed that Trp is not essential for the rapid photoyellowing of FWA-treated fibrous proteins to occur.

The same authors showed that wet FWA-treated wool irradiated with simulated sunlight produces significant amounts of H_2O_2 and superoxide radical anion ($O_2^{-\cdot}$), and that this process also occurs in aqueous FWA solutions (Millington and Maurdev, 2004). However, in solution the rate of production of $O_2^{-\cdot}$ and H_2O_2 was far more rapid in the presence of an electron donor amino acid, such as Trp or arginine (Arg). This strongly suggested that a photoinduced electron transfer (PET) reaction occurs in irradiated FWA-treated wool, where electrons are transferred from the wool to the FWA. Electron-rich amino acid groups in the wool act as the electron donors. The PET reaction leads to high concentrations of $O_2^{-\cdot}$ and H_2O_2 at the fabric surface. Irradiation of H_2O_2 near the surface subsequently leads to the formation of hydroxyl radicals which are highly reactive and cause rapid yellowing of both the wool and the FWA (Fig. 9.9).

This PET mechanism has previously been observed between *trans*-stilbene and tertiary and secondary amines (Lewis, 1986), which suggests that amino acid residues with secondary or tertiary amino groups in wool (arginine, proline, possibly also histidine and tryptophan) may be the electron donor species. It has recently been shown that arginine can act as an electron donor to the excited state of a distyryl biphenyl FWA by fluorescence quenching (Millington and Maurdev, 2005). Removal of electrons from wool is facile and it is interesting that a wool cloth is often used as a good electron donor system in electrostatic experiments. Photoreduction of the FWA by wool is also in accord with a previous study many years earlier. This study showed that for most dyed fibres irradiated in air, the dye is photo-oxidised and the fibre is photoreduced, whereas for the proteinaceous fibres wool and silk, the dye is photoreduced and the fibre is oxidised (Cumming *et al.*, 1956). Since FWAs can be considered as colourless fluorescent dyes, this theory

$$\text{FWA} \xrightarrow{h\nu} {}^1\text{FWA}^* \xrightarrow{\text{wool}} \underset{\text{Exciplex}}{\overset{\text{Non-polar environment}}{[\text{wool}^{\delta+}\ldots\ldots\text{FWA}^{\delta-}]}} \leftrightarrow \underset{\text{Oxidised wool Reduced FWA}}{\overset{\text{Polar environment}}{\text{wool}^{+}\cdot + \text{FWA}^{-}\cdot}} \quad (14)$$

photoinduced electron transfer from wool to FWA

$$\text{Wool}^{+}\cdot + \text{FWA}^{-}\cdot \xrightarrow{\text{Proton transfer}} \text{Wool}\cdot\ \cdot\text{FWAH} \xrightarrow{\text{Combination}} \text{Wool} - \text{FWAH} \quad (15)$$

covalent wool–FWA bonding

$$\text{FWA}^{-}\cdot + H^{+} \longrightarrow \text{FWAH}\cdot \rightarrow \text{FWAH}_2 \quad (16)$$

Reduced FWA products

$$\text{FWA}^{-}\cdot + O_2 \longrightarrow \text{FWA} + O_2^{-}\cdot \quad (17)$$

$$O_2^{-}\cdot + H^{+} \longrightarrow \cdot\text{OOH} \quad (18)$$

$$O_2^{-}\cdot + H^{+} + \cdot\text{OOH} \longrightarrow H_2O_2 + O_2 \quad \text{Dismutation of superoxide} \quad (19)$$

$$H_2O_2 + M^{n+} \longrightarrow M^{(n+1)+} + \text{OH}\cdot + \text{OH}^{-} \quad (20)$$

$$M^{(n+1)+} + O_2^{-}\cdot \longrightarrow M^{n+} + O_2 \quad (21)$$

$$H_2O_2 \xrightarrow{h\nu} 2\ \cdot\text{OH} \quad \text{Generation of}\ \cdot\text{OH} \quad (22)$$

$$\text{FWA} + n\ \cdot\text{OH} \longrightarrow \text{mono and dihydroxylated FWAs} \quad (23)$$

$$\text{Dihydroxylated FWA} \xrightarrow{-2H} \text{yellow stilbene quinones} \quad (24)$$

Yellow FWA photoproducts

$$\text{Wool} + \cdot\text{OH} \longrightarrow \text{yellow wool oxidation products} \quad (25)$$

$$\text{Wool}^{+}\cdot + O_2^{-}\cdot \longrightarrow \text{yellow wool oxidation products} \quad (26)$$

Yellow wool photoproducts

M^{n+}, $M^{(n+1)+}$ = catalytic metal ions

9.9 Photoreduction mechanism of FWA on wool and superoxide generation via photoinduced electron transfer.

explains very well the observed differences in the photostability of FWA-treated wool and silk and other FWA-treated fibres. It is also supported by a previous study showing the formation of reduced FWA photoproducts on irradiated FWA-treated wool, which would otherwise be highly unlikely in an oxidising environment (Holt and Milligan, 1974).

Millington and Maurdev also showed that irradiation of aqueous FWA solutions doped with H_2O_2 gave rise to yellow FWA oxidation products (Millington and Maurdev, 2005). Recent HPLC/MS/MS analysis of an aqueous distyrylbiphenyl (DSBP) FWA (Uvitex NFW) irradiated in the presence of H_2O_2 has confirmed the presence of mono- and dihydroxylated DSBPs and a yellow stilbene quinone chromophore (Dyer *et al.*, 2008). Stilbene quinones are highly coloured and have also been implicated in the photoyellowing of cellulosic materials (Ruffin and Castellan, 2000). It is likely that the dihydroxylated DSBP initially arises via attack of hydroxyl radicals on the FWA (Fig. 9.9, Equation 23) followed by oxidation and loss of H_2 (possibly by H atom abstraction by •OH), to form the stilbene quinone as shown in Fig. 9.10.

It is not possible from mass spectrometry to determine whether the stilbene quinone adopts an *ortho*- or *para*-geometry or a mixture of isomers, but both

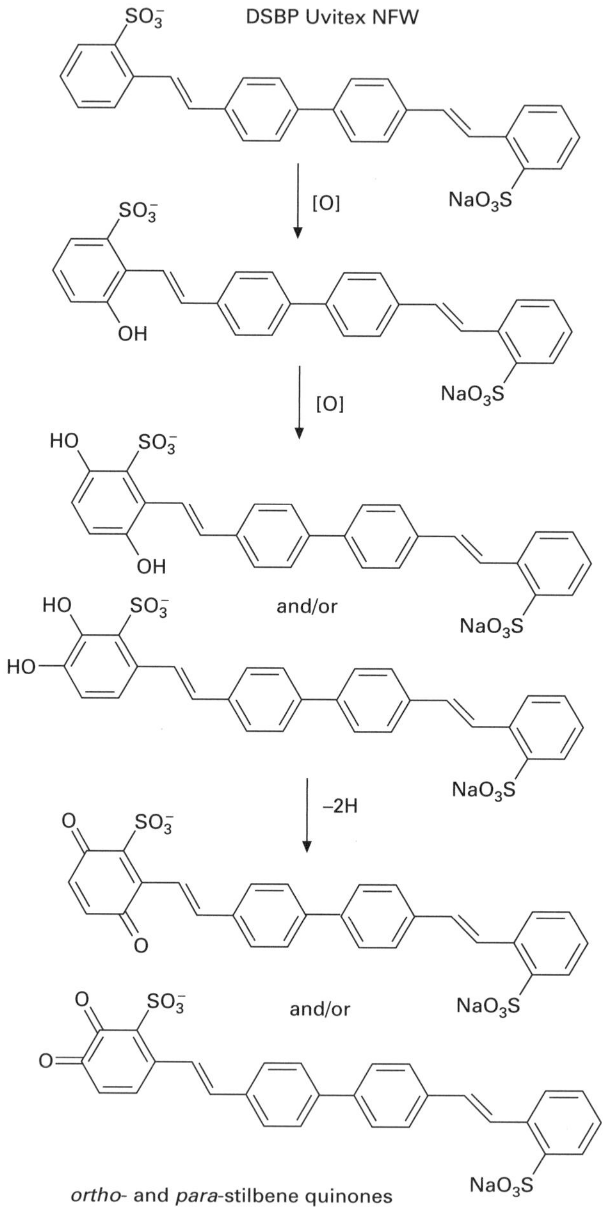

9.10 Formation of a coloured stilbene quinone from an aqueous DSBP FWA irradiated in the presence of H_2O_2.

forms would be highly coloured. It is interesting that *ortho*-quinones, in particular dopaquinone, are involved in the formation of melanin.

9.5 Methods for improving the whiteness and photostability of wool

There are currently a number of effective ways in which wool whiteness can be improved on the farm. As discussed in Section 9.2.1, heritability data show that selection of sheep for high clean wool brightness and low yellowness would result in improved wool colour characteristics. In addition, the rugging or shedding of sheep, or the annual application of UV absorbers to the backs of sheep would lead to reduced weathering damage and improved colour.

Several treatments have been described in the literature that can improve the photostability of wool, but so far only one has been found suitable for commercialisation, and mainly as a treatment against phototendering rather than photoyellowing. A UV-absorber with a sulphonated 2-hydroxyphenylbenzotriazole structure was developed by CSIRO in Australia that offers wool good protection against the reduced tear strength, abrasion resistance and dye fading usually experienced following prolonged sunlight exposure, with some reduction in photoyellowing for undyed wools (Mosimann *et al.*, 1990). Intramolecular hydrogen-bonded 2-hydroxyphenylbenzotriazoles are commonly used as UV-absorbers in other polymers, and they dissipate the absorbed UV energy as heat through conversion to a tautomeric form in the excited state (Fig. 9.11).

Hydroxyphenylbenzotriazoles used commercially for stabilisation of other polymers are highly hydrophobic materials that have no affinity for wool. Introduction of the sulphonate group imparts both water solubility and substantiveness for application to wool under slightly acid conditions, similar to acid dyes. However, for wool the key to development of an effective additive was in the nature of the R group. A bulky hydrophobic substituent is required to shield the internal hydrogen bond from interaction with polar groups in wool (Mosimann *et al.*, 1990). This chemistry was commercialised in 1990 for use as Cibafast W (Ciba), mainly for use on upholstery and curtaining fabrics. It cannot be used on FWA-treated wool since it absorbs the UV wavelengths necessary to excite the FWA.

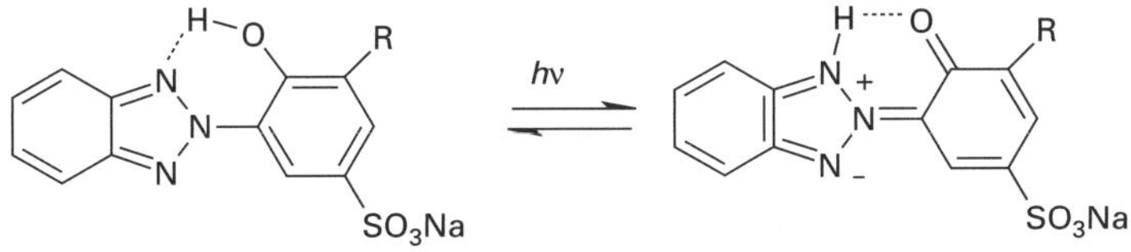

9.11 Tautomerisation of a 2-hydroxyphenylbenzotriazole UV absorber.

Improved bleaching processes are an obvious way to improve wool colour. Working on crossbred wools for use in wool carpets, New Zealand workers at Canesis found that performing an acid extraction during scouring produces a much brighter fibre (i.e. a higher *Y* tristimulus value) (McKinnon *et al.*, 1999). It is claimed that the acid extraction, performed at pH 2–3 using sulphuric acid and ethylenediaminetetraacetic acid (EDTA), removes absorbed iron from the wool. This agrees with previous work by Simpson who detected iron (10 mg/kg) and copper ions (1 mg/kg) after extraction of scoured wool fabric with molar sulphuric acid and with phosphate buffer at pH 7 (Simpson, 1987). Although acid extraction improves the brightness of treated wool, its perceived yellowness (*Y–Z*) also increases since the *Z* tristimulus value is not significantly improved. The new scouring process, marketed as Glacial® wool, therefore includes a peroxide bleaching stage and an optional reductive bleach. Incorporating acid extraction and an oxidative and reductive bleaching stage into the scouring process produces wool at least 7*Y* tristimulus units brighter than conventionally scoured wool. The *Z* value is also improved by ~8 units, making the wool appear significantly less yellow.

Another recent development in wool bleaching has been the development by Australian workers at CSIRO of a new reductive bleaching technology for wool, ColorClear™ WB (Australian Wool Innovation, 2006). This method utilises the reaction between sodium borohydride and sodium bisulphite to produce the active bleaching species, sodium dithionite, *in situ*. This technique was already in use by the textile industry to strip excess dyestuffs from synthetic fibres. Large-scale commercial trials on wool and wool blends have confirmed that the developed procedures are technically robust and give significant benefits over the use of conventional 'hydros' (sodium dithionite) and 'thiox' (thiourea dioxide) reductive bleachings in terms of improved whiteness. This is illustrated in Figure 9.12 for natural and machine-washable (chlorine/Hercosett-treated) wool fabric.

ColorClear™ WB can also be used for the whitening of wool blends. A polyester disperse FWA is first applied to a wool/polyester blend, followed by a sequential double bleaching procedure using H_2O_2 followed by reductive bleaching with a mixture of sodium borohydride and sodium bisulphite. In this process, the ColorClear™ WB carries out a dual role. It not only acts as a reductive bleach to whiten the wool component, but also strips any disperse FWA from the wool component. This is necessary since any FWA remaining on the wool component would lead to rapid photoyellowing of the wool on exposure to sunlight. ColorClear™ WB is produced and marketed by Rohm and Haas.

On the basis of the metal-catalysed oxygen free radical yellowing mechanism discussed in Section 9.4.4, a combination of a water-soluble antioxidant and a metal chelator was expected to be a more effective treatment for preventing the rapid photoyellowing of FWA-treated wool than an antioxidant alone.

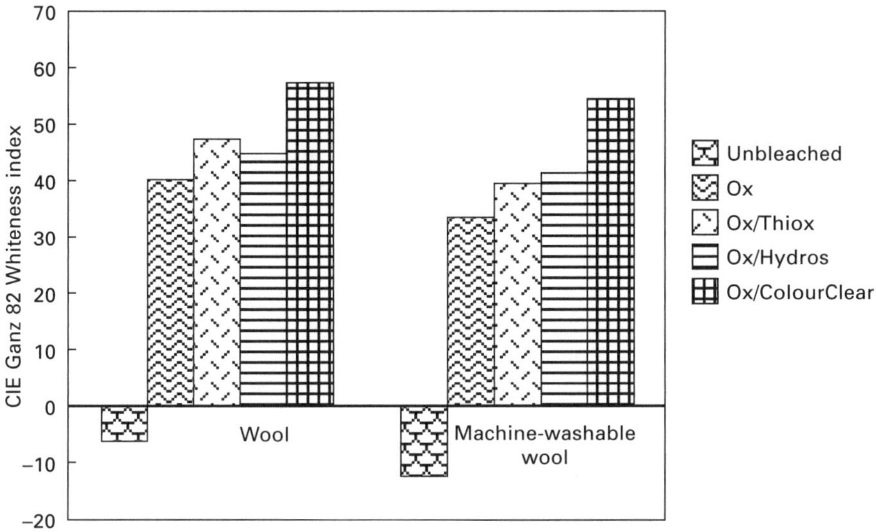

9.12 Plot of the CIE Ganz 82 whiteness index of wool fabric after various bleaching treatments. Ox = H_2O_2 bleaching, Thiox = thiourea dioxide, Hydros = sodium dithionite.

Rinsing FWA-treated wool with low concentrations of an antioxidant such as *N*-acetylcysteine (NAC) was highly effective against photoyellowing when combined with a metal chelator such as oxalic acid, and provided much better photoprotection than the antioxidant alone, particularly under wet conditions (Millington, 2006a). Unfortunately the benefits of the rinse treatment are lost on laundering, and efforts to identify a substantive antioxidant/metal chelator combination that could be applied during wet finishing and provide effective protection against photoyellowing under both wet and dry conditions have so far been unsuccessful.

Improving the photostability of wool to photobleaching can also be important for certain applications. When wool carpets are exposed to sunlight through window glass they undergo photobleaching soon after the carpet is laid during a period when consumers are particularly sensitive to product performance. The effect is particularly apparent when the yellowness of the wool is fairly high and the carpet is dyed to a pale shade. New Zealand workers at WRONZ developed a dyebath additive that photoyellows at the same rate as photobleaching occurs, so that the overall colour is unchanged throughout exposure. This is now marketed by Clariant as Lanalbin APB™.

9.6 Future trends

Selective sheep breeding for improved wool colour characteristics has already been shown to be highly feasible, although its implementation to date has been very limited. The current rapid progress in genetics and molecular biology suggests that application of these techniques to the genes that control wool colour could be very useful in the longer term for producing whiter wools with improved photostability. Selection of sheep for improved wool traits by conventional breeding has been limited by the genome and through each selected generation the genetic profile of the original strains is altered at an unknown number of loci. In contrast, transgenesis and cloning methodologies allow the valuable genetic background of pre-existing strains to be preserved. Moreover, these methods can now be used to alter single or multiple loci and to add or remove genes. Transgenic studies on sheep have already shown that the properties of the wool fibre can be significantly changed and that such changes are heritable (Powell *et al*., 1994; Bawden *et al*., 1998). However, in order to clearly define the experimental options for sheep transgenesis and wool fibre modification, a better knowledge of the molecular biology of the wool follicle and the molecules which control initiation of wool follicle growth is required. Realistically, one would expect transgenic studies to focus initially on traits such as wool yield, growth rate and fibre diameter, which potentially offer a higher return to the grower than improved colour.

It has been suggested that creating wool flocks where albinism could be used as the basis for pigmentation control would be a novel area of research (Fleet, 2002). Transgenics may also offer the opportunity to create small air cavities within fine wool fibres that are similar to the medulla found in coarser wool fibres, such as the Scottish Blackface. If this could be accomplished without significantly affecting the strength of the wool fibre, then the increased reflectance and reduced light penetration through the fibre would produce a brighter, whiter visual effect.

The origin and identity of the visible yellow chromophores in wool remains an area where further fundamental research is essential. This would not only allow geneticists to identify the genes responsible for their production, offering the possibility of control, but might also allow chemists to design more effective bleaching processes to remove them from the fibre during commercial processing.

Although a great deal of work has been done to characterise wool cortical proteins and define their structure and function, knowledge of the cuticle proteins remains very rudimentary. Increased knowledge of the biology and biochemistry of cuticle proteins in wool is an important issue, as most of the high-energy UVB radiation in sunlight which causes photoyellowing will be absorbed within the cuticle cells. One of the key questions is whether chromophores form preferentially near the fibre surface in the cuticle cells during photoyellowing by sunlight, as suggested by Simpson for UVB exposure (Simpson, 1999), or whether they are more evenly distributed throughout the cuticle and cortex.

Zinc oxide is a UV-absorbing white pigment commonly used in sunscreens to protect against skin damage. Currently it is not clear whether the application of inorganic nanoparticulates to wool, capable of penetrating into the fibre, might offer similar protection against photoyellowing. This would be unlikely if most yellowing occurs in the highly crosslinked regions of the exocuticle. Schäfer has claimed that the white inorganic pigments titanium dioxide and barium sulphate retard the photoyellowing of FWA-treated wool, but they also impact on its whiteness by reducing the effectiveness of the FWA (Schäfer, 1990). The materials used were not nanoparticles and presumably were simply deposited on the fibre surface.

A great deal of research has already been carried out on wool photoyellowing. The complexity of the chemistry involved suggests that an empirical approach to improving wool's photostability is most unlikely to be successful. Detailed knowledge of the photoyellowing mechanism(s) is therefore essential. Such knowledge could then be used to develop a photostable FWA system for use on wool and other protein fibres. This is urgently required to allow wool to compete on a level playing field with other textile fibres and to produce photostable brilliant whites and pastel shades.

9.7 Sources of further information and advice

The evolution of sheep and wool colour is described in detail in two papers by Ryder (1987, 1980) and wool colour genetics is reviewed in Sponenberg's chapter in *The Genetics of Sheep* (Sponenberg, 1997). There are also a number of useful websites on current research into sheep and wool genetics, including:

http://www.sheepgenomics.com/ (*Sheep Genomics*)
http://www.sheepgenetics.org.au/ (*Sheep Genetics Australia*)
http://www.agwine.adelaide.edu.au/animal/skin/ *(University of Adelaide)*
http://www.sardi.sa.gov.au (*South Australian Research and Development Institute*) http://www.sheepcrc.org.au/ (*Cooperative Research Centre (CRC) for Sheep Industry Innovation).* This CRC commenced a seven-year project aimed at improving wool colour and photostability in July 2007.
http://www.wool.com.au/ (*Australian Wool Innovation*)

Useful online information on wool colour measurement is available from http://www.awta.com.au/en/Home/Our_Services/Average-Yellowness/ (Australian Wool Testing Authority) and from http://www.wooltesting.sgs.com (SGS Wool Testing Services, New Zealand).

There is a helpful and comprehensive *'Whiteness Primer'* available for download, to assist understanding the concepts of colour measurement in general and whiteness in particular. http://www.axiphos.com/Reports/reports.html

There are a large number of excellent reviews on wool photoyellowing. There is a comprehensive recent (2006) two-part review that gives a very detailed description of the chemistry, instrumental methodology and mechanisms involved, and cross-references most of the extensive literature in this field (Millington, 2006b,c).

9.8 References

ADALSTEINSSON, S (1977), 'Albinism in Icelandic sheep', *J Hered*, **68**, 347–349.

ADALSTEINSSON, S (1978), 'Note on an Icelandic albino ram', *J Hered*, **69**, 430

Australian Wool Innovation (2006), Superwhite Australian Merino using ColorClear WB Bleaching Technology. Sydney: AWI.

Australian Wool Testing Authority (2008), 'Dark & Medullated Fibre Contamination', http://www.awta.com.au/en/Home/Our_Services/Dark—Medullated-Fibres/, accessed 5 May 2008.

BAWDEN, C S, POWELL, B C, WALKER, S K and ROGERS, G E (1998), 'Expression of a wool intermediate filament keratin transgene in sheep fibre alters structure', *Transgenic Res*, **7**, 273–287.

BAXTER, P (2001), 'Stability of wool colour', *Massey Wool Association Conference*, Napier, NZ,

BENAVIDES, M V and MAHER, P (2003), 'Genetic parameters of wool colour and skin traits in Corriedale sheep', *Genetics Mol Biol*, **26**, 267–274. DOI: 10.1590/S1415-47572003000300009.

BROWN, D J (2006), 'Genetic aspects of greasy wool colour assessments in Merino sheep', *Int J Sheep Wool Sci*, **54**, 1–10.

CAMERON, B A and STOBART, R H (2006), 'The colour of scoured and carded wools: a comparison of US, Australian and New Zealand wools', *Sheep Goat Res J*, **21**, 17–23.

CUMMING, J, GILES, C H and MCEACHRAN, A E (1956), 'A study of the photochemistry of dyes on proteins and other substrates', *JSDC*, **72**, 373–381.

DEAN, R T, FU, S, STOCKER, R and DAVIES, M J (1997), 'Biochemistry and pathology of radical-mediated protein oxidation', *Biochem J*, **324**, 1–18.

DUFFIELD, P A (1996), Review of wool bleaching processes. IWS Report.

DUFFIELD, P A and LEWIS, D M (1985), 'The yellowing and bleaching of wool', *Rev Prog Coloration*, **15**, 38–51.

DYER, J M, BRINGANS, S D, PLOWMAN, J E and BRYSON, W G (2005), 'Chromophores in photoyellowed wool fabric characterised by mass spectrometry', *Proc 11th Int Wool Text Res Conf*, Leeds, UK,

DYER, J M, BRINGANS, S D and BRYSON, W G (2006a), 'Characterisation of photo-oxidation products within photoyellowed wool proteins: tryptophan and tyrosine derived chromophores', *Photochem Photobiol Sci*, **5**, 698–706. DOI: 10.1039/b603030k

DYER, J M, BRINGANS, S D and BRYSON, W G (2006b), 'Determination of photooxidation products within photoyellowed bleached wool proteins', *Photochem Photobiol*, **82**, 551–557. DOI: 10.1562/2005-08-29-RA-663

DYER, J M, CORNELLISON, C D, MAURDEV, G and MILLINGTON, K R (2008), 'The photoyellowing of stilbene-derived fluorescent whitening agents – mass spectrometric characterisation of yellow photoproducts', *Photochem Photobiol*, **84**, 143–153.

FLEET, M R (2002), 'Pigmentation prevention in sheep: complex or simple ?', *Wool Tech Sheep Breed*, **50**, 410–416.

FLEET, M R, SMITH, D H and POURBEIK, T (1991), 'Age-related changes in pigmentation traits of adult Merino sheep', *Wool Tech Sheep Breed*, **39**, 24–34.

FORREST, J W, FLEET, M R and ROGERS, G E (1985), 'Characterisation of melaninocytes in wool-bearing skin of Merino sheep', *Aust J Biol Sci*, **38**, 245–257.

GANZ, E (1979), 'Whiteness formulas: a selection', *Appl Optics*, **18**, 1073–1078.

GARROW, C, LHUEDE, E P and ROXBURGH, C M (1971), 'An application of blue-light bleaching of wool', *Text I Ind*, 9, 286–288.

GEBICKI, S and GEBICKI, J M (1993), 'Formation of peroxides in amino acids and proteins exposed to oxygen free radicals', *Biochem J* , **289**, 743–749.

GUGUMUS, F (1993), 'Light stabilisers', in Gacher, R and Muller, H, *Plastics Additives Handbook*, 4th edn. Munich, Hanser, 129–270.

HOLT, L A and MILLIGAN, B (1974), 'Fluorescent whitening agents. III. Photodecomposition of disodium 4,4-diacetamidostilbene-2,2-disulphonate', *Text Res J*, **44**, 181–183.

HOLT, L A, LAX, J and MOLL, L (1994), 'The effect of weathering and weathering control measures on the colour of scoured wool', *Wool Tech Sheep Breed*, **42**, 151–159.

IWTO (2003a), 'International Wool Textile Organisation Standard Test Method IWTO–35–03. Method for the measurement of colour of sliver.'

IWTO (2003b), 'International Wool Textile Organisation Standard Test Method IWTO–56–03. Method for the measurement of colour of raw wool.',

JACKSON, I J (1994), 'Molecular and developmental genetics of mouse coat colour', *Ann Rev Genetics*, **28**, 189–217.

JAMES, P J, PONZONI, R W, WALKLEY, J R W and WHITELEY, K J (1990a), 'Genetic and phenotypic parameters for greasy and scoured wool colour', *Proc Aust Assoc of Animal Breeding and Genetics*, **8**, 521–523.

JAMES, P J, PONZONI, R W, WALKLEY, J R W and WHITELEY, K J (1990b), 'Genetic parameters for wool production and quality traits in South Australian Merinos of the Collinsville family', *Aust J Agric Res*, **41**, 583–594.

KRON, A, STENBERG, B and REITBERGER, T (1997), 'Chemiluminiscence applied to oxidation of polyolefins', *Prog Rubber Plastics Technol*, Vol **13**(2), 81–107.

LAUNER, H F (1968a), 'Photobleaching – a common phenomenon', *Nature*, **218**, 160–161.

LAUNER, H F (1968b), 'Spectral conformity – a widespread phenomenon', *Nature*, **218**, 161–163.

LAUNER, H F (1971), 'Rapid bleaching of wool with extremely intense visible light', *Text Res J*, **41**, 311–314. DOI: 10.1177/004051757104100404

LEAVER, I H and RAMSAY, G C (1969), 'Studies in wool yellowing. Part XXVII. The role of water in the photoyellowing of fluorescent whitened wool', *Textile Res J*, **39**, 730–733.

LENNOX, F G and KING, M G (1968), 'Studies in wool yellowing. Part XXIII. UV yellowing and blue-light bleaching of different wools', *Text Res J*, **38**, 754-761. DOI: 10.1177/004051756803800711.

LEWIS, F D (1986), 'Proton transfer reactions of photogenerated radical ion pairs', *Acc Chem Res*, **19**, 401–405. DOI: 10.1021/ar00132a004.

MACLAREN, J A and MILLIGAN, B (1981), *Wool Science: The chemical reactivity of the fibre*, Marrickville NSW, Science Press.

MAHAR, T J and OSBORNE, L (1996), 'The relationship between the clean colour of commercial consignments of greasy wool and the colour of the resulting top or noil', *Wool Tech Sheep Breed*, **44**, 219–237.

MAIJALA, K (1997), 'Genetic aspects of domestication, common breeds and their origin', in Piper, L and Ruvinsky, A, *The Genetics of Sheep*, Wallingford, UK, CAB International, 13–49.

MCKINNON, A J, MCLAUGHLIN, J R, TAYLOR, M E, RANKIN, D A, MIDDLEWOOD, P G, LE PINE, P, MESMAN, P G R and MANSON, S B (1999), A method for significantly enhancing the quality of scoured wool and machinery for achieving those enhancements. WO 99/16942, to WRONZ.

MILLINGTON, K R (2005), 'Continuous photobleaching of wool', *Proc 11th Int Wool Text Res Conf*, Leeds, UK.

MILLINGTON, K R (2006a), 'Improving the photostability of whitened wool by applying an antioxidant and metal chelator rinse', *Color Technol*, **122**, 49–56. DOI: 10.1111/j.1478–4408.2006.00008.x

MILLINGTON, K R (2006b), 'The photoyellowing of wool. Part 2: Photoyellowing mechanisms and methods of prevention', *Color Technol*, **122**, 301–316. DOI: 10.1111/j.1478–4408.2006.00045.x

MILLINGTON, K R (2006c), 'The photoyellowing of wool. Part I. Factors affecting photoyellowing and experimental techniques.', *Color Technol*, **122**, 169–186. DOI: 10.1111/j.1478–4408.2006.00034.x

MILLINGTON, K R and KIRSCHENBAUM, L J (2002), 'Detection of hydroxyl radicals in photoirradiated wool, cotton, nylon and polyester fabrics using a fluorescent probe', *Color Technol*, **118**, 6–14. DOI:10.1111/j.1478–4408.2002.tb00130.x

MILLINGTON, K R and MAURDEV, G (2004), 'The generation of superoxide and hydrogen peroxide by exposure of fluorescent whitening agents to UVA radiation and its relevance to the rapid photoyellowing of whitened wool', *J Photochem Photobiol A: Chem*, **165**, 177–185. DOI: 10.1016/j.jphotochem.2004.03.017.

MILLINGTON, K R and MAURDEV, G (2005), 'The mechanism of photoyellowing of fluorescent whitened wool', *Proc 11th Int Wool Text Res Conf*, Leeds, UK,

MILLINGTON, K R, MAURDEV, G and JONES, M J (2007), 'The thermal chemiluminescence of fibrous proteins', *Polym Degrad Stab*, **92**, 1505–1512. DOI: 10.1016/j.polymdegradstab.2007.05.014.

MOSIMANN, W, BENISEK, L, BURDESKA, K, LEAVER, I H, MYERS, P C, REINERT, G and WILSHIRE, J F K (1990), 'A new commercial UV absorber for the protection of wool and wool dyeings', *Proc 8th Int Wool Text Res Conf, Christchurch*, NZ, **IV**, 239–249.

MULLER, U, RATCH, M, SCHWANNINGER, M, STEINER, M and ZOBL, H (2003), 'Yellowing and IR-changes of spruce wood as a result of UV irradiation', *J Photochem Photobiol B: Biol*, **69**, 97–105. DOI:10.1016/S1011–1344(02)00412–8

NICHOLLS, C H and PAILTHORPE, M T (1976), 'Primary reactions in the photoyellowing of wool keratin', *J Text Inst*, **67**, 397–403.

OZEKI, H, ITO, S and WAKAMATSU, K (1996), 'Chemical characterisation of melanins in sheep wool and human hair', *Pigment Cell Res*, **9**, 51–57. DOI:10.1111/j.1600–0749.1996.tb00088.x

POWELL, B C, WALKER, S K, BAWDEN, C S, SIVAPRASAD, A V and ROGERS, G E (1994), 'Transgenic sheep and wool growth: possibilities and current status', *Reprod Fertil Dev*, **6**, 615–623. DOI: 10.071/RD9940615

ROWETT, M A and FLEET, M R (1993), 'Albinism in a Suffolk sheep', *J Hered*, **84**, 67–69.

RUFFIN, B and CASTELLAN, A (2000), 'Photoyellowing of peroxide-bleached lignin-rich pulps: a photochemical study on stilbene-hydroquinone chromophores issued from β-5 units of lignin during refining and(or) bleaching', *Can J Chem*, **78**, 73–83. DOI:10.1139/cjc–78–1–73

RUSSELL, J E and TUMLINSON, R (1996), 'Comparison of microstructure of white winter fur and brown summer fur of some Arctic mammals', *Acta Zool – Stockholm*, **77**, 279–282.

RYDER, M L (1980), 'Fleece colour in sheep and its inheritance', *Animal Breed Abstr*, **48**, 305–324.

RYDER, M L (1987), 'The evolution of the fleece', *Sci Amer*, **256**, 100–107.

SCHÄFER, K (1990), 'Investigation to stabilise optically brightened wool against light damage', *Proc 8th Int Wool Text Res Conf*, Christchurch, NZ, **IV**, 250–259.

SHATKAY, A and MICHAELI, I (1970), 'EPR study of wool irradiated by UV and visible light', *Radiat Res*, **43**, 485–498.

SIMPSON, W S (1987), 'The influence of pH on the reflectance and photostability of wool to sunlight', *J Text Inst*, **78**, 430–438.

SIMPSON, W S (1992) Photobleaching process for natural fibres. WO 92/15744 to WRONZ.

SIMPSON, W S (1999), 'Physics and chemistry of wool yellowing', *Wronz Report* R217,

SIMPSON, W S (2002), 'Wool chemistry', in Simpson, W S and Crawshaw, G H, *Wool: Science and Technology*, Cambridge, Woodhead, 130–159.

SPONENBERG, D P (1997), 'Genetics of colour and hair texture', in Piper, L and Ruvinsky, A, *The Genetics of Sheep*, Wallingford, UK, CAB International, 51–86.

SUMNER, R M W, YOUNG, S R and UPSDELL, M P (2003), 'Wool yellowing and pH within Merino and Romney fleeces', *Proc NZ Soc Animal Production*, **63**, 155–159.

SUTTON, R, WARD, W G, LEISH, Z and FAWCETT, A A (1998), 'Effect of anti-sense transgenes on coat colour and skin tyrosinase levels', *Proc 6th World Conf on Genetics Applied to Livestock Production*, Armidale, Australia, **26**, 489–492.

THOMPSON, B and TEASDALE, D C (1986), 'The colour of commercially scoured Australian wool', *Wool Tech Sheep Breed*, **33**, 127–128.

VELDSMAN, D P and SWANEPOEL, O A (1971), 'The influence of the rate of chlorination on the shrinkproofing of wool with DCCA', *Appl Polym Symp*, **18**, 691–700.

WESTMORELAND, D J, SCHLINK, A C and GREEF, J C (2006), 'Factors affecting wool scouring performance, yield and colour measurements of Western Australian fleece wools', *Aust J Exp Agr*, **46**, 921–925. DOI: 10.071/EA05352

WOOD, E (2002), 'The basics of wool colour measurement', *Wool Tech Sheep Breed*, **50**, 121–132.

10

Enhancing wool products using nanotechnology

R. DENNING, CSIRO Materials Science and Engineering, Australia

Abstract: This chapter describes the application of nanotechnology to textile fibres, with an emphasis on wool fibres. The chapter includes a description of nanotechnology, explores some common misconceptions and reviews current uses of nanoparticles on textile fibres. The use of nanotechnology on wool fibres is then reviewed and current commercial applications described.

Key words: nanotechnology, nanostructured films, nanoparticle safety, stain resistance, self-cleaning textiles.

10.1 Introduction

Nanotechnology is a rapidly emerging interdisciplinary field that is having a significant impact on materials including plastics and textiles; electronics; optics; medicine; and energy. It is often referred to as 'bottom up' technology (Holme, 2008), as materials are constructed from the smallest building blocks of atoms, molecules and nanoparticles, in contrast to 'top down' technologies that start with the bulk material.

The Royal Society and The Royal Academy of Engineering (2004) in their report *Nanoscience and Nanotechnologies: Opportunities and Uncertainties* defined nanoscience and nanotechnology as:

> ***Nanoscience*** is the study of phenomena and manipulation of materials at atomic, molecular and macromolecular scales, where properties differ significantly from those at a larger scale.
> ***Nanotechnologies*** are the design, characterisation, production and application of structures, devices and systems by controlling shape and size at nanometre scale.

Nanotechnology involves the creation or manipulation of structures at the nanoscale. At this scale the characteristics of matter can be dominated by quantum effects, molecular recognition, confinement effects and increases in relative surface area. This can result in significant differences in behaviour of materials at the nanoscale when compared to the bulk material (Renn and Roco, 2006). Gold nanoparticles for example range in colour from deep red at 7–20 nm to blue at approximately 300 nm and finally gold in the bulk material.

For textile applications, nanotechnology is more than applying very thin coatings or small particles to fibre substrates. It is changing the properties of the fibre in a way that cannot be achieved by applying bulk materials or thin films to the fibre surface.

10.1.1 Principles of nanotechnology

The nanometre scale is used to describe objects having at least one dimension between 1 and 100 nm. The Greek word 'nano', which means dwarf, is used as a prefix meaning one billionth or 10^{-9}, hence one nanometre (nm) is one billionth of a metre or approximately 20 000 times smaller than a 20 μm wool fibre. A cubic nanometre (nm^3) can contain approximately 20 atoms. Some typical objects in the nanoscale range are shown in Figure 10.1.

Nanotechnology development has been described in terms of four generations (Roco, 2006) as listed in Table 10.1. The current generation, and most applicable to wool textiles, is the first generation, involving passive nanostructures in which the nanomaterials perform a single function. Later generations of nanotechnology will perform multiple functions and involve several nanotechnologies integrated into a single system.

Although first generation nanotechnology is usually the application of nanoparticles, the emerging consensus is that nanotechnology involves the control and restructuring of matter at the nanoscale leading to engineered functional systems at the molecular scale.

10.1.2 Introduction to nanoparticles and nano-emulsions

Nanoparticles are the building blocks of nanotechnology. They are available in a range of common shapes including spheres, rods, tetrahedrons, octahedrons and discs. Nanoparticles are characterised by a high surface to bulk atom ratio, which often results in enhanced chemical activity. Qi *et al.* (2004)

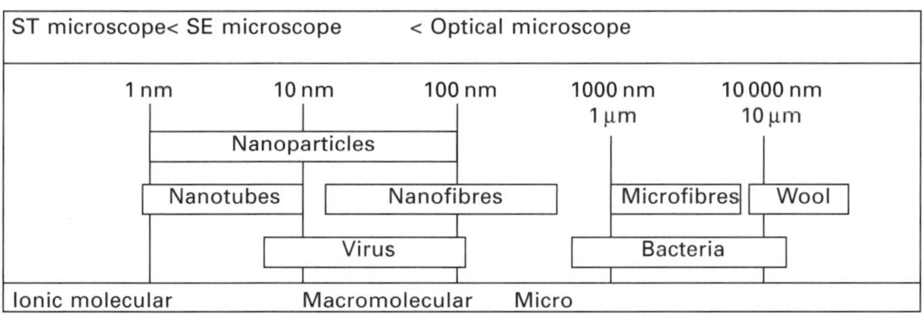

10.1 Some typical objects in the nanoscale range (ST = scanning transmission, SE = scanning electron).

Table 10.1 The generations of nanotechnology as defined by Roco and examples

Generations	Description	Examples
First	Passive nanostructures • Dispersed and contact nanostructures • Products incorporating nanostructures	 Colloids, nanoparticles, aerosols, nanotubes Coatings, nanoparticle reinforced composites, nanostructure metals and polymers
Second	Active nanostructures • Bioactive materials that change function during use • Physico-chemical active	 Targeted drugs and delivery systems, biodevices Amplifiers, actuators, adaptive structures, sensors and control systems
Third	Integrated nanosystems • Passive and/or active nanostructures integreated into systems • Convergence of nanotechnology, biotechnology, information technology and the cognitive sciences	 Artificial organs built from the nanoscale, evolutionary nanobiosystems, hybrid systems, 3D networking
Fourth	Molecular nanosystems • Engineered nanosystems and architectures created from individual molecules or supramolecular components each with a specific structure and role • The behaviour of applications being based on that of biological systems	Self-assembling products, nanoscale genetics, nanosystems integrated with industrial applications, atomic design

calculated the shape factor for several nanoparticle shapes, which can be used to calculate the ratio of surface to total atoms using equation 10.1:

$$N/n = 4\alpha d/D \qquad 10.1$$

where N/n is the ratio of surface to total atoms in the particle, d is the atomic diameter and D is the particle diameter. The shape factor, α, was calculated for a number of shapes. Examples of shape factors and percentage surface atoms for a number of nanoparticle shapes are shown in Table 10.2 and Fig. 10.2. The data were calculated for the disc-shaped particle, assuming the radius of the disc was twice the height.

The very high percentage of surface atoms of nanoparticles, particularly those less than 20 nm in diameter, results in many properties that differ from

Table 10.2 Calculated shape factors and percentage surface atoms for nanoparticles of different shapes with nominal diameters in the range 5 to 100 nm, assuming an atom diameter is 0.32 nm (silver)

Shape	α	*N/n* (%) for nominal diameter (nm)				
		5	10	20	50	100
Sphere	1	26	13	6	3	1
Rod	1.167	30	15	7	3	1
Tetrahedron	1.49	38	19	10	4	2
Disc*	2	37	18	9	4	2

*Radius to height ratio 2.

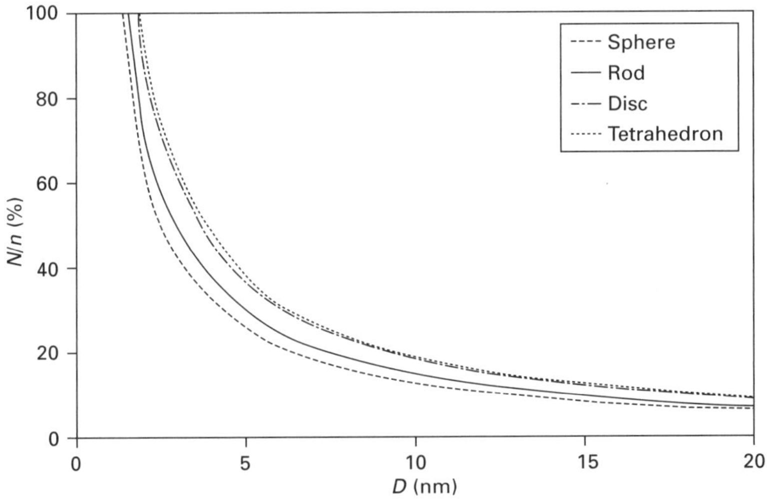

10.2 The percentage surface atoms of nanoparticles of different shapes calculated using equation 10.1 where *D* is the nominal diameter of the particle and *d* is the diameter of silver.

the bulk material. For example, the high curvature of nanoparticles results in an increase in surface energy that affects the interior bulk properties of the particle (Jiang *et al.*, 2006), giving a decrease in melting point. Buffat and Borel (1976) reported melting points for gold nanoparticles as low as 600 K for a 2 nm particle compared with the bulk melting point of 1336 K. Similarly catalytic activity of gold, normally considered an inert metal, has been reported for nanoparticles (Grisel *et al.*, 2002; Hurata, 2004) supported on metal oxide or activated carbon surfaces.

An emerging technology that may have application to wool is the nano-emulsion (Mason *et al.*, 2006). Nano-emulsions are conventional emulsions on the nano-scale, that is with radii in the range 1–100 nm. Nano-emulsions differ from micro-emulsions, which can have similar dimensions, in a number of important characteristics outlined in Table 10.3. Formation of nano-emulsions is usually by shear-induced rupture of conventional emulsions. Very high shear is applied to an emulsion containing excess surfactant. As the droplets rupture, they are stabilised by a coating of surfactant. As droplets approach each other in the emulsion, the surfactant films repel each other, preventing contact and coalescence. To achieve stable films, an equilibrium surface density of surfactant is needed. Given the very high surface area to bulk ratio of nanoparticles, surfactant concentrations are usually in excess of the critical micelle concentration. Nano-emulsions are therefore mixtures of emulsion droplets and surfactant micelles.

10.1.3 Introduction to nanostructured films

Nanostructured films are a new technology based on self-assembled monolayers (SAMs) where a single layer of a chemical or molecule is absorbed onto the surface of the fibre. These films differ from traditional textile coatings in that their thickness is in the nanometre range, they are often more even and develop higher film densities. Films can be built in multi-layers, each a mono-layer of the materials being applied as represented by Fig. 10.3. Often the layers contain different materials to achieve the final desired textile property.

The production of SAMs on textiles usually required a pre-treatment of the fibre to produce a surface suitable for self-assembly. For example to achieve a continuous monolayer of a charged molecule, an evenly distributed, dense, opposite ionic charge is needed on the textile fibre surface. Electrostatic interactions then build the monolayer on the surface. Techniques that have been used to produce the charge in the fibre surface include plasma, ion beam or chemical treatment such as oxidation or reduction.

Table 10.3 Comparison of nano-emulsion and micro-emulsion properties

Property	Nano-emulsion	Micro-emulsion
Droplet radius	2–100 nm	2–300 nm
Formation	Mechanical dispersion	Self-assembled thermodynamic phases
Shape	Spherical	Various including spherical, laminar sheet, hexagonal packed columns
Surfactant	Soluble in bulk phase Lower concentration	Soluble in both phases High concentration
Solubility of dispersed phase in bulk phase	Low	Relatively high
Surface tension	High	Low

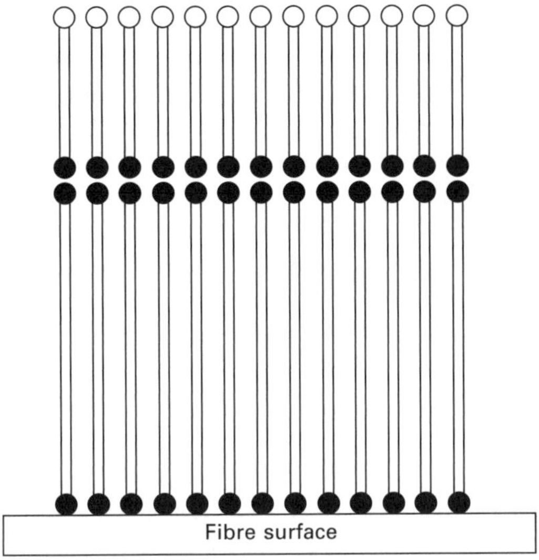

10.3 Representation of a two-layered self-assembled monolayer fibre coating.

Nanostructured films can be made self-repairable by engineering a storage source within the nanolayers. When material in the top layer is removed by, for example, abrasion, electrostatic interactions will try to neutralise the charge on the surface where the material was removed. These interactions will result in material moving from the storage source to cover the exposed charge. Unlike traditional coating that loose their effectiveness as the coating material is removed, the self-repair effect of SAMs can extend the life of the coating and maintain the finish properties throughout the garment life.

10.1.4 Advantages of nanotechnology on textiles

Nanotechnology has been applied to textiles at all stages from fibre production to consumer products. This section will summarise general applications of nanotechnology to textiles. The next section will describe specific applications to wool.

Fillers have been used for some time to produce composite fibres with improved properties such as increased strength. The high surface area and strong interactions of nanoparticles with polymers can be used to further enhance composite fibre properties. In particular, nanoparticles have been used to significantly improve fibre strength, stiffness and solvent resistance and reduce shrinkage and flammability.

The most common nanoparticle in use is clay. Clay nanoparticles are sheets of hydrous aluminosilicate that typically have a negative charge on the face and positive charge on the edge. Clay has been used to introduce dye sites into polypropylene (Fan *et al.*, 2003), to improve the flame resistance of nylon and to improve the heat resistance of polyester and nylon fibres. The use of nanoparticles in these applications reduced the mass of clay required to achieve the effect when compared to larger filler particles.

Self-cleaning cotton fabrics have been developed using titanium dioxide nanoparticles bound to the surface of the fibres (Bozzi *et al.*, 2005a; Meilert *et al.*, 2005). When titanium dioxide is exposed to light in the presence of oxygen and water vapour highly oxidative radical species such as $HO_2^{\cdot-}$ and $HO^{\cdot}$ along with H_2O_2 are produced. These species are able to decolour most stains. The use of a chemical spacer, a relatively unreactive poly-carboxylic acid, to bind the titanium dioxide to the cotton resulted in better removal of coffee and wine stains (Meilert *et al.*, 2005). Titanium dioxide and zinc oxide nanoparticles have also been used to protect textiles from UV damage and enhance the SPF rating of fabrics (Karst and Yang, 2006).

Nanoparticles and nano-emulsions have been used to protect reactive dyes and colour fibres. Microencapsulation techniques have been demonstrated (Baptista *et al.*, 2004) to increase the levelness of reactive dyes on cotton and polyamide textiles. Their use on wool will be discussed later in this chapter. Another method for colouring textiles has been the development of nanoparticle pigments that are small enough to diffuse into textile fibres. Carbon Black nanoparticles (Li and Sun, 2003) and self-dispersible Carbon Black nanoparticles (Li and Sun, 2007) have been used to colour textiles. It was found that only particles near 8 nm diameter diffused into the fibres using conventional dyeing conditions.

The fastest growing commercial use of nanoparticles in textiles is producing antimicrobial textiles using nanoparticle silver or silver-containing nanoparticles. Lee *et al.* (2003) found 2–5 nm silver nanoparticles applied to cotton and polyester fabrics gave good antimicrobial effect and had acceptable durability to laundering. Other applications of silver-containing nanoparticles on textiles include production of burn dressings (Atiyeh *et al.*, 2007) and the application of silver-containing sol–gels to cotton (Xing *et al.*, 2007).

Carbon nanotubes are becoming an increasingly important source of nanoparticles for textile applications, particularly as production rates increase and prices drop. Holme (2008) reports several laboratory-scale uses of carbon nanotubes, including producing conducting fibres by the application of nanotubes to the surface of polyester fibres using strong binders to ensure stability of the coating and increasing the strength of nylon and polyester by including carbon nanotubes in the bulk material. Other emerging technologies, not considered in detail in this chapter, include electrospun nanofibres and the use of nanofibres to produce yarns.

10.1.5 Risks of nanotechnology on textiles

Much has been written about the environmental and health risks of nanotechnology. Karst and Yang (2006) have reviewed much of this literature and concluded that 'along with these benefits come risks to health and the environment. ... As the textile industry continues to be one of the leaders in nanotechnology research and manufacturing, the health and environment risks will need to be evaluated.'

The toxicity of nanoparticles is not fully understood, particularly as the particles can behave differently from the bulk material. The very large surface to bulk ratio of nanoparticles and their enhanced surface chemistry can lead to increased toxicity. The International Risk Governance Council state in their White Paper on Nanotechnology Risk Governance (Renn and Roco, 2006):

> 'There is only a limited understanding of the potential EHS risks of nanomaterials and further studies are required for both: (i) hazard characterisation, in areas such as toxicity, ecotoxicity, carcinogenicity, volatility, flammability, persistence and accumulation in cells; and (ii) exposure, including the potential for oral, cutaneous and inhalative uptakes of nanomaterials during production, transport (in air, water, soil and biosystems), decomposition and/or waste disposal.
>
> - Human health risks. Several studies have shown that: (i) due to the high surface-area-to-volume ratio and higher reactivity of nanostructures, large doses can cause cells and organs to demonstrate a toxic response (in particular inflammation) even when the material itself is non-toxic; (ii) some nanosized particles are able to penetrate the liver and other organs and to pass along nerve axons into the brain; (iii) nanomaterials may combine with iron or other metals, thereby increasing the level of toxicity and presenting unknown risks; (iv) engineered nanomaterials raise particular concerns because of the unknown characteristics of their new properties and their potential use in concentrated amounts; and (v) some nanomaterials may have similar characteristics to known high-risk materials at the microscale.
> - Explosion risks. The higher surface reactivity and surface-area-to-volume ratio of nanopowders increases the risk of dust explosion and the ease of ignition.
> - Ecological risks. The impact of nanostructures on the environment may be significant because of the potential for: (i) bioaccumulation, particularly if they absorb smaller contaminants such as pesticides, cadmium and organics and transfer them along the food chain; and (ii) persistence, in effect creating non-biodegradable pollutants which, due to the small size of the nanomaterials, will be hard to detect.'

No reports or publications are available that specifically consider the potential risks of nanotechnology in textiles, although the potential risk is being recognised (Holme 2008). Studies have shown that some nanoparticles are able to pass through the blood/brain barrier of largemouth bass (Oberdorster, 2004) and cause lung inflammation in rats (Tetley, 2007). There are three risks that need to be considered when using nanoparticles on textiles:

1. The accumulation of nanomaterials in the environment from wastewater streams and the effect that this could have on the food chain. Some have speculated that nanoparticles in the environment could be absorbed at the lower end of the food chain and move up to eventually enter humans. These studies often use C-60 fullerenes (Brumfiel, 2003).
2. The transfer of nanoparticles from the textile through the skin into the wearer's bloodstream. Such materials could accumulate in the body with toxic effects.
3. Hazards to mill workers handling nanomaterials from inhalation or skin contact with the material. Many nanoparticles will pass through conventional dust masks.

At present there are no government regulations concerning nanomaterials, rather the industry is self-regulating. Many people believe nanomaterials are safe because the bulk material is safe; however, as stated above, the high surface area of nanoparticles can result in significant changes in the particle chemistry and therefore may increase the hazard of the material.

10.2 Nanotechnology application to wool

Applications of nanotechnology to wool at both research and commercial levels are emerging phenomena and are relatively few in number. Examples of the key applications are given below.

10.2.1 Antimicrobial applications

There is a growing interest by the consumer in health and well-being. With this growth has come a demand for antimicrobial textiles. The most common application of nanotechnology to wool at present is antimicrobial treatments using the application of silver nanoparticles during fabric finishing operations. Silver nanoparticles themselves do not inhibit bacterial growth on textiles, rather they are a source of silver ions that provide the biocidal activity. This presents a difficulty for wool, which contains a significant concentration of basic and thiol groups that strongly bind silver ions and therefore potentially reduces the availability for antimicrobial effects. Freddi *et al.* (2001) showed wool could absorb 0.7mmol/g silver ions under alkaline conditions and after soaking for 120 hours in an acidic solution released only 9% of the absorbed silver. Silver has, however, been successfully used as an antimicrobial agent on wool.

Silver nanoparticles can be applied to the fibre surface or allowed to diffuse into the fibre. The most common application method is padding of a suspension of silver nanoparticles in a suitable carrier such as ethanol onto the fibre surface. The relatively high solubility of silver in water renders aqueous suspensions relatively unstable and severely limits the storage time of the suspension. Silver binds strongly to the fibre surface, so binders and curing agents are often not required. Figure 10.4 illustrates the durability of silver nanoparticles applied to machine-washable (chlorine/Hercosett treated) wool knitwear. Approximately 1% silver by weight of fabric was applied by padding, the fabric was dried and then washed using Woolmark TM31 machine wash test method. The scanning electron micrographs show the presence of clumps of silver nanoparticles on the surface before and after washing.

Some other techniques that have been used in laboratory applications for applying silver-containing nanoparticles to wool or protein fibres include direct deposition of the silver onto the fibre surface from a silver solution, sonochemical deposition and the use of sulphur-silver colloids. Direct deposition uses redox reactions to reduce silver ions (Ag^+) to elemental silver (Ag) on the fibre surface. Dong *et al.* (2005) used silk fibroin fibres as a biotemplate to reduce silver ions to nanoparticles on the fibre surface. Tyrosine in the fibre surface, which has strong electron donating properties, was shown to reduce the silver ions to elemental silver, producing 5 nm nanoparticles on the fibre surface. Wool contains approximately 3.2% tyrosine and in the right environment could also reduce silver ions.

Ki *et al.* (2007) applied sulphur–silver nanoparticles to wool as an antimicrobial, antistatic and mothproofing agent. The particles, supplied by NP-Tech, were described as a nano-silver particle containing a few sulphur compounds. They were found to have very good antimicrobial and moth proofing effects at an application of 20 ppm silver particles and a small reduction in static developed on the fabric above 50 ppm silver particles.

10.2.2 Stain blocking and water repellence

Stain blocking using nanotechnology is based either on the 'lotus leaf' effect or on nanostructured films. However, specific applications to wool are very limited and many products, developed for use on cotton and synthetic fibres, have not been tested on wool.

Many stain-resist applications claim to use nanotechnology; however, closer examination of these technologies shows that some are simply the application of a thin film of a hydrophobic finish. Using the definition of nanotechnology above, these cannot be considered nanotechnology so have not been included here.

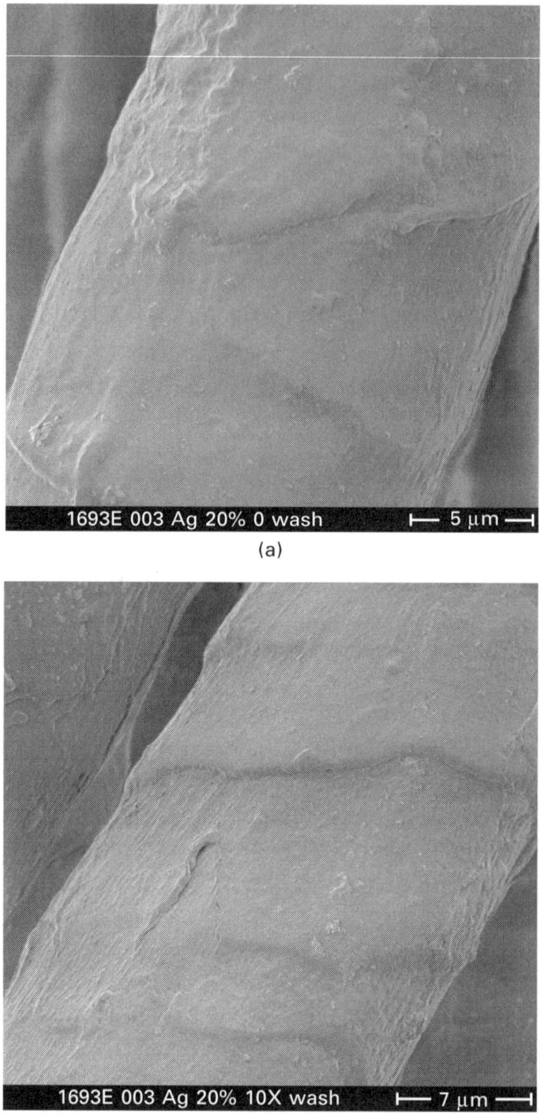

10.4 Silver nanoparticles applied at 20% product (1% silver) applied to chlorine/Hercosett treated machine-washable wool knit fabric: (a) as applied before washing, (b) after 10 ISO 5A wash cycles using ECE standard detergent.

The lotus leaf is well known to possess a natural super-hydrophobic surface. The leaf is able to shed water and remain clean due to the wax nodules on the leaf surface. Each micron-sized nodule is covered with nano-sized hairs that trap air under water drops that fall on the leaf, increasing the apparent contact angle and preventing the leaf from wetting. Artificial super-hydrophobic surfaces are often described as the lotus effect, but most treatments have been applied to flat substrates. Some textile finishes, for example Ruco-Nanoguard-GOR™ from Rudolf Chemie, apply nano-crystallites in a fluorocarbon matrix to the fibre surface to produce the lotus effect on fabrics.

Another approach to achieve water repellence or stain resistance is to use nanostructured films on the fibre surface. Rudolf Chemie's Bionic Finish™ range of products, Advanced Teflon™ from DuPont and Resist Spills™ from Nano-Tex form such nanostructured films on fibre surfaces. The films consist of self-assembled layers of polymers and may contain nanoparticles. The outermost layer is often a fluorocarbon to increase both the oleophobic and hydrophobic nature of the surface.

10.2.3 Self-cleaning wool textiles

Titanium dioxide, particularly in the anatase crystal form, is an excellent photocatalyst, able to oxidise most organic compounds. As the particle size is reduced, the effectiveness of the photocatalyst increases.

Bozzi *et al.* (2005b) applied titanium dioxide in the rutile form directly to plasma-modified wool–polyamide fabrics and found significant photo-oxidative activity under visible light and ambient conditions. The titanium dioxide was exhausted directly onto the modified fibre surface with good dry cleaning fastness. It was found that the effectiveness of the photocatalyst was decreased after dry cleaning due to the presence of chloride ions that interfere with the self-cleaning process.

10.2.4 Novel coloration effects

Nanoparticle pigments have been used to colour textiles. Many ink-jet printing inks contain nanoparticle pigments that are bound to the surface of the fibres with suitable binders. One unusual application is the use of self-dispersible Carbon Black pigments to dye wool fibres (Li and Sun, 2007). Good exhaustion was observed; however, the high water dispersibility of the particles and ease of exhaustion resulted in poor wash fastness.

Liposome microencapsulation (Gomes *et al.*, 1997, El-Zawahry *et al.*, 2007; De La Maza *et al.*, 1998) has been used to improve wool dyeing with a range of dyes. Improved exhaustion and lower energy requirements resulted in a cleaner dyeing technology. While it is usual in this process to encapsulate the dye in the liposome before addition to the dye bath, Coderch *et al.* (1999) found adding the liposome and dye at the start of the dyeing cycle had the same effect and good exhaustion could be achieved at 80 °C. The fastness of the dyeing to rubbing was not described.

10.2.5 Other applications

Cyclodextrins are unique ring molecules with a hydrophobic interior cavity and hydrophilic exterior. They are able to hold hydrophobic compounds in the interior cavity and bind to textile substrates via the interactions with the exterior. Two textile applications for cyclodextrins are to:

1. trap odours during wear that are released during washing; and
2. apply compounds, particularly fragrances, to fabrics preloaded into the cavity.

A common use of cyclodextrins preloaded with fragrance is the freshness sheets added to tumble dryers where heat from the dryer releases the fragrance.

'Nano-wool', (nano-sized keratin particles produced by grinding wool fibres) has been used to modify cotton fabrics (Hu *et al.*, 2007). The treated fabrics show increased thermal resistance, increased 'warm touch' and decreased wettability by water. Such particles have also been added to polypropylene films (Xu *et al.*, 2007) and polyurethane films (Xu *et al.*, 2006). The modified polyurethane films showed decreased mechanical properties, with reduced stress at break, elongation and breaking energy, but a significant increase in water vapour permeability and moisture regain. The polypropylene films also showed a significant decrease in mechanical properties.

10.3 Current commercial applications of nanotechnology to wool

At the time of writing, commercial applications of nanotechnology to wool were limited to antimicrobial treatments using nanoparticle silver and stain resist treatments. A number of companies produce nanoparticle silver antimicrobial finishes for wool. Treatment conditions for the application of these finishes are available from the supplier. These products include Nano-Tex Silver™ (NanoTex), nanoHorizons SmartSilver™ (NanoHorizons Inc), Ultra-Fresh Silpure™ (Thomson Research Associates) and Nano Tech AB™ (Australian Wool Innovation).

Most nanotechnology stain-resist finishes claim to be effective on wool, however recent investigation funded by Australian Wool Innovation suggested that many products were not as effective as claimed. Many new generation stain blockers do not require high-temperature reactivation after washing and, for example, Advanced Teflon™ (DuPont), Oleophobol 7713™ (Huntsman) and Rucostar EEE™ (Rudolf Chemie) are effective on wool. However some similar products do not retain their oil repellence unless heat reactivated. One interesting application of nanotechnology stain resist finishes is QuickDry Merino™ (Australian Wool Innovation). By reducing the surface energy of the wool fibres the mass of water retained by the fabric after washing is reduced, giving a significant reduction in drying time.

10.4 Future trends

Nanotechnology offers unique opportunities to bring value-added performance to textiles. Many of these technologies will be based on new materials such as carbon nanotubes and synthetic nanofibres. The challenge for wool is to be at the forefront of nanotechnology development or risk being left behind with an ever-decreasing market share.

Future developments are likely to be in two directions, improving product performance and in the development of smart or intelligent materials such as fibres with integrated sensors, conducting fibres and shape-memory fibres. Some new applications of nanotechnology that improve product performance (principally in stain resistance and antimicrobial treatments) are in the early stages of realisation as commercial products. However, other potential treatments remain as laboratory concepts and as 'thought experiments'.

Nanotechnology is one of the 'hot research topics' of the 21st century that will bring a proliferation of new products and processes. The scope for the application of nanotechnology to textiles is extensive, covering all areas of textile finishing. These include (Singh *et al.*, 2006):

- wrinkle, stain and static resistant finishes;
- nano-composite materials with superior performance;
- smart textiles that support climate control and enhanced comfort;
- UV protection of fabrics and wearers;
- odour absorbing finishes; and
- protection against chemical and biological threats.

While the focus of Singh's paper was on cotton, many of these developments will be applicable to wool. New dyeing technologies using nanoparticles and nano-emulsions are emerging that will reduce the energy requirements of the dyeing process or produce novel coloration effects. New lubricants and processing aids using nanoparticles to improve fibre handling are being released by major textile auxiliary manufacturers.

Smart textiles will include integrated electronics where the electronic components are part of the fabric, not attached to the garment as is the case for present electronic textiles. Flexible power sources and power storage devices; physical, chemical and environmental sensors and information systems are under intense research that have the potential to change the way we think about textiles.

10.5 Sources of further information and advice

The number of journals dedicated to nanotechnology is growing rapidly with the current count at 32. *Nano Letters* is the leading journal and a very good source of information on current nanotechnology developments. *Nanotoday* is a free quarterly journal published by Elsevier that contains article of general interest as well as latest developments in nanotechnology. Several web sources offer free email abstracting and news services highlighting recent developments. A simple Google search on nanotechnology received approximately 12 million responses. Good general-knowledge services include Nanotechweb.org/cws/home and Nanotechnology Now (www.nanotech-now.com). While these are not directly related to textiles or wool, both offer good background information.

Nanoparticle safety is an area of concern, particularly as many of the properties of nanoparticles are unknown. Two good sources of information are the reports from the Royal Society found on their website (http://royalsociety.org) and the Inter-national Risk Governance Council, Geneva (e.g. Renn and Roco, 2006). 'The Project on Emerging Nanotechnologies' (http://www.nanotechproject.org/) is also a good source of information on nanoparticle safety and nanotechnology applications.

10.6 References

ATIYEH BS, COSTAGLIOLA M, HAYEK SN, DIBO SA, 'Effect of silver on burn wound infection control and healing: Review of the literature' *Burns* 2007 **33** 139–148.

Australian Wool Innovation, http://wool.com.au/

BAPTISTA ALF, GOUTINHO PJG, REAL OLIVEIRA MECD, ROCHA GOMES JIN, 'Lipid interactions with textile fibres in dyeing conditions' *Progr Colloid Polym Sci* 2004 **123** 88–93.

BOZZI A, YURANOVA T, GUANAQUILLO I, LAUB D, KIWI J, 'Self-cleaning of modified cotton textiles by TiO_2 at low temperatures under daylight irradiation' *J Photochem Photobiology A: Chem* 2005a **174** 156–164.

BOZZI A, YURANOVA T, KIWI J, 'Self-cleaning of wool-polyamide and polyester textiles by TiO_2-rutile modification under daylight irradiation at ambient temperature' *J Photochem Photobiology A: Chem* 2005b **172** 27–34.

BRUMFIEL G, 'A little knowledge…' *Nature* 2003 **424** 246–248.

BUFFAT PH, BOREL J-P, 'Size effects on the melting temperature of gold particles' *Phys Rev A* 1976 **13**(6) 2287–2298.

CODERCH L, MANICH AM, MARTI M DE LA MAZA A, PARRA JL, SERRA S, 'Complementary study of optimizing a wool dyeing process with commercially available liposomes' *Text Res J* 1999 **69**(10) 789–790.

DE LA MAZA A, CODERCH L, MANICH AM, MARTI M, PARRA JL, 'Optimizing a wool dyeing process with an azoic 1:2 metal complex dye using commercially available liposomes' *Text Res J* 1998 **68**(9) 635–642.

DONG Q, SU H, ZHANG D, '*In situ* depositing silver nanoclusters on silk fibroin fiber supports by a novel biotemplate redox technique at room temperature' *J Phys Chem B* 2005 **109** 17429–17434.

EL-ZAWAHRY MM, EL-SHAMI S, HASSAN M, EL-MALLAH MH, 'Optimizing a wool dyeing process with reactive dye by liposome microencapsulation' *Dyes Pigments* 2007 **74** 684–691.

FAN Q, UGBOLUE SC, WILSON AR, DAR YS, 'Nanoclay-modified polypropylene dyeable with acid and disperse dyes' *AATCC Rev* 2003 **3**(6) 25–28.

FREDDI G, ARAI T, COLONNA GM, BOSCHI A, TSUKADA M, 'Binding of metal cations to chemically modified wool and antimicrobial properties of the wool–metal complexes' *J Appl Poly Sci* 2001 **82**, 3513–3519.

GRISEL R, WESSTRATE K, GLUHOI A, NIEUWENHUYS B, 'Catalysis by gold nanoparticles' *Gold Bulletin* 2002 **35**(2) 39–45.

GOMES JINR, GENOVEZ MC, HRDINA R, 'Controlling exhaustion of reactive dyes on wool by microencapsulation with liposomes' *Text Res J* 1997 **67**(7) 537–541.

HOLME I, 'Textile nanotechnology upgrades performance', *Int. Dyer* 2008 **193**(4) 8–10.

HU JY, LI Y, CHANG YF, YEUNG KW, YUEN CW, 'Transport properties of fabrics treated with nano-wool fibrous materials', *Coll Surfaces A: Physiochem Eng Aspects* 2007 **300** 136–139.

HURATA M, 'Gold as a novel catalyst in the 21st century: preparation, working mechanism and applications' *Gold Bulletin* 2004 **37**(1) 27–36.

JIANG H, MOON K, DONG H, HUA F, WONG C P, 'Size-dependent melting properties of tin nanoparticles', *Chem Phys Lett* 2006 **429** 492–496.

KARST D, YANG Y, 'Potential advantages and risks of nanotechnology for textiles', *AATCC Rev* 2006 **6**(3) 44–48.

KI HY, KIM JH, KWON SC, 'A study on multifunctional wool textiles treated with nano-sized silver' *J Mater Sci* 2007 **42** 8020–8024.

LEE HJ, YEO SY, JEONG SH, 'Antimicrobial effect of nanosized silver colloidal solution on textile fabrics' *J Mat Sci* 2003 **38** 2199–2204.

LI D, SUN G, 'Coloration of polyester and acrylic fabrics with carbon black nanoparticles', AATCC Rev 2003 **3**(12) 19–22.

LI D, SUN G, 'Coloration of textiles with self-dispersible carbon black nanoparticles' *Dyes and Pigments* 2007 **72** 144–149.

MASON TG, WILKING JN, MELESON K, CHANG CB, GRAVES SM, 'Nanoemulsions: formation, structure and physical properties' *J Phys Condens Matter* 2006 **18** R635–R666.

MEILERT KT, LAUB D, KIWI J, 'Photocatalytic self-cleaning of modified cotton textiles by tio_2 clusters attached by chemical spacers' *J Molec Catal* 2005 **237** 101–108.

OBERDORSTER E, 'Manufactured nanomaterials (fullerenes, c-60) induce oxidative stress in the brain of juvenile largemouth bass', *Environmental Health Perspectives* 2004 **112** 1058–1062.

RENN O, ROCO M, 'White Paper No 2 Nanotechnology Risk Governance' International Risk Governance Council, Geneva, June 2006.

Royal Society and The Royal Academy of Engineering, *Nanoscience and Nanotechnologies: Opportunities and uncertainties* 2004, London.

ROCO M, 'Global strategy for the risk governance of nanotechnology: an introduction to the IRGC Project on Nanotechnology', at IRGC 2006 Conference 'The Risk Governance of Nanotechnology: Recommendations for Managing a Global Issue', Zurich', 6–7 July, published at http://www.irgc.org/IMG/pdf/Mike_Roco_Risk–Governance_for_Nanotechnology/pip.php?page=irgc&id_ rubrique=24, accessed 1 July 2008.

QI WH, WANG MP, LIU QH, 'Shape factor of nonspherical nanoparticles', *J Mat Sci*, 2004 **40**(9–10) 2737–2739.

SINGH KV, SAWHNEY PS, SACHINVALA ND, LI G, PANG S, CONDON B, PARACHURU R, 'Applications and Future of Nanotechnology in Textiles' Beltwide Cotton Conference, San Antonio, Texas 3–6 Jan 2006 2497–2503.

TETLEY TD, 'Health effects of nanomaterials' *Biochem Soc Trans* 2007 **35** part 3 527–531.

XING Y, YANG X, DAI J, 'Antimicrobial finishing of cotton textile based on water glass by sol–gel method', *J Sol-Gel Sci Technol* 2007 **43** 187–192.

XU W, FANG J, CUI W, HUANG J, 'Modification of polyurethane by superfine protein powder' *Poly Eng Sci* 2006 **46**(5) 617–622.

XU WL, WANG X, LI WB, PENG X, LIU X, WANG XG, 'Characterization of superfine wool powder/poly(propylene) blend film' *Macromol Mater Eng* 2007 **292**(5) 674–680.

11
Wool performance apparel for sport

B. HOLCOMBE, University of Wollongong, Australia

Abstract: Although initially slow to capitalise on the explosive growth in the sports apparel sector and hampered by an 'old-fashioned' image, wool's complex morphological structure means that it is uniquely positioned for tailoring into high-performance products suitable for a wide range of sports applications. Indeed wool can be considered as one of the most high-tech of all apparel fibres. In this chapter the benefits of wool in relation to apparel for active sports will be examined against a background of the physiological and sensory processes that create the environment in which wearer comfort must be achieved.

Key words: thermal and subjective comfort, moisture buffering, sweat wicking, Sportwool, intelligent textiles.

11.1 Introduction

Wool was once commonly used in a wide variety of sports apparel. It was the product of choice for participants in soccer, rugby, cricket, cycling, sailing and outdoor pursuits and until the 1950s it was even commonly used for swimming costumes. Since the advent of mass-produced synthetic fibres in the 1940s, wool's position in this apparel sector has been steadily eroded, to the point where today all that remains is a very tenuous share of niche markets such as outdoor apparel and thermal underwear.

The reasons for wool's demise are complex. Although synthetic fibre manufacturers initially strove to copy the properties of natural fibres, in the early 1980s they began to move in new directions, developing and expanding the attributes of their own products. Growing competition from low-cost countries in commodity fibre markets further encouraged this. Producers in Europe, Japan and the United States moved into high value-added fibres with engineered aesthetic or mechanical properties.

Active sportswear was a natural outlet for many of these developments. The availability of textile materials with technological origins provided the catalyst for a rapid evolution in fabric and garment design and led to the emergence of specialist active sportswear manufacturers. This group exploited the technical attributes of fabrics in garment design to differentiate their products in a highly competitive market and establish for themselves credibility as market leaders in their sports. Differentiation was linked to identifiable end-product requirements, now commonly known as functional attributes and supported by substantial investment in marketing.

Objective measurement of clothing 'thermophysiological' properties by specialised laboratories such as the Hohenstein Institute (Hohenstein, 2008) became popular as manufacturers sought to establish unique advantages for their products related to their target markets. The increasing technical awareness of consumers was exploited through the introduction of advertising based on product differentiation and the inclusion of swing tags in garments to draw attention to the technical basis of product benefits.

Over the same period there was an enormous increase in the level of participation in active sport, fuelled by growing consumer awareness of the importance of healthy, more active lifestyles, and the general increase in available leisure time. There was also a steady shift in the balance between casual and formal apparel as casual dress styles move out of the home and into all but the most formal of occasions. One of the most significant consequences of these two events has been the growth in sports apparel, trickling down from the developments of the specialist manufacturers to an explosion in look-alike sportswear worn as casual and leisurewear and becoming a major segment of the apparel market in its own right.

By the 1990s wool was seen as the embodiment of all that the serious sportsperson should avoid. Wool products were regarded as too heavy, slow drying, unable to handle moisture efficiently and altogether old-fashioned – an image that dogs wool to this day. Although much of the negative sentiment can be attributed to clever marketing by competing fibre manufacturers, until recently wool marketing authorities held the view that wool's future lay in other sectors of the apparel market and little was done to counter this perception.

There is no good reason why wool should not be a successful competitor in the active sportswear market. Its morphological structure is extraordinarily complex, giving rise to a wide range of physical and chemical properties that form a unique base for beneficial exploitation. While competitors have copied some of these properties, none can match wool's versatility, nor is this likely to happen in the foreseeable future. In many ways, wool is the most high-tech of all apparel fibres. Most importantly, many of the treatments and processing technologies that are used to add functionality to synthetics can equally be applied to wool.

In this chapter the benefits of wool in relation to apparel for active sports will be examined against a background of physiological and sensory processes that create the environment in which wearer comfort must be achieved.

11.2 The human system

Sport imposes some rather severe demands on clothing over and above the basic requirements of general comfort, modesty and appearance. Sports apparel must contend with a much wider range of heat and moisture loads than conventional apparel and is often subject to far greater physical abuse and put through more cleaning cycles during its life cycle. Perhaps the most challenging requirements are the maintenance of thermal and subjective comfort.

11.2.1 Metabolic heat loss and thermal comfort

The human body is an exothermic system, constantly generating and releasing heat to the surroundings. In a healthy state the temperature of its internal organs, known as the core temperature, is maintained at around 37 °C. This varies by a degree or so depending on the time of the day, the level of physical activity and the emotional state. Greater changes occur during illness, in extreme environmental conditions and during vigorous activity. When heat is generated more rapidly than it can be dissipated, the core temperature rises. An increase of around 2 °C can be safely tolerated but further sustained rises can lead to heat-related illness (Nielsen, 1928).

Work done to move the body around during exercise is derived from biochemical reactions that occur within the muscles. These processes are accompanied by the release of heat, the rate of heat production depending largely on the level of physical activity involved. Muscles are very inefficient at converting stored energy into physical work done and waste energy released into the tissue as heat is typically as much as five times the actual energy expended in moving the body around (Coyle *et al.*, 1992; De Prampero *et al.*, 1971).

Highly trained athletes can maintain work rates of around 500 watts for periods as long as 30 minutes with a total heat output of the order of 2000 watts – more than 20 times greater than the resting metabolic output (Padilla *et al.*, 2000). Such a wide range of heat generation requires a very sophisticated temperature control system in order to maintain thermal or physiological comfort.

Heat is transferred from the muscles to the skin via the cardiovascular system. An increase in body core temperature stimulates expansion of the blood vessels near the skin and increases the blood flow to the skin surface. During sport, blood flow is further accelerated by an increased heart rate. The result is elevation of the skin temperature. Heat loss from the skin to the surroundings normally occurs by a combination of infrared radiation, air convection and as latent heat from the evaporation of moisture. Each of these modes relies on a difference in either temperature or moisture concentration between the body and the environment to drive the heat transfer process.

The skin loses moisture continually, sourced from evaporation within the tissue and diffusing as moisture vapour through the dermal structure. During sleep and very low activity conditions, this amounts to about 30 millilitres per square metre per hour (Hendrie *et al.*, 1997). When the heat lost by radiation and convection is not sufficient to prevent the core temperature from rising, sweat glands within the skin are activated, releasing moisture for evaporation directly at the skin surface. The dissipation of energy as latent heat of vaporisation in this way significantly increases the rate of body heat loss. In extreme conditions sweat evaporation becomes the major route for heat loss by the body. With sweat glands fully activated, the moisture output may be one to two litres per hour or more (Hendrie *et al.*, 1997), occurring both as moisture vapour and if it cannot evaporate quickly enough, liquid sweat.

11.2.2 Skin sensory behaviour and subjective comfort

The physiology of the skin plays an important role in the subjective comfort of clothing. Just below the skin surface are sensitive nerve endings that respond to movement, temperature and pain. The nerve endings that detect temperature are very much more sensitive to changes in the temperature of the skin (or more correctly, rate of change of temperature) than they are to its actual temperature.

The lowest rate of change of temperature that the skin is able to detect is about one hundredth of a degree Celsius per second (Hensel, 1981). A rise in ambient temperature of one degree Celsius over a period of an hour is unlikely to be noticed because the rate of change is very low. The same change in just 1 second produces an instant reaction – something akin to the experience encountered when first entering a coolroom or a sauna.

The sensations we understand as warmth or coolness to the touch are the result of physical processes that occur at the skin and trigger responses in the skin's thermal sensors. When an object contacts the skin, the resulting heat flow between the two depends on how much warmer or cooler the temperature of the object is than the skin and how well it conducts heat. The faster the heat flow into or out of the skin, the greater the rate of change of skin temperature and the warmer or cooler the object feels.

A brushed knit fabric with a hairy surface feels warmer than the smooth surface of a dense, tightly woven sateen because for the same temperature difference, the rate of heat flow from the skin into the knit is lower and produces a weaker thermal sensation. Smooth fabrics have more fibres in contact with the skin than hairy fabrics, giving greater contact area and more rapid heat flow, cooling the skin faster and producing a stronger coolness sensation.

Human skin has no known nerve endings to detect moisture directly, and yet discomfort associated with clamminess and stickiness is invariably attributed to moisture. It is likely that these sensations result from indirect changes in the skin that the brain associates with moisture although the actual mechanisms involved are yet to be identified.

Coarse fibres tend to indent the skin rather than buckle when under external pressure, triggering pain receptors and causing irritation (Naylor *et al.*, 1992). Thus parameters such as fibre diameter and fabric stiffness are important considerations in the comfort of active sportswear. Like wool, the skin contains proteins that make it hygroscopic so that, like wool, it absorbs and desorbs moisture as the humidity around it changes. As its moisture content increases it becomes softer, making it easier for textiles on the skin surface to activate the underlying nerve receptors. As a consequence, the skin is more sensitive during active sports or in warm climates than when dry and cold.

For wool products to be worn in skin contact during active sports the issue of fibre diameter is quite important. Wool is commercially available in a wide range of diameters and diameter distributions and the fine wools required to ensure universal acceptance in active sportswear (typically 18.5 μ and finer) come at a significant cost disadvantage. This is one of the factors that have discouraged the use of wool fabrics in active sportswear, particularly in the high-volume look-alike sector where base fabric cost is a critical issue.

11.3 Textile properties relevant to sports apparel

11.3.1 Thermal resistance

The physical quantity associated with heat transfer through textiles is thermal resistance, which is a measure of the ability of a material to resist heat flow. Heat flow between any two points in a material occurs in response to a gradient in temperature between those points. Thermal resistance is generally measured in steady-state conditions, that is, once the heat flow resulting from the applied temperature gradient has stabilised.

Fibres typically make up less than 20% of the total volume of textile materials, even for dense structures such as woven fabrics. This falls to about 5–10% for knitwear and can be as low as 0.1% in insulating webs. For knitted fabric structures of the type used in active sportswear, air contributes about half the total thermal resistance (Holcombe and Hoschke, 1983). The remainder of the resistance can be attributed to a combination of fibre thermal properties and fabric packing density. The thermal conductivities of conventional apparel fibres are all of a similar order of magnitude and much greater than that of air (see Table 11.1).

Figure 11.1 shows fabric thermal resistance plotted against thickness for a large number of fabrics of various structures. The thermal resistance of conventional textile materials largely reflects their thickness – the thicker the material, the greater its thermal resistance, which implies that there is no intrinsic fibre 'warmth' property (Holcombe and Hoschke, 1983). Thermal resistance is mainly determined by fibre characteristics that influence the resulting fabric thickness and hence the amount of air trapped in the fabric.

Table 11.1 Thermal conductivities of textile fibres

Fibre	Conductivity mW/(m K)
Cotton	461
Polyacrylonitrile	200
Polyamide	243
Polyester	141
Polypropylene	117
Polyurethane	126
Polyvinylchloride	167
Viscose rayon	289
Wool	193
Air	26
Water	600

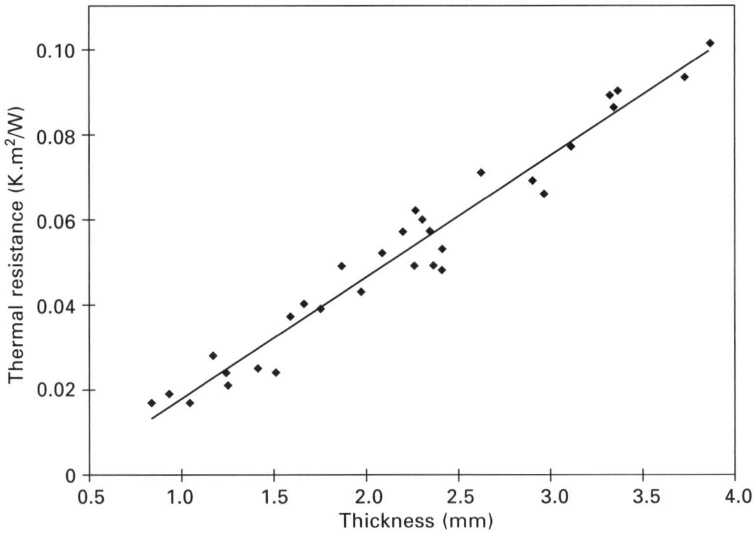

11.1 Relationship between fabric thickness and thermal resistance.

Wool's traditional reputation as a warm fibre is firmly grounded in its natural properties. The level of technology of earlier times was limited to converting the natural crimp, elasticity and bulk of wool into fabrics that were characteristically hairy, conferring a warm, soft touch and thickness for good insulating properties. These fabrics were ideally suited for garments needed to provide warmth in times when domestic heating was a luxury rather than taken for granted as it is today.

11.3.2 Vapour resistance

The mechanism of moisture vapour transport through textile materials has some parallels with heat flow. In steady-state conditions, in this case with the rate of vapour flow constant, moisture movement is opposed by a resistance that is largely a function of fabric thickness. The rate of diffusion of moisture vapour through air is very much greater than that through fibres, so most of the vapour flow takes place through the air spaces between fibres and yarns. As with thermal behaviour, fibre properties have little direct influence on steady-state resistance to moisture vapour flow.

Rain protective shell products are an integral part of the clothing system of many outdoor sports participants. They incorporate coatings or waterproof breathable membranes for which the vapour and wind resistance are significantly higher than that of a conventional textile of equivalent thickness.

Measurements of steady-state thermal and vapour resistance are useful for quantifying the performance of clothing worn for protection against the climatic extremes encountered in outdoor activities such as skiing and climbing. They are far less important in the lighter and thinner fabrics used for active sportswear where the air gaps between clothing layers can have greater resistance than the textiles themselves. In addition, bellows ventilation of air inside the clothing induced by body activity and external air movement can contribute significantly to heat and moisture loss from the body independently of heat and moisture losses through the textiles themselves (Bouskill *et al.*, 2002).

11.3.3 Fibre moisture sorption

Water vapour moves through the air spaces between fibres and yarns as the result of either forced convection caused by external wind and body movement, or by diffusion from regions of high vapour concentration to regions of low concentration. Similarly, water vapour moves in and out of fibres and through their internal molecular structure under the influence of moisture concentration gradients between regions. As mentioned earlier, this process is slower because the diffusion coefficient of water vapour inside fibres is less than the diffusion coefficient of water vapour in air.

The amount of water vapour inside the fibre is determined by the relative humidity and, to a lesser extent, the temperature of the adjacent air (Watt and D'Arcy, 1979). An equilibrium exists between fibre moisture content and relative humidity. If the moisture content of the fibre is below its equilibrium level for the relative humidity of the surrounding air, water molecules diffuse from the air into the fibre. Conversely, if the moisture content of the fibre is above equilibrium for the external conditions, water molecules are released from within the fibre and diffuse back into the air. This behaviour, known as hygroscopicity, is common to all textile fibres.

An important difference between natural fibres such as wool and synthetic fibres is that the chemical structure of natural fibres enables them to store much more water vapour internally. Table 11.2 shows that wool has a sorption capacity of about 35% at saturation whereas for fibres such as polyester and polypropylene, the corresponding figure is of the order of 1% (CSIRO, undated).

Steady-state measurements take no account of the responses of fibres to changing temperature and humidity conditions. In practice the vapour pressure in and around clothing is never constant, particularly at the skin, and fibre properties play a very important role in the management of moisture in the microclimate adjacent to the skin. The extent of that role is dependent on fibre dynamic moisture vapour sorption characteristics, that is, the speed and extent of their response to changes in the moisture in the air around them.

11.3.4 Heat of sorption

An important feature of the behaviour of hygroscopic materials is the energy associated with changes in moisture content. Water molecules absorbed by such materials associate with the relatively polar polymer chains within the structure, dropping to a lower energy state and thereby releasing heat. The reverse of the absorption process requires energy (Hearle, 2002).

Thus moisture absorption by fibres as humidity rises causes the fibre temperature to rise, and moisture release following a decrease in humidity lowers the fibre temperature. For fibres such as wool that can absorb a large amount of moisture, the amount of heat involved is quite significant. A kilogram of dry wool placed in an atmosphere of air saturated with moisture releases about the same amount of heat as that given off by an electric blanket running for 8 hours! A good demonstration of this is to take a loose handful of wool fibre that has been oven dried and lightly spray it with water from an atomiser. The heat released is very obvious – sufficient to cause the temperature of the wool to rise by as much as 10–12 °C.

Table 11.2 Saturation moisture regain of some common sports apparel fibres

Fibre	Saturation moisture content (% regain)
Wool	35
Cotton	24
Polyamide	7
Polyester	1
Polyolefin	0.05
Polyacrylonitrile	7

11.3.5 Moisture vapour transport

In most everyday wear conditions moisture is present in clothing only as vapour or as adsorbed water within the fibre, that is, in the form of individual gaseous water molecules. However during active sport where the metabolic heat loss from the athlete is typically quite high, liquid water and water vapour often coexist. The physical processes that control the behaviour of these two states of water in textile materials are largely independent.

The diffusion of moisture vapour into and out of fibres, known as dynamic moisture sorption, is a time-dependent process. The main rate-determining factors are fibre diameter, saturation moisture regain and the diffusion coefficient of moisture vapour inside the fibre (Li, 2001). The finer the diameter, the higher the sorption capacity or the higher the diffusion coefficient, the more rapid the change in fibre moisture content.

Dynamic moisture vapour sorption behaviour is responsible for a property known as buffering. This is only significant in fibres such as wool where the moisture sorption capacity is high. As the rate of moisture evaporated from the skin increases, the relative humidity of the microclimate adjacent to the skin increases and the fibres respond by increasing their moisture vapour content. This slows (or buffers) the rate of rise in humidity within the microclimate. As moisture levels decrease, the fibres give up some of their stored moisture, again slowing the rate of humidity change.

Wool has a favourable combination of the key rate-determining factors and so its buffering effect is far more pronounced than that of other fibres. By slowing the rate of change in humidity, wool worn next to the skin makes the wearer less conscious of changes in humidity and reduces sensations of discomfort that might otherwise occur. This ability to buffer humidity change is responsible for wool's widely recognised 'breathable' reputation.

Buffering is in effect a form of skin-clothing microclimate humidity control that is most important at low to moderate activity levels or in the early stages of vigorous activity. It sets wool apart from other materials used in sport and is one of the cornerstones of the product known as 'Sportwool™'. This is described in Section 11.5.

11.3.6 Liquid moisture transport

The behaviour of textile materials in contact with liquid sweat is a critical factor in their ability to influence wear comfort at high levels of physical activity. Water is picked up and moved either laterally in the plane of the fabric or transversely from one face of the fabric to the other by the process of wicking.

Unlike moisture vapour, the driving force for liquid sweat movement is the surface energy of the fibre, that is, the attraction between the liquid and the fibre surface, and the size and configuration of the spaces within the fabric (Das *et al.*, 2007). The mechanisms underlying wicking in textiles closely parallel those responsible for the wicking of liquids in capillaries. The two key factors that influence the pressure driving liquids along a capillary are the radius of the capillary and the surface energy (wettability) of its solid surfaces. The smaller the radius and/or the higher the surface energy, the more readily wicking occurs. Table 11.3 shows the surface energies of some typical fibres used in active sportswear.

Although the spaces between fibres and yarns do not have the continuous circular form on which wicking theory is based, close proximity between fibres and their parallel alignment in yarns create discontinuous capillaries of irregular shape through which liquids can migrate. Capillaries are effectively formed both between fibres within yarns and between the outer fibres of adjoining yarns. The apparent capillary radius of spun yarns depends on the relative alignment and proximity of the fibres, which in turn depend on properties such as:

- fibre cross-sectional shape and diameter;
- yarn structure (ring, open-end or vortex spun);
- yarn twist factor and count; and
- knit or weave structure and tightness factor.

The terms hydrophilic, meaning water loving or attracted to liquid water, and hydrophobic, meaning water hating or repelling liquid water, are widely used to describe the properties of the surface of fibres. In general, the more hydrophilic the fibre, the higher its surface energy and the more readily a fabric made from it will wick liquid water. These terms are often wrongly used interchangeably with the term hygroscopicity. The former relate to the interaction of liquid water with the surface of fibres, the latter to the exchange of water vapour with the internal structure of the fibre. These two properties are quite independent.

Table 11.3 Surface energy of some common sports apparel fibres

Fibre	Critical surface energy (mJ/m^2 @ 20 °C)
Cotton	200
Polyamide (nylon 6)	46
Wool (natural)	29
Wool (chlorinated)	45
Polyester	43
Polyvinylchloride	37
Polypropylene	29

Fibre surface chemistry has become quite an advanced area of textile science and is today an important aspect of many textile products. It is possible to chemically modify the surface energy of a fibre so that its surface energy is anywhere between the extremes of strongly hydrophobic and strongly hydrophilic without changing the underlaying hygroscopic behaviour significantly.

A key point here is that although it is relatively easy to modify the surface properties of fibres to change them from hydrophilic to hydrophobic and vice versa, modification of their internal structure to increase or decrease hygroscopic moisture vapour capacity is extremely difficult. Although the synthetic fibre industry has tried very hard to increase the moisture vapour capacity of its products to match that of wool or cotton by processes such as hydrolysis in the case of polyester, synthetic fibres are still well short of the performance of most natural fibres in this regard.

11.3.7 Drying of textiles

An important requirement of sports apparel is that it is able to dry as quickly as possible after absorbing liquid sweat or after washing. Apart from the discomfort aspect, the replacement of the air in a textile by water dramatically increases its thermal conductivity. As Table 11.1 shows, the thermal conductivity of water is more than 20 times greater than that of the air that makes up the bulk of a textile. In cold or windy conditions where the rate of evaporative heat loss from the wet textile is high, post-exercise chill can result.

The rate of moisture evaporation from wet textiles is a function of the surface area available for evaporation and the surrounding climate conditions. Any two fabrics of the same area that contain the same amount of interstitial water will take the same time to dry in a given environment, regardless of the type of fibre involved. This is illustrated by Fig. 11.2. In each case the rate of loss of moisture, represented by the slope of the line, is about the same. The total time taken for the fabric to dry depends on the amount of water present in the fabric at the start of the drying process, indicated by the intersection of each line with the vertical axis.

Hydrophobic fabric manufactured from fibres such as polypropylene retain very little water after complete immersion when compared with strongly hydrophilic fibres such as cotton and appear to dry more quickly as a result. Yet polypropylene films used for plastic wrapping are commonly treated to make them hydrophilic so that they retain the printing inks used for labelling. A strongly hydrophilic fibre such as cotton can be treated with a hydrophobic agent so that it picks up little or no water after immersion.

The key factors that determine the amount of water retained in fabrics after wetting are the thickness or physical volume of the fabric and as with wicking, fibre surface energy and effective capillary size. Treating a textile

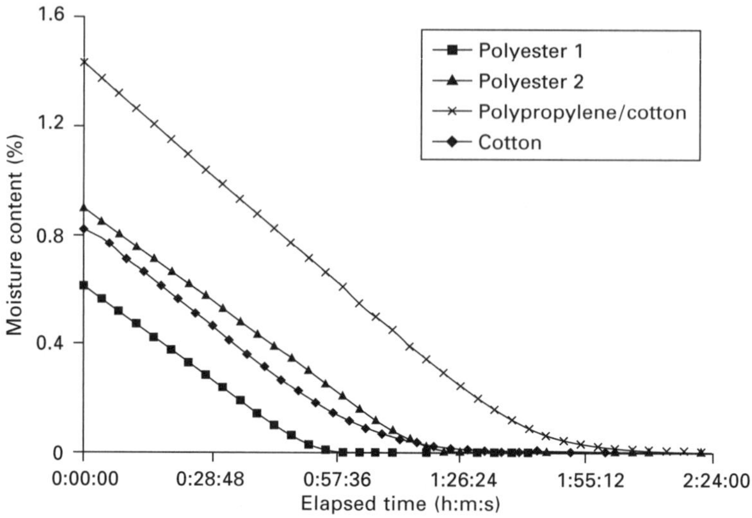

11.2 Drying behaviour of typical textiles.

with a hydrophobic agent will shorten its drying time but at the expense of wicking behaviour and vice versa.

Speed of drying is considered a negative for wool in the active sports and outdoor areas. Although the surface of dry wool fibres is relatively hydrophobic due to fatty acids in the outer scale structure, in the presence of moisture the surface inverts to a 'protein out/fatty acid in' arrangement (Huson *et al.*, 2008) that is more hydrophilic and leads to significant water retention after immersion. Similarly, machine wash treatments for wool generally leave the fibre surface quite hydrophilic and increase water retention.

The recently development of the QuickDry Merino concept (Denning, 2006) is a means of shortening drying time in applications where speed of drying is an issue. In this process, hydrophobic, low-temperature activating polymers are applied to the wool fibre surface, reducing drying time by as much as 70% in the case of machine-wash treated wool. These treatments have negligible influence on moisture vapour absorption by the fibre and therefore are not detrimental to buffering behaviour. They have the added advantage of improving the stain resistance properties of wool fabric.

11.3.8 Odour management

The high levels of moisture generally present in clothing worn during active sports and the presence of bacteria originating from the skin provide the ideal environment for bacterial and fungal growth. This is particularly the case if the garments are worn and then left damp for a while before washing.

One side-effect of the growth of these micro-organisms is the development of unpleasant odours.

The use of antimicrobial agents that inhibit the growth of micro-organisms is becoming more widespread in an effort to add a 'hygienic' or 'fresh' image of active sports apparel. Wool has its own natural way of dealing with this issue. Athletes such as long-distance runners and mountaineers who are obliged to wear the same clothing for long periods have reported less odour build-up in their clothing when wearing wool than with synthetic fibres.

In addition to the ability to buffer and thereby reduce moisture levels in the clothing microclimate that facilitate the growth of micro-organisms, the wool fibre has a very complex internal chemical structure that includes many side-chains. These side-chains are able to bind the acidic, basic and sulphurous compounds that are important components of body odour and thus capture them inside the fibre. When wool garments are washed, the odour compounds are readily removed.

11.4 Textiles for sports apparel

For a clothed athlete, body heat loss occurs both from exposed skin and from the outer surfaces of clothing. Some convective and evaporative losses also occur at the inner surface of clothing and from covered skin through the process of forced or bellows ventilation via openings such as those at the sleeves and neck. Heat lost from the outer clothing surfaces must first pass through the clothing itself. Similarly, some of the moisture evaporated from the skin must diffuse through the clothing and most of the liquid sweat picked up by the clothing must wick to the outside for evaporation to occur.

Clothing interferes with metabolic heat loss. Reducing fabric thickness increases the rate of heat and moisture flow through clothing but there is a lower limit to thickness determined by the point at which the garment becomes too transparent to be viable, or basic structural characteristics such as strength and elasticity are below practical limits.

The ideal clothing for active sports where sweat is an issue should maximise the rate of evaporation of sweat while minimising the discomfort associated with wet clothing against the skin. It should pick up sweat from the skin, transport it to the outer fabric face, expose the maximum wetted surface area for evaporation and draw the latent heat required for evaporation from the body. During periods when the rate of sweat output from the skin is greater than the rate of evaporation from the clothing, it should store this excess liquid in such a way that the clothing neither feels wet to the skin nor clings to it.

Unfortunately these attributes cannot all be achieved simultaneously and the most practical sports apparel will inevitably be a compromise. As an example, a simple knit manufactured from fibres that have a very high surface energy will readily pick up sweat from the skin and give good evaporation from its external surface. However, the sweat is held uniformly through the fabric and so feels damp when in contact with the skin. If its inner surface is dense enough and hydrophilic enough it may form liquid 'bridges' across to the skin that cause 'wet cling' or drag and interfere with physical movement.

In order to avoid skin contact issues while maintaining good evaporation performance, some active sportswear manufacturers use double-knit fabric structures, also known as double layer knits or double-faced knits. These consist essentially of two discrete fabrics joined during the knitting process, either by interlacing the two faces, or by the addition of a third binder yarn crossing from one to the other. An example of an interlaced double-knit is shown in Fig. 11.3.

The almost complete separation of the two faces enables different physical characteristics to be achieved at the internal and external surfaces of the fabric. Manufacturers have sought to use the separation available with this structure to build functional characteristics into textiles to improve sweat transport and evaporation. If the two surfaces of the fabric have different sweat attracting characteristics, then by putting the less wettable face in contact with the skin, sweat picked up from the skin is transported to the outside of the fabric where it is more readily exposed to air movement for evaporation into the surrounding atmosphere.

Early versions of double-faced fabrics comprised fibres such as polypropylene or polyester on the skin side, with cotton or viscose on the outside. Current products tend be 100% synthetic, consisting of textured continuous filament yarns or spun staple yarns (the latter made from filaments cut up into individual fibres of specific length) on the inner face, and a textured, low bulk continuous filament yarn on the outer face, both in the same fibre type.

Another option is to use density gradient structures – fabrics in which the diameter of the fibres on the inside face is greater than that on the outside. In such structures the packing density of the fibres, that is, the effective capillary size, increases from inside to outside, encouraging liquid migration away from the skin.

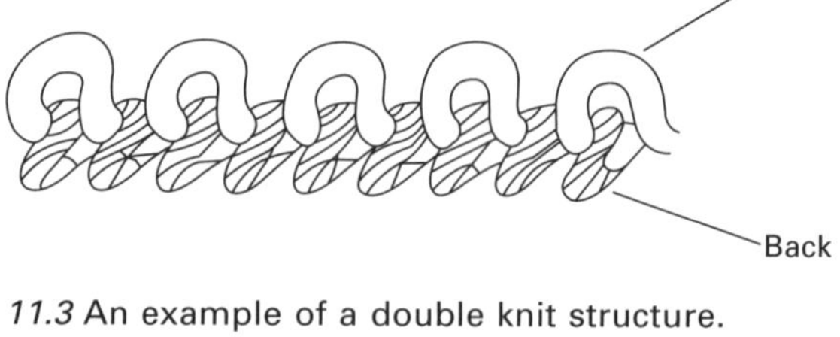

11.3 An example of a double knit structure.

At the other extreme is the apparel worn in some of the more technically aware sports such as climbing and skiing where sweat is not such an issue. These are often treated with hydrophobic agents to prevent liquid pick-up from snow or rain, avoiding the enormous loss in insulating properties that results from water in the fabric structure.

Most synthetic fibres tend to be relatively hydrophobic (water repellent) in the raw polymer state. Over the past decade or so a range of treatments has been developed to make synthetic fibres more hydrophilic. Many treatments and treated fabrics have proprietary brand names that are widely recognised by active sportswear users, such as Akwatec™, Climalite™, HydroMove™, H2-Out™, Dri-FIT™ and Coolmax™. Most consist of a hydrophilic polymer coating applied to the fabric after manufacture so that sweat wicked up by the fabric is uniformly distributed throughout the structure.

The incorporation of natural fibres such as cotton and wool to active sportswear fabrics adds another dimension to moisture management behaviour that cannot be matched by all-synthetic products – that of vapour buffering. This is particularly the case if the natural fibre is directly adjacent to the skin. Sportwool™, wool's first serious attempt at a high-performance textile suitable for active sports, emerged in 1994 from a need for a cycling product with the buffering capabilities of wool with the ease of transfer printing of polyester.

11.5 The Sportwool™ concept

The original Sportwool™ concept was a joint development by CSIRO and the former International Wool Secretariat. It emerged from a major programme of research into the behaviour of textiles during wear to investigate the role of fibre properties in the interaction between skin and clothing.

In its earliest form, Sportwool™ was a double-knit structure in which superfine, machine-wash treated wool formed the skin contact face, and polyester the outer face. The superfine wool ensures skin comfort, while the polyester provides a platform for transfer printing of complex designs in bright colours with good definition. Sportwool™ combines the comfort of a natural fibre with the durability and ease of coloration of synthetics. This juxtaposition of dissimilar fibres gives the added advantage of unique sweat transport performance.

The properties of the two component yarns were engineered to ensure that the inner wool face readily picks up sweat without causing wet cling, and the outer face strongly attracts sweat. The use of two quite dissimilar fibres creates a density gradient structure, with different surface properties on the inner and outer faces of the fabric. The resulting sweat transport behaviour is illustrated in Fig. 11.4.

On the left, a drop of coloured water (simulating sweat) has been placed

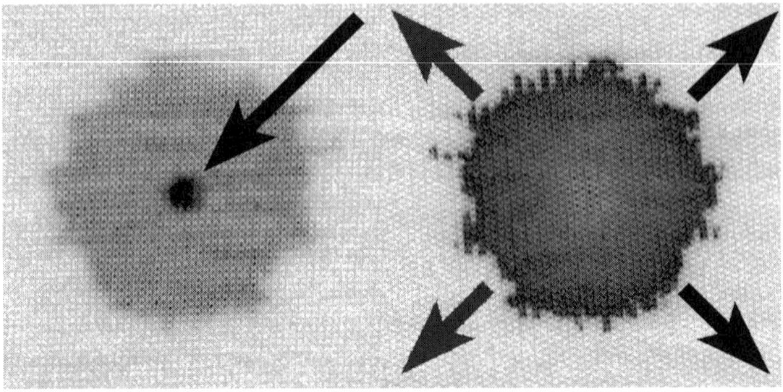

11.4 The sweat absorption behaviour of Sportwool.

on the wool face of the fabric. The dark spot in the centre shows how the wetted area on the skin side remains about the same size as the original drop. This minimises contact between skin and wet fabric and helps maintain skin comfort. The paler area around the spot is the outer face visible through the back of the fabric. The photo on the right shows the other side of the fabric and how effectively sweat spreads on this face to maximise evaporation.

In the early stages of exercise before liquid sweat appears, the buffering ability of the wool inner layer slows the increase in the humidity of the skin-clothing microclimate. The ability of Sportwool™ to deal pro-actively with moisture vapour is a major point of difference in the highly competitive active sportswear market. During sustained high levels of physical activity where sweat is present, the powerful wicking behaviour ensures that sweat is removed from the skin and held at the outer face of the fabric adjacent to the surrounding atmosphere to maximise evaporative cooling. This unique combination of behaviours is best described as two-stage moisture management.

11.6 Future trends

At the high-performance end of the active sportswear market the development of new products continues apace and will do so well into the foreseeable future. Manufacturers continue to innovate, seeking an edge over their competitors. The key areas are fabric structure, mechanical properties, durability of chemical treatments and targeting garment design to end use. Others include the incorporation of elastomers for muscle compression, antimicrobials for odour control, welding and bonding of seams and the use of three-dimensional anatomical mapping to design products with better fit or localised pressure relief.

Perhaps one of the more significant advances in recent years has been the introduction of whole body knitting machinery with selection and cuff/collar/hem capabilities that had its origins in the sock industry. These machines have substantially expanded the palette of designers, enabling one-piece seamless garments tailored for individual sports to be manufactured in one step. It is now possible to bring different functionality to specific locations within garments rather than using much more clumsy cut and sew techniques.

Another recent advance rapidly finding acceptance among sports participants is that of wearable electronic devices, some of which have already been directly incorporated into sports apparel and footwear. These devices include music players, GPS systems that give information about pace, speed, distance covered and even calories burned, heart rate monitors, communications systems and more.

At the moment, apparel-based devices are achieved by deconstructing existing systems, micro-miniaturising the components and rearranging them to be more compatible with the clothing environment. The resulting garments tend to be stiff and bulky in order to support the weight of the components and are still reliant on conventional metal wire technology and batteries that change the handle of the underlying fabric. Such systems will not find universal acceptance until they are able be withstand the everyday rigours of traditional garments and can be crumpled, ironed, thrown on the floor or into the washing machine without compromising functional performance.

Until very recently, functionality added to sports apparel has generally been passive. Wearers have no influence over the magnitude of the functionality or when it occurs. Functionality is generally associated with the response of one or more physical properties of the material to a particular set of circumstances. Benefits that might be highly desirable in one situation can be very uncomfortable in another. These benefits are usually small in real terms and often barely noticeable to the wearer. This situation is on the threshold of change that will have significant implications for the textile industry as a whole.

Recent developments in polymer technology have achieved properties in polymeric materials previously only possible with metals or solid-state devices. Properties such as exceptional strength, electrical conductivity, the ability to store energy or convert it into light or mechanical action, to convert light into energy and to sense chemicals have all been achieved at levels that have the potential to be exploited in some beneficial way, particularly in the area of sports apparel.

The incorporation of mechanical or electronic functionality into textiles using these materials, sometimes at the nano-scale, will require the development of new forms of sensors, actuators, energy sources and methods of interconnection with mechanical properties that are more compatible with textiles than current technologies. The key to compatibility is to achieve similar levels of conformation in the components as the underlying substrate. In some instances this will be achieved by developing components based around fibres and yarns, in others by the application of flexible coatings. Most will involve polymeric materials with specific properties assembled or applied within the textile-processing pipeline rather than discrete devices retro-fitted to the finished clothing.

The terms 'intelligent' and 'smart' clothing have been coined to describe the products now beginning to emerge from these developments. In the textile sense, intelligent implies that the garment has the ability to sense and proactively respond in some beneficial or desired way. The possibilities that this sort of functionality could bring to sports apparel are endless. Mechanical changes in fibre orientation and arrangement could be used to vary the thermal resistance and air permeability of textile fabrics, enabling a single garment or combination of garments to be worn by those involved in outdoor and winter sports rather than adding or removing layers as weather and activity levels change. Switchable fibre surface energy that is hydrophilic when sweat pick-up is needed but could be reversed to become hydrophobic as desired to make the garment repellent to rain would find a ready outlet in a wide range of applications.

Active functionality generally requires some form of energy source together with associated electronics and circuitry. Over the past few years the development of polymeric electronic circuitry and conducting elements has attracted the interest of many research organisations around the world. In the apparel context, such materials are known as electronic textiles. Power sources based on the capture and storage of solar energy directly within the garment are a key to truly stand-alone devices and this is an area of research that is currently expanding rapidly.

11.7 The role for wool

Most of these developments can be utilised in conjunction with wool substrates just as easily as any other fibre. Indeed value-adding in this sector is far greater than with conventional products so the cost disadvantage of fine wools is much less of an issue. The chemical versatility of wool creates an endless variety of avenues for interaction with intelligent polymeric materials that can add their own form of functionality, modify existing properties of wool in some way or create new benefits that are quite unique.

Irrespective of the form of functionality added to the garment, it is the textile substrate that is either in contact with the skin or seen from the outside or both. Garment appearance and comfort will always be important aspects of consumer appeal and wool has much to offer in creating that appeal. With modern processing technology, wool can be made into cool, lightweight fabrics suitable for active sportswear and offers significant benefits that synthetics cannot match, particularly in the area of moisture management and odour control. It can also be used in combination with other fibres to produce synergistic products with benefits not possible from the individual components.

As products such as Sportwool™ and the success of niche manufacturers such as Smartwool™ and Icebreaker™ have shown, the right wool products have a ready market. There is a strong consumer interest in things natural, and high-performance products that meet all the technical requirements of this market and retain a natural focus have great potential appeal.

11.8 References

BOUSKILL LM, HAVENITH G, KUKLANE K, PARSONS KC, and WITHEY WR (2002), 'Relationship between clothing ventilation and thermal insulation', *AIHA J.*, **63** (3) 262–268.

COYLE E, SIDOSSIS I, and HOROWITZ J (1992), 'Cycling, efficiency is related to the percentage of Type 1 muscle fibres', *Med. Sci. Sports Exercise*, **24** (7) 782–788.

CSIRO (undated), 'Moisture buffering', http://www.csiro.au/files/files/p50w.pdf, accessed 5 May 2008.

DAS B, DAS A, KOTHARI VK, FANGUIERO R, and DE ARAUJO M (2007), 'Moisture transmission through textiles. Part 1: Processes involved in moisture transmission and the factors at play', *AUTEX Research Journal*, **7** (2) 100–110.

DENNING R (2006), *QuickDry Merino; Save Time, Save Energy. Technical Marketing Manual*, CSIRO Textile and Fibre Technology.

DE PRAMPERO PE, CORTILI G, CELENTANO F and CERRETELLI P (1971), 'Physiological aspects of rowing', *J. Appl. Physiol.*, **31** (6) 853–857.

HEARLE JWS (2002), 'Physical properties of wool', in Simpson WS and Crawshaw GH, *Wool: Science and Technology*, Cambridge, Woodhead, 80–129.

HENDRIE AL, BROTHERHOOD JR, BUDD GM, JEFFERY SE, BEASLEY FA, COSTIN BP, ZHIEN W, BAKER, CHENEY NP, and DAWSON MP (1997), 'Project Aquarius 8: Sweating, drinking and dehydration in men suppressing wildland fires', *Int. J. Wildland Fire*, **72** 207–218.

HENSEL H (1981), *Thermoreception and Thermal Regulation*, Monographs of the Physiological Society No. 38, Academic Press, London.

HOHENSTEIN (2008), http://www.hohenstein.de/en/content/content1.asp?hohenstein=46-0-0-0-0, Accessed 5 May 2008.

HOLCOMBE BV and HOSCHKE BN (1983), 'The dry heat transfer characteristics of underwear fabrics', *Text. Res. J.*, **53** (6) 368–374.

HUSON M, EVANS D, CHURCH J, HUTCHINSON S, MAXWELL J, and CORINO G (2008), 'New insights into the nature of the wool fibre surface', *J. Struct. Biol.*, in press.

LI Y (2001), 'The science of clothing comfort', *Textile Progress*, **31** (1/2) 1–138.

NAYLOR GRS, VEITCH CJ, MAYFIELD RJ, and KETTLEWELL R (1992), 'Fabric evoked prickle of some wool fabrics', *Text. Res. J.*, **62** 487–493.

NIELSEN M (1928), 'Die' Regulation der Korpertemperatur bei Muskelarbeit', *Skand. Arch. Physiol.*

PADILLA S, INIGO M, ANGULO F, and GOIRENA JJ (2000), 'Scientific approach to the one hour cycling world record: a case study', *J. Appl. Physiol.*, **89** 1522–1527.

WATT IC and D'ARCY RL (1979), 'Water-vapour adsorption isotherms of wool', *J. Textile Inst.* **70** 298–307.

12
High-performance wool blends

M. MIAO, CSIRO Materials Science and Engineering, Australia

Abstract: Wool is blended with other fibres to improve comfort, protection, durability and easy-care characteristics, and to impart special functionality. This chapter first reviews products made from wool blends, their beneficial characteristics and applications. It then discusses the developments of wool blend yarn structures and processing technology.

Key words: wool blend, blending, wool blend yarn spinning.

12.1 Introduction

Wool is blended with other fibres to utilise the respective advantages of the fibres. These advantages may be either unattainable, deliver undesirable side effects or be too costly if only one type of fibre is used. The advantages may relate to comfort, appearance or style, durability, easy care and specialised functionalities, such as heat and flame resistance, conductivity, chemical and biological behaviours. From a commercial point of view, blends are formed to reduce the total costs of production and to increase the sales prices of the products by utilising the differences between available fibres. In this chapter, we will focus on the technical benefits that can be achieved by blending wool with other fibres and the technologies that are used to process wool blends.

Wool is highly regarded for a wide range of properties. It absorbs water vapour, yet has a water-repellent surface. Wool can absorb up to 30% moisture to its own weight without feeling wet. The natural crimp and resilience of wool fibres helps the maintenance of high loft and thermal insulation of fabrics. The high absorption heat it releases when absorbing moisture makes it an ideal material for cold and humid climate conditions. Although not a very strong fibre, wool is highly extensible, flexible and resilient. These properties give fabrics the special wool handle, drape and fashion style. Wool is naturally flame resistant and does not melt or drip polymer when it is burnt. Because of these desirable properties, wool has been considered for blending with practically all traditional fibres and new fibres.

Other types of fibres, such as polyester and nylon, offer high strength and abrasion resistance. A relatively small percentage of these fibres in a wool blend may improve the processing performance of wool and the durability of the final products. Many new fibres that have emerged in recent years offer high functionalities, such as flame retardance (for example, Nomex), softness to handle (microfibres), high stretch (elastane fibres, such as Lycra®), wicking (for example, Coolmax®) and electrical conductivity (conductive polymers and metal fibres). The combination of functionality of these high-performance fibres with the comfort of wool provides endless opportunities for product development.

12.2 Wool blend products, their benefits and applications

We will first review how different fibres are combined with wool to achieve optimum properties in the final products. While we discuss the benefits of blends under different categories of product properties, it should be kept in mind that one wool blend product often offers a range of benefits.

12.2.1 Felting shrinkage

Natural wool fibre demonstrates differential friction due to the presence of surface scales. The wool fibre slides over other wool fibres more easily when it is pulled by the root than when it is pulled by the tip. This differential friction induces a ratchet effect as wool fibres move relative to each other, which causes wool fabrics to felt under the action of machine washing or tumble drying. The effect is strongest when the glass transition temperature (T_g) of wool is exceeded, i.e. when fibre is wet or warm and loses some of its stiffness. Fabric shrinkage is an apparent result of felting. Felting shrinkage can happen both in hand washing and in machine-washing conditions. The conditions in the tumble drying machine, that is, high temperature, low liquid ratio and high degree of mechanical action, promote felting more readily than in a washing machine. Wool knitted garments that are not shrink-resist treated must be handled gently in washing, or alternatively can only be cleaned in solvents using specialised dry cleaning equipment. Wool stays below T_g in most dry cleaning conditions. However, if the water content is too high, wool also shrinks in dry cleaning.

The popularity of domestic washing machines and dryers created a consumer demand for garments that do not require dry cleaning. A widely adopted technology to eliminate the felting shrinkage of wool fabrics is the chlorine–Hercosett shrink-resist treatment, which modifies the wool fibre surface and masks the scales on the fibre surface with a polymer.

Long before the chlorine–Hercosett treatment, wool was often blended with cotton to control shrinkage and to soften the fabrics. The mixing of non-felting fibres (cotton in this case) with wool reduced the opportunity of direct contact between wool fibres, which reduced the ratchet mechanism operating between wool fibres and reduced, or even eliminated, felting shrinkage. The best known example of wool/cotton blend is the Viyella fabric (www.viyella.co.uk), a twill woven fabric made from a blend of 55% lambswool and 45% cotton. The fabric was developed in the 19th century by James and Robert Sissons of William Hollins & Company. Viyella fabrics have a fine flannel appearance and have been used in many kinds of children's and baby's wear, sportswear, men's and women's tailored shirts and dresses. A cheaper version of the Viyella fabric, Clydella, was made by the same company at a later date. Clydella was a 20/80 wool/cotton blend fabric used especially for babies' clothes. After more than a century, a wide range of fabrics broadly categorised as Viyella are still sold around the world. Colana™ was a 30/70 wool/cotton blend yarn produced from stretch broken wool top and cotton by the Australian spinner Rocklea in the 1990s.

The degree of felting shrinkage in wool/cotton blends is largely determined by the blend ratio. Byrne (1998) produced fabrics at three blend ratios and tested the felting shrinkage due to machine washing. He considered that 20% wool had little influence on the washing performance of wool/cotton intimate blend products, but fabrics with 30 to 40% wool content could shrink during machine washing unless the wool component was stabilised.

Polyester is the most commonly used fibre for blending with wool to achieve easy-care properties in woven fabrics. Wool-poor polyester blend fabrics (wool ratio < 50%) require no special finishing to prevent felting shrinkage for machine washing. Wool-rich garments (wool ratio > 60%) in general require special finishing and garment setting procedures. The maximum wool percentage possible for shrink resistance can be further increased when very fine fibres (for example, microfibres) are used (de Boos *et al.*, 1996). This is because the large number of fine fibres in the yarn further reduces the opportunity of direct contact between wool fibres. Felting shrinkage can also be reduced or eliminated by adopting high-twist yarns and tight weave structures to restrict fibre movement. In tightly twisted plain weave structures, wool-blend fabrics containing as much as 80% wool can resist felting shrinkage (de Boos *et al.*, 1996). Some singles yarn structures (such as SoloSpun yarns) made from 100% wool can achieve machine washability without shrink-proof treatment when tight weave structures and suitable finishing treatments are used.

12.2.2 Wrinkle recovery and easy care

Fabrics wrinkle or crease during wear in the presence of moisture and heat. This is especially a problem for travellers who sit in a constrained space for a prolonged period of time. It is generally considered that improving wrinkle recovery is more desirable than increasing fabric resistance to creasing through increasing fabric stiffness. The viscoelastic properties of the wool fibre, and to some extent the coercive couple in the bending hysteresis curve of the woven fabric due to fibre to fibre friction, are the main factors responsible for fabric wrinkle recovery (Smuts, 1989).

Natural cotton fabrics wrinkle badly largely because cotton is a high-friction fibre. Wool is a highly elastic and low-friction fibre and under conditions where moisture and temperature are constant, wool fabrics have excellent recovery from deformation. It is only under severe wear conditions (where the wearer is hot and sweating) that the glass transition of wool is exceeded and the wrinkle recovery of wool deteriorates. The wrinkling conditions in Fig. 12.1 were designed to simulate difficult wrinkling conditions for wool, with creases inserted above the glass transition temperature and recovery below. Under these conditions polyester does not absorb moisture and its recovery properties are not affected.

Polyester is the only common type of fibre that provides higher wrinkle recovery than wool. In general, the wrinkle recovery of wool/polyester blend fabric improves with increasing polyester content, but the relationship is not a linear one (de Boos *et al.*, 1975).

Hayes *et al.* (1980) noted that the wrinkles in trousers form mainly across the leg, in other words, the warp threads form the wrinkles. Union blend fabrics consisting of a pure polyester warp or 20/80 wool/polyester warp and 100% wool weft demonstrated similar wrinkle recovery to a pure polyester fabric in wear (Hayes *et al.*, 1980; Phillips, 1982). However, as with pure polyester fabrics, repeated deformations led to 'mussiness' behind the knees and this was found to be as objectionable as sharp wrinkles to some observers.

An additional benefit of adding at least 20–30% polyester to wool is that the polyester component, with appropriate heat setting, can provide sufficient set to stabilise pleats to survive repeated laundering. It is generally considered

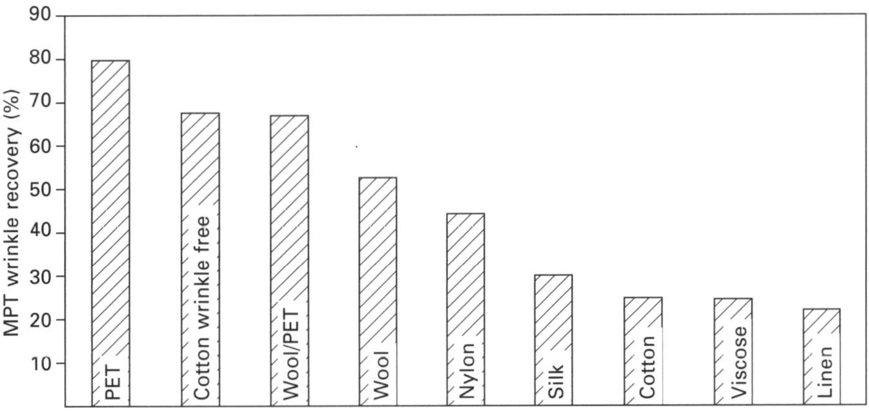

12.1 Wrinkle recovery (multiple pleat test) of various fibres (Leeder, 1984).

that about 50% polyester is required to provide easy-care properties, i.e. adequate resistance to felt shrinking as well as smooth drying and flat set properties. This is a very important property as consumers now demand 'total easy-care' garments that can withstand repeated machine washing and tumble drying, while maintaining their 'just pressed' appearance. Garments must remain wrinkle-free after laundering, garment seams must remain flat and without pucker, and pleats must stay sharp.

Many garment manufacturers now offer machine washable men's suits made from wool/polyester blends. Vertical worsted textile and garment company Heilan in China manufactures a drip-dry, machine-washable men's suit using a 50/50 wool/polyester blend. Bagir introduced an easy care (tumble-dry, machine-washable) men's suit (Anon, 2005a) by using a shrink-resist treated wool in a blend with more than 50% polyester and around 3% Lycra. The suit remained both shrink-resist and wrinkle-free after 30 cycles of machine wash and tumble dry. Another product offered on the Internet (www.eshopone.com) claims wrinkle-free, stain protection, two-way stretch and machine washability using a special Ultimate Performance washbag.

12.2.3 Prickle reduction

Some fabrics worn in contact with the skin can evoke an unpleasant sensation of prickle. The effect is caused by mechanical excitation of small diameter nerve fibres of the thermal or pain groups in the skin. When the force exerted by a fibre pressed against the skin is greater than a certain threshold, it can trigger the pain receptors. The density of prickly protruding fibre ends on a fabric could be correlated with the magnitude of estimated prickle intensity (Garnsworthry *et al.*, 1988). Various methods were used to determine the threshold diameter of fibre for causing prickle. In general, fibres coarser than 32 μm were considered responsible for prickle (Naylor, *et al.*, 1997). In relation to wool knitwear, it has been suggested that the percentage of fibres coarser than 30 μm could be used as an indicator of the level of skin discomfort that the final fabric could cause. If less than 5% of the fibres that were coarser than 30 micron were in a blend, the blend could be judged as non-prickly.

It is important to recognise that the prickle effect is not specific to wool. When pure synthetic fabrics were made with fibres above 30 μm, fabrics were equally prickly. Wool is more susceptible to the effect because of the natural variability of fibre diameters that occur on a sheep. The variability increases within a flock and eventually within a processing lot made up of wool from different sources.

Prickle in wool fabrics can be reduced or eliminated by blending with fine, non-prickly fibres. This is achieved by diluting the concentration of prickly fibres in fabrics to a level below the prickle threshold. Natural fibres such as cotton and most synthetic apparel fibres are finer than their respective prickle threshold diameters, with lower variability of diameters. Most wool blends achieve some degree of prickle reduction. The manufacturers of Viyella-type wool/cotton fabrics claimed their garments to be prickle-free long before the physiology of fabric-evoked prickle had been understood.

Another strategy to eliminate or reduce fabric evoked prickle is physically keeping prickly wool fibres out of direct contact with the skin. Bilayer fabrics can be produced, in which the wool layer is shielded from direct skin contact by a non-prickly layer. The main disadvantages of such bilayer fabrics are their heavy weight and limitations to fabric design. Experiments have also been carried out to produce core yarns whereby coarse wool fibres are wrapped inside fine fibres, such as cotton, fine wool and synthetic fibres. The difficulty to fully cover wool fibres in the yarn core and the increase of processing costs are some of the obstacles for the commercial application of such a prickle reduction technology. A further question raised on the shielding strategy for solving the prickle problem is that the coarse fibres may gradually migrate to the next-to-skin layer during wear.

Because prickle is caused by mechanical pressure from protruding fibres, fabric structures and yarn tightness can also affect prickle. The prickle threshold mean fibre diameters for woven fabrics was considered to be about 3 μm finer than that for knitted fabrics (Naylor and Phillips, 1995), and reducing fabric cover factor was found to be helpful for prickle reduction (Naylor and Phillips, 1997).

A composite wrap spun yarn structure was evaluated for prickle reduction for winter-weight knitwear (Miao *et al.*, 2005). The composite wrap spun yarn consisted of a core of wool fibres and a nylon multifilament yarn wrapped by another nylon multifilament. The wrap yarn was then back-twisted to about 50% of the initial wrap twist so that the resultant yarn had a low-twist core consisting of wool fibres and a multifilament, wrapped by another multifilament. The composite yarn structure offered higher yarn bulk than a conventional ring spun two-fold yarn structure without sacrificing its pilling performance. The apparent micron shift achieved by the modified structure depended on the mean fibre diameter of the wool used: a 3 μm shift was achieved for a 31 μm wool but it reduced to about 1 μm for a 25 μm wool, and no micron shift was detected for wools finer than 22 μm.

12.2.4 Softness to handle

Softness of a fabric can be assessed subjectively by rubbing and pressing the fabric between the thumb and fingers, with or without visual assistance. There are a wide range of descriptors that are used to describe the sensory reactions from fabric handling, including smoothness, fullness, coarseness, warmth and stiffness. Despite enormous research efforts, opinions still vary widely on what objectively measured properties should be used to represent fabric softness (Li, 2001). Softness was considered to be the opposite of stiffness that can be measured by bending length. Softness was also considered to be the ease of yielding to pressure that can be measured by a compression test. Fabric surface smoothness plays an important part when describing fabric softness. Bishop (1996) considered that fabric softness was associated with bending, compression and tensile properties, shear stiffness and hysteresis, areal density and friction. Fibre diameter, crimp and surface friction are some of the key fibre attributes that influence fabric softness. These fibre attributes, together with appropriate yarn and fabric structures, determine the fabric compliance to touch, bulk, resilience to compression, smoothness to the stroke of fingers.

In the context of lightweight wool and wool blend knitwear, 'cashmere-like handle' is often considered to be the target. Fine wool is often blended with cashmere down to achieve such 'luxury softness'. The blending of wool improves the processing and wear performances of cashmere, and reduces material costs. Many specialty animal hair fibres, such as camel, alpaca, rabbit, yak and possum, are blended with wool to achieve softness, smoothness and lightweight warmth. Wool/possum fibre blends are spun on the woollen system and made into lightweight warm knitwear in New Zealand. The straight and smooth possum fibres tend to end up on the outer layer of the yarn and in surface hairs, giving the end product a luxury smooth handle, while the merino wool component improves the processing and wear performance of the fabric. Manufacturers claim that possum fibres have a low apparent density due to their foam- or honeycomb-like structure, similar to a highly medullated wool fibre.

Natural cellulose fibres such as cotton and linen, as well as synthetic fibres, are also blended with wool to achieve soft handle. Wool is often blended with these low-crimp fibres to improve bulk and softness. Microfibres (fibres finer than 1 denier) can be made into fabrics that have a special soft handle, commonly referred to as peach skin effect. Normal denier Tencel fibres can also provide a microfibre-like handle by subjecting the fabric to a fibrillation treatment. Wool/Tencel blends (Roberts, 1996) acquire their chamois softness from the Tencel and their bulk and drape from the wool.

12.2.5 Stretch and fit

An inextensible garment can impose considerable strain on the skin when body parts move. The skin strain imposed can be reduced by adopting loose fit garment design, by reducing fabric-skin friction, and by increasing fabric 'give' or stretch (Kirk and Ibrahim, 1966). Research sponsored by DuPont carried out by Penn State University showed that all types of fatigue (strength, endurance and power) could be significantly reduced by means of reducing the degree of muscle oscillation as a consequence of wearing stretch garments made from Lycra® Power fabrics (Voyce *et al.*, 2005). It was estimated that nearly 50% of all fabrics made in recent years had some degree of stretch (Anon, 2004).

Most fabrics obtain their stretch by including elastane fibres. Elastane fibres can be integrated into wool fabrics in many ways, such as feeding elastane yarns and wool yarns together in the fabric forming stage, forming plied or bicomponent yarns by twisting together a wool yarn and an elastane yarn, elastane fibre wrapped wool yarns, and wool wrapped elastane core yarns. Fabrics with different levels (or power) of stretch are made in both woven and knitted forms to meet a wide range of consumer demands. IWS/DuPont jointly launched the Wool Plus Lycra® and Woolblendmark Wool Plus Lycra® brands in the mid-1990s to assist product development in the wool processing industry. Wool Plus Lycra® and Woolblendmark Wool Plus Lycra® have been used for many different men's and women's wear, not only in activity garments to provide freedom of movement, but also in formal dresses to enhance garment shape retention and wrinkle recovery.

Wool stretch fabrics have also been developed for technical applications. A wool/Lycra knitted fabric has recently been developed by CSIRO for the prevention of skin tears for the elderly people (Douglas, 2006). As a result of age-related changes, the skin layers become less well anchored together, and the skin becomes dry and inelastic. Minor bumps and knocks can cause open wounds or skin tears. The wool/Lycra® fabric provides a layer of protection to the skin and is comfortable to wear 24 hours a day. The fabric is easy to stretch-fit to the limbs and shrink-resist wool is used so that the garment is machine washable and can be tumble-dried.

12.2.6 Lightweight fabrics

The demand for lightweight fabrics has been an important driving force for wool spinning technology development over a long time. Since it is impossible to arrange fibres end to end in a staple fibre yarn, the yarn irregularity that can be achieved by modern spinning machinery with a commercially acceptable spinning efficiency is largely governed by the mean number of fibres in the yarn cross-section. This sets the limit for the finest count yarn that can be spun from fibres of given linear density. The minimum area weight of the resultant fabric is in turn determined by the achievable yarn count. Depending on the labour cost involved to repair yarn breaks during spinning and weaving, 35 to 40 fibres are generally considered to be the minimum mean fibre number in a singles yarn cross-section for commercial worsted spinning.

Blending synthetic filaments with wool at the spinning stage can produce finer count composite yarns. SiroFil is an extension of the SiroSpun™ technology by replacing one of the two wool rovings with a filament. A further development along this direction is the so-called Treotek, or LincSpun, yarn technology, in which two fine filaments are fed to the sides of the roving. The lightweight fabrics made from Treotek yarns have high abrasion resistance and have been used as upholstery fabrics in passenger airlines.

With the availability of low-cost water-soluble fibres, another approach to lightweight fabrics becomes available. Water-soluble PVA (polyvinyl alcohol) fibres, in either filament or staple form, are spun in a blend with wool to a normal count yarn and woven into a normal weight fabric. Wet finishing treatment of the fabric dissolves the water soluble fibres to obtain a lighter weight, soft handle and highly breathable pure wool fabric. The percentage of the water-soluble fibres used in the blend is normally between 10% and 20% to achieve a meaningful weight reduction while fabric performance is still acceptable. PVA filaments or pre-spun yarns can also be used to spin a wool covered core yarn. If appropriately constructed, a pure wool sleeve-like 'hollow yarn' is left when the PVA is dissolved in fabric stage. The result is a lightweight, high bulk and soft handle fabric.

12.2.7 Bulk, cover and warmth without weight

Traditional cold weather garments are heavy and baggy. The consumer's desire for greater mobility demands that thermal insulation garments provide 'warmth without the weight'. This concept also appeals to manufacturers because of the implied material savings.

The thermal conductivities of most common textile fibres are remarkably similar (Morton and Hearle 1993). In fact, the thermal conductivity of constituent fibres has little influence on the thermal insulation of the final fabrics. Dry, still air provides much better thermal insulation than any textile fibres. Fibres form only a skeleton of a fabric and the spaces between the fibres are filled by still air. So long as convection air flow does not dominate the heat transfer, fabric density is only of second order importance. Fibrous structures with as little as 1% fibre by volume, such as duvet inners, are capable of trapping still air. The thermal insulation of fabrics is mainly determined by their thickness.

Bulk (volume per unit weight, mL/g) is used to measure the power of a wool fabric to hold still air. Fibre crimp is the main fibre attribute that determines fabric bulk. Crimp is also the major factor that determines the resistance to compression of a fabric, and possibly the single most important contributor to fabric softness. High bulk structures provide higher fabric cover with the same material, or the same fabric cover with less material. The ability of fabric to recover its bulk after compression is also very important. Wool fibre, with its high crimp and resiliency, is highly valued for its thermal insulation. High-crimp wool varieties are especially sought after for producing lightweight thermal insulation products such as duvet and futon inners.

An effective method to increase the bulk of yarns and fabrics is blending retractable fibres with non-retractable fibres. Retractable and non-retractable fibres (usually in the ratio of 40/60, 50/50 or 60/40) are blended during the spinning preparation stage. When the yarn made from such a blend undergoes an activation process, such as steaming or a wet process, the retractable fibres shrink and cause the yarn to shorten its length, which forces the non-retractable fibres to buckle and the yarn to expand its volume. The activation of the retractable fibres can also be carried out in the fabric stage, usually as a part of the finishing routine. Retractable acrylic fibres have been commercially used to produce high bulk wool/acrylic blended yarns for many years. In recent years, retractable pure wool tops have been produced by stretching wool tops without applying a permanent setting. Optim Max® developed by CSIRO and lincLITE® developed by WRONZ are two commercial technologies for producing temporarily stretched wool tops. These stretched tops are used in high bulk machine- and hand-knitting yarns.

Water-soluble PVA filaments and pre-spun yarns have been used to produce wool covered core yarns which are later dissolved to produce 'hollow yarns' to create high bulk yarns and fabrics. Besides limiting the thickness of the PVA core used, positioning the core at the centre of the yarn and correctly choosing twist level during core yarn spinning are some of the key requirements to maintain the hollow yarn structure after the PVA is dissolved and to optimise the bulk of the resultant yarns and fabrics.

Synthetic fibres can be made with very high crimp. These high-crimp synthetic fibres can be blended with wool to achieve bulk levels that natural wool cannot achieve alone. High-crimp synthetic fibres have been produced specifically for the wool carpet manufacturing industry to produce high-bulk pile yarns for improved carpet cover and resilience (Fisher, 2002).

12.2.8 Fibre security

Fibres need to be securely locked into yarn and fabric structures so that they do not migrate out during processing and end use. Two major problems associated with the lack of fibre security are pilling and fibre shedding. Pilling is a widespread problem for the textile industry, escalated by the fashion trend towards soft handle fabrics, and it is especially serious for knitwear. A comprehensive review on pilling and its control was carried out by Ukponmwan *et al.* (1998). Several seemingly contradicting approaches may be used to control fabric pilling, including increasing fibre security to stop fibre migration out of the fabric and accelerating fibre fall off. Each of these approaches has its own drawbacks and a 'silver bullet' solution to pilling has not been found (Miao, 2004).

The traditional wisdom on the pilling of wool/synthetic blend fabrics is that the strong synthetic fibres form anchors for the wool fibres to gather around to form pills. The strong anchor fibres are difficult to break off so that pills remain on the fabric for a longer time during which more pills accumulate on the fabric. Most published work supports this theory although contrary reports appear from time to time. The synthetic fibre industry has developed anti-pilling synthetic fibres by introducing intermittent weak spots along the fibre (Salvio and Stibal, 2002) that allow the pills to break off instead of clinging to the fabric. Anti-pilling synthetic fibres should be considered when constructing a wool/synthetic blend for use in soft handle products.

Fibre shedding happens in practically every process from yarn spinning to fabric finishing, and it becomes a concern only when excessive amounts of fibres are lost. Fibre shedding can be a serious problem in cut pile carpets and fabrics, especially in the initial stage of carpet wear. Bursting of cut ends of the pile yarn causes the loss of tuft definition that gives the new carpet appearance. Melt bonding fibres have been used in blend with wool to increase fibre security. The melt-bonding fibres, upon activation by steaming or wet treatment, adhere with themselves and other synthetic fibres to form a network through the spaces between the wool fibres in the yarn. However, melt bonding fibres in general do not bond well with wool fibres. 80/10/10 wool/nylon/bonding fibre is a popular blend for wool-rich cut pile carpets.

The melt-bonding technique has also been experimented for pilling reduction in wool-rich woollen knitting yarns. The key in this case is to strike a balance between the deterioration in fabric soft handle and the improvement in pilling performance.

12.2.9 Flame retardance

Wool is regarded as a naturally flame-resistant fibre and wool fabrics were the traditional material for flame and heat protection. A survey carried out in the early 1980s (Mehta and Norman, 1983) indicated that wool and wool blends were widely used materials in firefighter tunics in Europe (Table 12.1). Wool has a high ignition temperature, high limiting oxygen index and a low heat release from combustion. Unlike thermoplastic fibres, wool does not melt or drip, but forms a self-extinguishing char in a flame. For a higher degree of protection, a Zirpro treatment developed by IWS can be used to meet regulatory fire retardancy requirements in specialised applications such as aircraft interiors.

Table 12.2 compares the major thermal properties of wool with some common and flame-resistant (FR) fibres. Because wearers of thermal protective clothing work in extreme conditions, good moisture transport and thermal insulation are important requirements. Natural flame retardance, nondripping, moisture absorption and relatively low price are the main attractions for wool as a component in high-performance thermal protection applications,

Table 12.1 European firefighter tunics (Mehta and Norman, 1983)

Country	Tunic				
	Mass (kg)	Thickness (mm)	No. of layers	Vapour	Type of outer fabrics
Austria	1.3	5.7	3	No	Wool/FR viscose/FR polyester fibre (50/25/25)
	0.9	0.7	1	No	Zirpro wool
Belgium	1.9	5.1	3	Yes	Aramid fibre (coated)
Denmark	1.6	2.5	2	No	Wool
Finland	2.5 overall	2.3	2	Yes	Wool/acrylic fibre (80/20)
France	2.1	2.3	2	No	Leather
Germany	1.0	1.05	1	No	Zirpro wool
Netherlands	2.3	3.9	1	No	Wool
	2.1	2.2	2	No	Wool/nylon (80/20)
	2.2	4.9	3	Yes	Aramid fibre
Norway	1.8	1.8	1	No	Wool
Sweden	1.3	5.8	2	No	Zirpro wool
	1.8	2.2	2	No	75/25 wool/FR viscose
	2.2	2.6	2	No	60/40 wool/FR viscose
UK	1.9	4.5	3	No	Wool/nylon (80/20)
	1.7	3.2	4	No	Aramid fibre

Table 12.2 Thermal transition temperatures of some fibres (Bajaj and Sengupta, 1992)

Fibre	T_m (°C) melts	T_p (°C) pyrolysis	T_c(°C) combustion	ΔH (kJ/g)	Limiting oxygen index (LOI)
Wool	–	245	600	27	25
Cotton		350	350	19	18.4
Viscose		350	420	19	18.9
Nylon 6	215	431	450	39	20–21.5
Nylon 6.6	265	403	530	32	20–21
polyester	255	420–477	480	24	20–21.5
Acrylic	>220	290	>250	32	18.2
Polypropylene	165	469	550	44	18.6
Nomex	375	310	500	30	28.5–30
Kevlar	560	590	>550	–	29
PBI		>500	>500	–	40–42

such as in blends with Nomex (an aramid fibre), Kermel (a polyamide-imide fibre) and PBI (polybenzimidazole) to provide primary protection for firefighters. On the other hand, pure wool, wool in blends with FR viscose and FR-treated cotton are used in base layer and secondary protection for firefighters.

Joint research between Lenzing AG and IWS (Mach, 1983) demonstrated that wool and FR viscose fibres had a synergistic effect when encountering fire. The wool/FR viscose blend offers several benefits:

- with 30% or more FR viscose, the wool component did not require flame retardant treatment;
- reduced breaking up of charred areas, resulting in higher tear strength after flame exposure;
- reduced felting shrinkage due to the blending of viscose;
- improved mechanical properties and wear comfort due to the fine viscose fibre content and the elimination of Zirpro finishing; and
- reduced material price due to the lower price of the FR viscose.

Many applications were developed from the wool/FR viscose blend, including workwear, uniforms for firefighters, pilot overall, army underwear, upholstery fabrics and seat cover for Airbus 310 (Bajaj, 2000). PR97™ made from wool/viscose FR is now a prominent international fabric brand for molten metal splash, electric arc and other heat protection.

Upholstery fabrics such as airline seat covers require frequent machine washing. 55/45 wool/FR viscose blend and 55/45 wool/modacrylic blend did not require flame retardant treatment and could meet shrink resistance requirement as their felting potential is reduced by the blending of the synthetic fibres. However, wool-rich upholstery fabric required both shrink-resist and flame-retardant treatments. Commercial wool shrink-resist treatments (such as chlorine-Hercosett, Synthappret BAP) are not compatible with Zirpro treatment. Benisek (1998) reported the development of a shrink- and flame-resistance (FRS) polymer for 100% wool and 90/10 wool/nylon upholstery fabrics.

12.2.10 Moisture transport

Natural wool fibre is well known for its moisture absorption. It can absorb up to 30% moisture without feeling wet. Synthetic fibres, on the other hand, can be made with certain surface hydrophilicities and cross-sectional shapes to greatly enhance their wicking property. An important development in this area, Sportwool™, as a wool polyester bilayer fabric is discussed in details in Chapter 11. A more recent example is a blend of wool and Coolmax® fibres to produce quick dry fabrics for use in sports, outdoor base layers and outerwear (Anon, 2007).

12.3 Processing systems for wool blend yarns

Wool blends can be formed at almost any stage of the process from fibre to fabric. If blending takes place before or at spinning, a blended yarn is produced.

A pure wool yarn can also be plied with a non-wool yarn to form a twist blend or union blend yarn. Pure wool yarns may be combined with other yarns during fabric making to produce a union fabric. We will focus on the production and quality aspects of wool blend yarns.

Fibre length in a wool blend is of primary importance in early stage textile processes up to spinning. The woollen spinning system is often chosen for its unique appearance and the bulky handle of its products. The system is known for its tolerance for processing blends with a wide range of fibre lengths for both long and short staple fibres. The woollen spinning system produces relatively coarse count yarns for a given average fibre fineness. When a smoother and less bulky yarn is required, high-draft spinning systems that straighten and parallelise the fibres are used. When wool and long staple synthetic fibres are blended, a long staple worsted spinning system is used. Short staple fibres, such as cotton, are usually blended with short wools (short shorn wool, combing noils, broken tops, etc.) and processed on short staple (cotton) spinning systems.

12.4 Spinning wool blend on the long staple worsted system

Long staple synthetic fibres are often processed into tops first and then blended with wool in top form on gilling machines. The number of gilling operations should be sufficiently large so that good fibre mixing is achieved. The blended slivers are then processed into rovings and yarns in much the same way as pure wool slivers.

Fibre diameter and length distributions of the blend are important for the spinners to determine the final yarn count that can be made from the blend and to set the roller drafting system during the process from top to yarn. Spinners are well aware of the relationship between the mean number of fibres in a yarn cross-section and the theoretical minimum mass irregularity that the yarn can achieve. The theoretical irregularity of a pure wool yarns can be predicted from the mean fibre diameter and its coefficient of variation. The fineness of synthetic fibres, on the other hand, is commonly measured in terms of linear density (denier, or decitex). The minimum mass irregularity of a two-component wool/synthetic fibre blend yarn (Rae and Bruce, 1973) can be calculated using

$$\mathrm{CV}\% = \frac{1}{\sqrt{T}}\sqrt{10\,[w_{\mathrm{w}}p_{\mathrm{w}}(1 + 4\mathrm{CVD}_{\mathrm{w}}^{2}) + w_{\mathrm{s}}p_{\mathrm{s}}]} \times 100\% \qquad 12.1$$

where T is the linear density of the blended yarn in tex, w_{w} and w_{s} are the ratios of the wool and synthetic fibres by weight in the blend yarn, p_{w} and p_{s} are the mean linear densities of the wool fibre and the synthetic fibre expressed in decitex, and $\mathrm{CVD}_{\mathrm{w}}$ is the fractional coefficient of variation of wool fibre diameter. The mean wool fibre diameter in micron, d, can be converted into mean linear density in tex using $p_w = 0.01\ 062\ 5d^2$.

The fibre length of long staple fibre tops may be expressed in terms of barbe or hauteur. Most woollen spinners prefer to use barbe (mm) while worsted spinners tend to use hauteur. If a number of wool batches with barbe values of $B_1, B_2, \ldots, B_n$ are blended according to their proportions by weight $w_1, w_2, \ldots, w_n$, the barbe value of the resultant wool batch will be the weighted average of the components:

$$B = w_1B_1 + w_2B_2 + \ldots + w_nB_n \qquad 12.2$$

Hauteur of a resultant wool sliver (H), on the other hand, follows a harmonic relationship with its components ($H_1, H_2, \ldots, H_n$) (Miao, 1998):

$$\frac{1}{H} = \frac{w_1}{H_1} + \frac{w_2}{H_2} + \ldots + \frac{w_n}{H_n} \qquad 12.3$$

When using the above formula on blends of wool and other types of fibres, the fibre weight-based blend ratio may need to be replaced by a fibre volume-based blend ratio or a dielectric constant-based blend ratio, depending on the fibre length measurement method adopted.

12.5 Processing wool and wool blends on the cotton spinning system

There are a number of reasons why short staple (cotton) spinning system is preferred for processing wool blends. Cotton has a much larger share of the total fibre market than wool. This means that cotton processing machines are lower in cost, more widely available and more automated. Major cotton processing machines, including carding and spinning machines, operate at higher speed than their worsted and woollen counterparts. In addition, the carded cotton spinning route involves fewer processing steps than the worsted spinning route. When short wools are spun on the cotton system, the mean number of fibres in the yarn cross-section can be set to about 60–80 fibres (Harry and Robinson, 1977a), while 100 to 120 fibres are typically required for the woollen spinning system.

The wide use of the cotton system for processing wool and wool blends has been mainly limited by the quantity and price of short wool available. Short-shorn wools and combing noils are much cheaper than short wool converted from wool tops. There have been large-scale installations of cotton processing machines for processing wool blends in China in recent years. This has raised the demand for short wools. A sector of the Chinese cashmere dehairing industry started to separate the fine fibre components (called 'wooldown') from wools grown on some Chinese traditional breed sheep. These traditional wools have very broad fibre diameter distributions and contain a significant quantity of fine fibres for commercial separation using low-cost dehairing machinery modified from cotton cards. Wooldown now makes up a considerable proportion of the short wool supplies in China.

12.5.1 Cleaning and carding of short shorn wool on cotton card

Owing to the high bulk of wool fibres, it has been found difficult to form all-wool laps on some old cotton blowroom machinery and to unroll the laps when being fed to the card. Addition of cotton or synthetic fibres could improve the situation. In modern cotton spinning mills, the cotton/wool blend is fed to the carding machine by a direct chute feeding system and the problems associated with forming and unrolling the lap are avoided. Hopper feeding to the cotton card, in a similar fashion as in woollen carding, has been an economical option for many of the short staple facilities in China.

The blending of scoured wool in the cotton blowroom can be problematic owing to inadequate cleaning, high fibre breakage and other processing difficulties. Woollen and worsted cards are installed with sophisticated cleaning systems to deal with vegetable matter (VM) in scoured wool. The cotton card, on the other hand, is not designed to deal with the vegetable matter in wool. Wools with a high VM content produce high waste, which increases with the wool content in the cotton/wool blend (Spencer and Taylor, 1979). On the other hand, blends of cotton and low VM wool (such as wool converted from tops) produced less waste in comparison with processing cotton alone.

Wool fibre breakage by the cotton card is difficult to accurately quantify when blends are processed. In one experiment on pure wool top fed to a cotton card, a mean fibre length reduction of 25% (from 40.1 to 30 mm) was observed when ignoring the effect of carding waste (Aldrich, 1975). Major fibre breakage took place at the feed roller/licker-in region, while the main carding elements (flats and cylinder) caused little fibre breakage. Fibre breakage was reduced by reducing the speeds and opening settings in the feed roller and licker-in region. Owing to the lower fibre-to-fibre cohesion, all-wool webs processed on cotton card were found to sag prior to its entry into the trumpet to form a sliver. This problem was resolved by installing a web supporting pan (Louis and Pardo, 1980). The processing performance of wool and wool blends was improved by some other minor changes on the cotton card, including a 5–6% reduction in tension draft between the doffer and the coiler (Harry and Robinson, 1977a), replacing the licker-in wires with wires that were normally used for processing synthetic fibres, increasing the flat-cylinder setting and by running the card at a lower production rate.

12.5.2 Square-cut and stretch-broken wool tops

Wool tops are converted into short wools to process on the cotton system for different reasons, including processing efficiency and product quality. Converted short wool also provides the option of sliver blending on the cotton drawframe, avoiding the difficulties of processing blends in blowroom and carding processes mentioned above.

Cutting of wool tops and card slivers can be carried out using a variety of highly productive guillotine and rotary cutters. Stretch breaking, on the other hand, is usually carried out on machines consisting of essentially a roller drafting system (Eley *et al.*, 1980). Stretch-breaking acts only on fibres longer than the effective roller ratch while cutting applies to all fibres. Consequently, the coefficient of variation of fibre length of cut wool is higher than that of stretch-broken wool. The required level of roller loading in stretch breaking largely depended on the size of the slivers fed to the stretch breaker. A minimum draft of 1.3 is required to break wool fibres, but breaking efficiency increases with increasing draft up to an optimum value between 2.7 and 3. The resultant mean fibre length (MFL) was approximately proportional to the setting of the last ratch (MFL = 0.55 × final ratch).

Lupton (1980) compared 60/40 wool/polyester blends composed of cut-top and 6-month shorn wools. The mean fibre length of the shorn short wool (30.2 mm) was appreciably longer than the cut-top (25.9 mm). It was concluded that fabrics containing cut-top wool in general exhibited higher strength, but lower resistance to flex abrasion, than their short-shorn wool counterparts. Other fabric properties such as tear strength and laundering shrinkage were similar. Short wools produced by stretch-breaking (51 mm setting) and square cutting (38 mm setting) were compared by Louis and Pardo. (1980). The stretch-broken top had greater mean fibre length and smaller amount of short fibres than the cut top. However, after blending with 40% or more cotton, the difference in wool fibre length did not transfer into any significant differences in fibre bundle strength and yarn strength.

Stretch-breaking of wool tops has also been carried out in steps using a combination of modified worsted gilling machines and modified cotton drawframes (Kim, 1997). The drawn sliver is then ready for roving and spinning. When woollen-type fibres (such as lambswool and cashmere) are used, the fibres are carded on a woollen card into a sliver, which is then fed to a series of modified cotton drawing frames to stretch-break the longer fibres (Kim, 2002). Three to four passages of drawing were recommended. The system, known commercially as AFAY (All Fibres for All Yarns), has been used to produce blends of wool and short staple fibres commercially. (Anon, 2003).

12.5.3 Drawing, roving and ring spinning

Slivers of short length wool and wool blend can be processed on cotton drawframes, speedframes and ringframes without major changes. Wools with a staple length of 40–45 mm were processed in this way without major alteration. Roller ratch in the drafting systems can usually be adjusted to accommodate the relatively long fibres from the wool component. In case such adjustment is not sufficient, a recess may be cut in the back top roller of the front drafting zone to allow for slip draft of the longer fibres (Harry and Robinson, 1977a). Slip drafting was used to process wools with a mean fibre length up to 53 mm. Wool slivers with a mean fibre length of 65 mm were processed by removing the top middle roller (Ellis and Robinson, 1979). A thicker apron spacer was used to accommodate the high bulk of wool at the speedframe and ringframe. A lower twist in roving (twist factor $= 950 - 1100\ T/m\sqrt{\text{Tex}}$) was required for a wool/cotton blend than for pure cotton (Harry and Robinson, 1977b). Yarn twist levels used on wool/cotton blends were set according to pure cotton yarns (Harry and Robinson, 1977b).

12.5.4 Open-end rotor spinning

The open-end rotor spinning system offers an even shorter processing route with higher levels of automation and productivity than the conventional cotton ring spinning system. The twist insertion rate of rotor spinning is five to ten times higher than that of the ring frame. Rotor spinning is especially suitable for producing medium to coarse count carded cotton yarns. The best known application of rotor spun cotton yarns is found in denim fabrics. The productivity advantage of rotor spinning reduces when longer fibres are processed because correspondingly large rotors are required, which limits the rotor speed. The presence of long fibres can also cause excessive spinning breaks due to the interference of these fibres with the yarn tail in the rotor groove. Special long staple rotor spinning machines fitted with large rotors were designed in the 1970s and used commercially for spinning wool (Landwehrkap, 1979). Spinning of short wools (lambswool, combing noils, etc.) using relatively small diameter rotors has been investigated in recent years. A higher twist factor is required to achieve optimum yarn strength in rotor spun yarns than in ring spun yarns, and even then, yarn and fabric strengths, abrasion performance and yarn evenness will be reduced (Louis and Pardo, 1980; Lehmann and Philippen, 1998; Schmidt, 2002).

12.6 Wool–filament composite yarns

Continuous filaments or a pre-spun yarn can be spun together with wool or wool blend to produce a composite yarn. With the availability of increasing number of high-performance synthetic filaments, composite yarn spinning has become a useful vehicle for developing wool blend products that are both highly functional and comfortable to wear. Composite yarns may fall into several types of structures, for example wrap yarn, bicomponent yarn, core yarn, and intermingled composite yarn.

Wrap yarns can be produced on a hollow spindle wrap spinning machine or by twisting a pre-spun singles yarn and a filament yarn in the opposite direction. Hollow spindle wrap spinning machines can offer cost advantages over ring spinning in some coarse count yarns. Owing to the parallel lying of the wool fibres, wrap yarns generally offer higher bulk and softer handle than conventional ring spun yarns.

There are several methods to produce wool/filament bicomponent yarns. The most commonly used method is the SiroFil method, which is a variation of the SiroSpun yarn system, by replacing one of the two rovings with a filament yarn. The filament appears as a helix (parallel to the staple fibres) on the surface of the final yarn. SiroFil provides the opportunity to spin a fine count, low hairiness and strong yarn for use in lightweight fabrics. However, the appearance of filament on the yarn surface can be a distraction for some products. A further development in this direction is the Treotek or LincSpun yarn produced from two fine filament yarns and a wool roving. The two filament yarns are fed to the sides of the wool strand at the front roller nip. Treotek yarns are used in high abrasion resistance and lightweight upholstery fabrics, especially for passenger airlines. Wool/filament bicomponent yarns can also be produced by twisting a wool yarn with a filament yarn.

12.6.1 Core yarns

Core yarns have a layered core-and-sheath structure. The core component fed to the yarn formation zone can be a monofilament, a multifilament yarn, a pre-spun staple fibre yarn, or another strand of staple fibres. The sheath or cover material is usually staple fibres, but it can also be in other formats, such as filaments and prespun yarns. Core yarns can be produced on a spinning machine or a twisting machine. Besides the ring spinning system, friction spinning, rotor spinning and air-jet spinning systems have been used to produce core yarns, but ring spinning dominates. Core yarns can be made on most ring spinning machines with relatively minor changes. Two types of roving feeding can be adopted, the conventional single roving feeding and the SiroSpun type of double-roving feeding. Besides the ratio of surface fibres used, the correct placement of the core filament relative to the rovings at the front roller nip, the tension applied to the core filament, and the yarn tension in the spinning zone are important conditions that influence how well the core is covered by the staple fibres in the resultant core yarn.

Although core yarns can be produced relatively easily, perfect coverage of the core component by the surface fibres is not always achieved, especially when a relatively heavy core component is used. In a poorly covered core yarn, the bare core is exposed on the yarn surface, leading to an undesirable barber's pole effect and poorer product performance. The level of yarn coverage can be assessed subjectively or objectively by using appropriate image analysis techniques. Maximum visibility of core exposure is obtained when the core and surface fibres are in highly contrast colours. Commercial flatbed scanners and image analysis techniques have been used to quantify the degree of core yarn coverage (Sawhney *et al.*, 1992; Miao and Barnes, 2006).

Even when complete yarn coverage is achieved, the cohesion between the covering fibres and the core material may still be low, and the surface fibres can slip along the core when the yarn is subjected to rubbing. This problem is caused by the lack of penetration between the surface fibre layer and the filament core. Different types of yarn abrasion tests could be adapted to test core yarn resistance to surface fibre stripping (Miao *et al.*, 1996). The surface fibre stripping resistance was found to improve with the level of twist inserted (Jou *et al.*, 1996) and the level of pre-tension applied to the core filament (Miao *et al.*, 1996). The use of a filament yarn pre-twisted in the opposite direction also brought improvement to core yarn strip-back resistance in friction spun core yarns (Miao *et al.*, 1996).

12.6.2 Elastane core yarns

Stretch fabrics can be made from either two separate yarns (an inelastic yarn and an elastic yarn), or from a composite yarn consisting of an elastic component and an inelastic component. The composite yarn route is suitable for both knitting and weaving, and it is the more convenient route for fabric manufacturers. Several types of wool/elastane composite yarns can be made, including twisted yarns produced from an elastane yarn and a wool spun yarn, wrap yarns produced using a hollow spindle wrap spinner, and core spun yarns produced from wool rovings and an elastane yarn on a worsted spinning frame. The core yarn type is preferred for its soft, smooth fabric handle and uniform fabric appearance that it produces.

In the spinning of elastic yarns, a positively driven creel is commonly used to maintain a constant feeding rate of the elastane core. The draw ratio between the positive feed roller and the front roller is normally between two and five and is decided according to the fabric stretch required. A roller with a V-shaped groove above the front top roller is commonly used as a guide to the elastane core. Su *et al.* (2004) found that increasing the wrap angle of the elastic core on the top front roller could improve yarn coverage.

Special attention is required in the winding and splicing of elastane core yarns. In the process of preparing yarn ends for splicing, the elastic core tends to retract from the cut ends. Longer joint length and higher splicing air pressure could be used to achieve a satisfactory result (Anon, 2005b). Winding of an elastane core yarn on a standard winding machine can cause the inner layers on the package flanks to bulge out due to pressure created by the yarn tension in the outer layers. Special winding and splicing machines that avoid these problems are now available (Schlafhorst & Co., 2002a,b).

12.6.3 Intermingling filaments with staple fibres

Core yarns are often found to possess low stripping resistance because of the lack of penetration between the filaments and the surface staple fibres. One solution to this problem is to intermingle the filaments and the staple fibres before twisting them together to form a composite yarn. Such intermingling can be achieved by spreading the filaments across the width of the staple fibre strand at the front roller nip of a ring spinning machine. Filament yarns can be spread mechanically (Hicks, 1958), electrostatically (Uchiyama *et al.*, 1972; Jou *et al.*, 1996) and aerodynamically (Arnold *et al.*, 2004). Up to 12 mm filament spread width could be achieved using an electrostatic spreader (Wu and Lee, 1995), but 6 mm was optimum, giving yarns with higher resistance to stripping back and lower surface hairiness than regular core yarns, but similar tensile properties.

12.7 References

ALDRICH D V (1975), The processing of blends of cotton and wool on the cotton system; Part 1: An introductory investigation, SAWTRI Technical Report No 248

ANON (2003), AFAY spins cost-cutting web, *Beyond the Bale*, Issue 4, March

ANON (2004), Elastic yarn technologies, *Melliand International*, **10** (March), 14–16

ANON (2005a), Bagir launches machine washable suit that can be tumble dried, *Wool Record*, March 2005, p11

ANON (2005b), EliCore and EliCoreTwist: production of compact core yarns, *The Indian Textile Journal*, **115** (issue 9) 113–116

ANON (2007), A new twist on wool, *WSA*, March/April 30–33

ARNOLD A, ARTZT P and BETZ D (2004), New process for hybrid yarn production from filament and staple fibres, *Technical Textiles*, **47**, (May) E59–E62

BAJAJ P and SENGUPTA A K (1992), Protective clothing, *Textile Progress* **22** (2/3/4), 6

BAJAJ P (2000), Chapter 10, Heat and flame protection, in *Handbook of Technical Textiles*, edited by Horrocks A R and Anand S C, Woodhead Publishing Ltd, pp223–265

BENISEK L (1998), Innovations in flame resistant wool transportation furnishings, *Textile Asia*, **29** (Issue 8), 36–42

BISHOP D P (1996), Fabrics: sensory and mechanical properties, *Textile Progress* **26** (3), 1–62

DE BOOS A G, JONES F W, LEEDER J D and TAYLOR D S (1975), The wrinkling of wool and wool/polyester blends, *Proceedings of 5th International Wool Research Conference*, Aachen, Sept, Vol 3, pp472–481

DE BOOS A G, FINCHER K W and WEMYSS A M (1996), 'Wrinkle-free' wool-rich trousers, Books of Paper, AATCC International Conference & Exhibition, 15–18 Sept, pp137–146

BYRNE K M (1998), Easy care wool/cotton blends, *Proceedings of the Beltwide Cotton Production Conference*, 5–9 January, **1**, pp771–775

DOUGLAS S (2006), CSIRO remains at the cutting edge of wool research, *Wool Record*, Oct, 40–41

ELEY J R, HARROWFIELD B V, HENSHAW D E and ROBINSON G A (1980), Stretch-breaking and cutting of wool for the short-staple system, *International Wool Textile Research Conference*, 26 August–3 September, *Proceedings* Vol. 3, pp349–362

ELLIS B C and ROBINSON G A (1979), Processing medium-length wools on the short-staple system, CSIRO Report No. G38

FISHER G (2002), Self-crimping fibre for wool-blend carpet yarns, *Wool Record*, Nov 82–83

GARNSWORTHRY R K, GULLY R L, KANDIAH R P, KENNIS P, MAYFIELD R J and WESTERMAN R A (1988), Understanding the causes of prickle and itch from skin contact of fabrics, *Australasian Textiles*, **4** 26–29

HARRY A A and ROBINSON G A (1977a), Processing of wool and wool-rich blends on the short staple system, I: A modified system for wools of average 40 mm length but with a non-square fibre length distribution, CSIRO Report No. G31

HARRY A A and ROBINSON G A (1977b) Processing of wool and wool-rich blends on the short staple system, II: Noils and carbonized lambs wools, CSIRO Report No. G32

HAYES R L, PHILLIPS D G and RUSSEL I M (1980), Wrinkling of wool and wool blends in wear, *Proceedings of 6th International Wool Research Conference*, Pretoria, Vol 4, pp199–216

HICKS J W (1958), US Patent 2,825,199

JOU G T, EAST G C, LAWRENCE C A and OXENHAM W (1996), The physical properties of composite ayrns produced by an electrostatic filament-charging method, *Journal of the Textile Institute (Part I)*, **87** 78–96

KIRK W J and IBRAHIM S M (1966), Fundamental relationship between fabric extensibility to anthropometric requirement and garment performance, *Textile Research J*, **26** 37–47

KIM K W (1997), Method of spinning all-wool yarn on cotton system, US Patent No. 5,638,580

KIM K W (2002), Manufacturing method for carded woollen yarn, US Patent No. 2002/0059702

LANDWEHRKAP H (1979), Preparation and spinning wool with the rotor-spinning system, *Textile Research Journal*, **39** 137–139

LEEDER J D (1984), *Wool – Nature's Wonder Fibre*, Australian Textiles Publishers

LEHMANN K H and PHILIPPEN H (1998), Production of yarns from blends of carbonised wool noils and cotton, *Melliand English* E31–E33

LI Y (2001), The science of clothing comfort, *Textile Progress*, **31** (1/2), 102–103

LOUIS G L and PARDO C E (1980), USDA natural fibre blend research, *Textile Research Journal*, **50** 130–136

LUPTON C J (1980), Wool processing on the cotton system: a comparison between cut-top wool and six-month shorn wool in a blend with polyester, *Textile Research Journal*, **50** 119–129

MACH D (1983), Flame resistant work and protective clothing with Lenzing viscose fibres, *Melliand Textilberichte* (English Edition) **12** May 321–327

MEHTA P N and NORMAN D L (1983), Development of a new fire service tunic, *Textile Research Journal*, **50** 153–159

MIAO M (1998), Combination of Hauteurs and CV_H's, unpublished WRONZ Report

MIAO M (2004), The knitwear paradox: handle and bulk versus pilling, where is the balance? *Textile Institute World Conference*, Sept, Shanghai

MIAO M and BARNES S (2006), Determination of core spun yarn coverage, *Processings of 2006 China International Wool Textile Conference and IWTO Wool Forum*, Nov Xi'an, China, pp16–22

MIAO M, HOW Y L and HO S Y (1996), Influence of spinning parameters on core yarn sheath slippage and other properties, *Textile Research Journal* **66** 676–684

MIAO M, COLLIE S C, WATT J D and GLASSEY H E (2005), Prickle and pilling reduction by modified yarn structures, paper presented to *International Wool Research Conference*, Sept, Leeds

MORTON W E and HEARLE J W S (1993), Chapter 23, Thermal properties and setting, in *Physical Properties of Textile Fibres*, The Textile Institute, pp585–610

NAYLOR G R S and PHILLIPS D G (1995), Skin comfort of wool fabrics, *Proceedings of the 9th International Wool Textile Research Conference*, Biella, Italy, Vol 2, pp203–209.

NAYLOR G R S and PHILLIPS D G (1997), Fabric-evoked prickle in worsted spun single jersey fabrics, part 2: the role of fibre length, yarn count and fabric cover factor, *Textile Res J*, **67** 354–358

NAYLOR G R S, PHILLIPS D G, VEITCH C J, DOLLING M and MARLAND D J (1997), Fabric-evoked prickle in worsted spun single jersey fabrics, part 1: the role of fibre end diameter characteristics, *Textile Res J*, **67** 288–295

PHILLIPS D G (1982), The assessment of wrinkling in the laboratory and in wear, *Proceedings of Japan–Australia Joint Symposium on Objective Specification of Fabric Quality, Mechanical properties and performance*, Kyoto, Japan, May, pp75–88

RAE A and BRUCE R (1973), *The WIRA Textile Data Book*, WIRA, pp B90–91

ROBERTS S (1996), Tencel's appeal is widening, *Wool Record*, May 59

SALVIO G and STIBAL W (2002), Terital MAP – a new antipilling polyester fibre, *Chemical Fibres International*, **52** 320–324

SAWHNEY A P S, ROBERT K Q, RUPPENICKER G F and KIMMEL A B (1992), Improved method of producing a cotton covered/polyester staple-core yarn on a ring spinning frame, *Textile Research Journal*, **62** 21–25

SCHLAFHORST A G & CO (2002a), Variopack FX: optimized package build for elastic yarn, *Melliand International*, **8**, (March) 41

SCHLAFHORST A G & CO (2002b), New Elastosplicer, *Melliand International*, **8**, (September) 183

SCHMIDT R (2002), Spinning pure virgin wool on Autocoro rotor spinning machines, *Melliand International*, **8** (March 23–25)

SMUTS S (1989), A review of the wrinkle of wool and wool/polyester fabrics, *Tex Report*, CSIR, July 1989

SPENCER J D and TAYLOR H (1979), Removal of vegetable matter from scoured wool during the processing of wool/cotton blends on the cotton system, SAWTRI Technical Report No. 440

SU C-J MAA M-C and YANG H -Y (2004), Structure and performance of elastic core-spun yarn, *Textile Research Journal*, **74** 607–610

UCHIYAMA S, KAKU E, KOAYASHI M and ZODA T (1972), US Patent 3,657,871

UKPONMWAN J O, MUKHOPADHYAY A and CHATTERJEE K N (1998), Pilling, *Textile Progress*, **28** (3), 1–44

VOYCE J, DAFNIOTIS P and TOWLSON S (2005), Chapter 10, Elastic textiles in *Textiles in Sport*, Edited by Shishoo R, Woodhead Publishing Limited, pp204–232

WU W-Y and LEE J-Y (1995), Effects of spread width on the structure, properties, and production of a composite yarn, *Textile Research Journal*, **65** 225–229

13
Intelligent wool apparel

S. R. COLLIE, AgResearch Limited, New Zealand

Abstract: This chapter describes how wool is currently being used in the field of intelligent textiles, and how it may be exploited in this field in the future. Initially, categories of intelligent or 'smart' textiles are defined, and what constitutes a smart textile is discussed. Specific applications of wool in the field are then reviewed, focusing on its combination with inherently conductive polymers and its use in wearable electronic systems. Finally, future opportunities for wool to be exploited in smart textiles are discussed.

Key words: intelligent wool textiles, categories of smart and intelligent textiles, wearable electronics, inherently conductive polymers, *in situ* polymerisation

13.1 Introduction

There is often a perception among consumers and even textile manufacturers that wool is a fibre best suited to, and most often used in, very traditional types of textiles, such as suits, knitwear and carpets. It is perhaps not seen as a fibre to use in technical or high-performance applications such as sportswear or industrial textiles. However, as is discussed in other chapters of this book, wool textiles are now becoming established in these less conventional application areas. Perhaps the most unconventional class of textiles undergoing development at present is that of the so-called 'smart' or 'intelligent' textiles, and as in the other high-performance sectors, wool has an opportunity for use. It is the purpose of this chapter to review the extent to which wool is being employed in this emerging sector, and to discuss how it will be further exploited in the future. It is beyond the scope of this chapter to review the full scope of intelligent textiles (there are excellent publications that do this; Mattilla, 2006, for example); rather the intention is to provide a brief introduction to the field and discuss the role that wool already has, and could have, in the future.

13.2 Definitions and categories of smart textiles

It is useful initially to consider what is meant by an 'intelligent' textile system. Vigo (1999) defines intelligent materials as 'materials that usually have a dramatic and reversible response to one or more external stimuli'.

With this definition, we can expect a smart textile to behave in a manner that goes well beyond our normal expectations of textiles. Similarly, Tao (2000) proposes that 'smart materials and structures can be defined as the materials and structures that sense and react to environmental conditions or stimuli, or respond to, or be activated to perform a function by manual operation or in a pre-programmed manner'. The implication, therefore, is that three components may be present in intelligent textiles: a means of detecting changes in the environment (sensors); a means of responding to that change (actuators); and controlling units. In a simple analogy to the human body, the sensors are our five senses detecting mechanical (touch, hearing), electromagnetic (vision) and chemical stimuli (taste, smell), while the actuators include (for example) muscles, sweat glands, and so on. The controlling unit is our brain – allowing us to process the information and control our response. Clearly some actuation responses seem to happen automatically (such as sweating in response to elevated temperature), while others can be consciously processed (such as deciding to run away when we see a potential threat approaching in the distance).

It is this author's preference to identify three categories of intelligent textiles, according to the type of features they have and the degree of sophistication of their intelligent response and behaviour. Similar classifications have been used by others. The categories are:

1. *Passive* intelligent textiles, which are capable of sensing environmental conditions or stimuli, and can supply that information to another system (human or technological). These types of sensors often utilise changes in the electrical properties of conductive fibres as a means of detecting stimuli. The textile-based strain gauges developed and used by De Rossi *et al.* (1999) and CSIRO/IPRI (Munro *et al.*, 2004) are good examples of these. The stimulus is a mechanical deformation, which brings about a change in the electrical resistance of a conductive fabric. Detection of changes in the light transmission properties of optical fibres in response to stimuli has also been demonstrated (Guan *et al.*, 2000; Tao *et al.*, 2000).
2. *Active* intelligent textiles would be capable of not only sensing environmental conditions or other stimuli, but also responding to them by adapting their properties in some advantageous way. Shape memory materials (Vigo, 1997; Marks, 2001) or electro-active actuators (Otero and Sansiñena, 1998) are examples of these. Sometimes the sensing and actuation functions cannot be separated, such as in (at a very simple level) thermo-chromic materials that change colour in response to a change in temperature. The colour change is an outcome of thermally induced changes to the molecular structure of the material.
3. The final category is intelligent textiles that include systems for processing

and/or transmission of data collected by sensor systems. This might be used to process sensor inputs and control actuator functions. This category also includes the field known as 'electronic textiles' or 'wearable electronics', and includes textile-based electrical devices. The textile-based control systems developed by Softswitch (www.softswitch.co.uk) and Eleksen (www.eleksen.com) are examples of these, as are the numerous heated textile technologies now available.

In this chapter the terms 'intelligent textile' and 'smart textile' are taken to be synonyms and are used interchangeably.

13.2.1 'Smart' textiles that are not really smart

To a large degree manufacturers and marketers are determining the boundaries of what constitutes a smart textile. In some circumstances this means that conventional technologies are being re-packaged as 'smart', when perhaps a better definition of them would simply be 'high performance'. For example microporous waterproof breathable membranes are sometimes referred to as smart, because they allow moisture vapour to pass through, while preventing liquid moisture from penetrating. This occurs because of the size of the pores in the membrane (vapour molecules are small, while droplets are large) and while this is a very useful function, the material cannot be considered to be smart because it does not sense any environmental change, nor does it respond to one. If this is a smart textile then so is the fishing net that allows small fish to escape while trapping big fish! The concept and the method of manufacturing the material may be sophisticated, and certainly offering a level of performance that exceeds that of conventional materials, but there is no smart behaviour or smart functionality in the textile itself.

Conversely, just because a material responds to an environmental stimulus it is not automatically smart either. Metal railway lines might respond to an increase in temperature by expanding in length, resulting in buckled tracks and derailed trains. This is clearly not a smart response.

13.3 Wool's intrinsic 'smartness'

It could be argued that wool is inherently 'smart' in its behaviour regardless of modifications to it, or the structures it is put into. As has been discussed in other chapters of this book, wool is a remarkably complex fibre (a comprehensive review of wool's performance attributes in technical applications was also provided by Johnson *et al.*, 2003). It is worth considering why wool has become so complex, and the functions it can perform in the context of intelligent textiles. It has evolved over millions of years to provide comfort and protection for a warm-blooded organism that lives in a wide range of climates from the heat and dryness of the Mediterranean to the cold and damp of Northern Europe. It has a unique structure with a hydrophilic interior and hydrophobic surface, meaning that it can allow the passage of water vapour while repelling liquid water. It also has the ability to release substantial amounts of heat when it becomes wet – a useful thermoregulation capability in damp, cold environments. All this suggests that it may not be unreasonable to consider wool an inherently smart material. Capabilities such as thermoregulation and selective porosity are typical of smart textile functionalities.

Another concept in intelligent textile development is the field of biomimetics. This term refers to the technique of imitating natural systems in the development of new technologies. With wool, biomimetics is somewhat irrelevant as the fibre is already a complex biological material produced in the skin of the sheep.

13.4 Current applications of wool in intelligent textiles

At the 1999 Tissu Premier exhibition in Lille, a speaker from the Institut Textile de France suggested that the properties of wool would endear it to developers of emerging intelligent textile technologies (Anon., 1999). However, the following years have seen only a trickle of wool-based intelligent textiles. This may be due to the overall slower-than-expected commercialisation of smart textile technologies (Wilkinson, 2004), with wool's prominence in the intelligent textiles field simply being a reflection of its prominence in textiles globally. Therefore it is perhaps not surprising that much of the research and product development involving wool in intelligent textile systems has taken place in the countries where wool has greater prominence: Australia and New Zealand.

13.4.1 Wool and inherently conductive polymers

Electrical conductivity is an important property of fibres used in many intelligent textile systems. Conductive fibres have the potential to be used in a number of ways in intelligent textile systems, including as sensors (where the sensor stimulus is detected as change in electrical properties, such as resistance) and in wearable electronics (discussed in the following section). Therefore, a key initial stage in the development of electronic sensing and responsive mechanisms for wool is to provide the fibre with useful electrical conductivity. The technique used by many researchers has been to incorporate a conductive coating onto wool by the deposition of inherently conductive polymers (ICPs) on the fibre surface. One particular reason for interest in the use of ICPs as a means of making textiles conductive is their potential to be chemical sensors (via interaction of the species to be sensed with the chemical structure of the polymer), and potentially even electro-active materials.

One question that should be addressed is the level of conductivity that is required to make a wool fibre useful in an intelligent textile system. This depends on the application being targeted. For example, in many sensing applications the actual level of conductivity is almost irrelevant, as the most important characteristics are how conductivity changes in response to the stimulus being detected, and how reproducible the electrical behaviour is over time and with repeated use. On the other hand, in a wearable electronic application it might be necessary to have a high conductivity to allow power to be efficiently transferred, or a signal to/from a device to be accurately transmitted/received. In the former case, resistance levels into the mega-ohm range may be acceptable, while in the latter case ohm-order levels or lower will be required. The level of conductivity achievable in ICPs can be very high, depending on their composition and the synthesis technique used. Generally though, the conditions of polymerisation onto a textile substrate are such that the polymer formation is imperfect, and the conductivity diminished well below theoretical maximums.

Alongside considerations of an appropriate level of conductivity is the consideration of suitable ICP loading on the fibres, i.e. the ICP 'add-on'. Generally, putting more ICP onto a substrate will make it more conductive (as would be expected), but this relationship is not universal. For example, in work carried out by this author, the conductivity of a fabric with a thiophene variant polymerised onto it did not increase with increasing ICP add-on (Collie, 2007). In fact there was an optimum add-on level, above and below which conductivity decreased. This was probably due to better-ordered polymer formation under conditions that led to lower add-on, and when multiple treatments at the optimum level were used there was good correlation between the amount of ICP present and conductivity. It should be noted that the ICP add-on is often not reported by researchers, but is critical to considerations of the economic efficiency of the deposition process, and in all likelihood influences the intrinsic textile mechanical properties, such as flexibility and extensibility.

As discussed elsewhere in this volume there is ongoing interest in exploiting the chemical functionality of the wool fibre surface for the attachment of functional finishes, and conductive polymers are no exception. Conductive polymers (most commonly polypyrrole and polyaniline) are deposited onto textile substrates by two main techniques: *in situ* oxidative polymerisation of the ICP on the fibre surface (and often also in the fabric interstices), or deposition of a soluble form of the ICP. 'Doping' of the ICP is an essential process to make it fully conductive, and is simultaneous with polymerisation when ionic salts are used as oxidants. A very commonly used reagent is iron(III) chloride, where the iron(III) cation is the oxidant species, and the chloride anion is incorporated into the ICP as the dopant. Also, aromatic sulphonic acids (such as *p*-toluene sulphonic acid or naphthalene disulphonic acid) are used as sources of dopant anions. Iodine vapour or hydrochloric acid are also commonly used to dope ICPs after they have been polymerised. A discussion of ICP deposition techniques as they have been applied to wool follows.

A well-known and widely used deposition method is an *in situ* polymerisation technique developed by Milliken Research Corporation (Gregory *et al.*, 1989; Kuhn and Kimbrell, 1989). In some publications Milliken researchers suggest that fibre type has minimal influence on the conductivity of the substrate (Kuhn and Kimbrell, 1989; Kuhn and Child, 1997), which of course means that wool would offer no advantage (or disadvantage) over other fibre types. This is somewhat surprising, as in an adsorption reaction (as this deposition technique is), it is logical that the chemical and physical properties of the substrate onto which reagents are being adsorbed will be relevant. Other researchers have examined the effect of substrate fibre type, and several have included wool in their comparisons. For example, the interaction between the wool fibre and the polymerisation environment was discussed by Kaynak *et al.* (2002), with the polymer–dopant system thought to attach to molecular groups on the wool fibre by ionic and hydrogen bonds. For this reason, they concluded that wool is an ideal substrate to use in the preparation of textile-conductive polymer composites. The research found that treated yarns showed increased tenacity, due to the reinforcing effect of the polypyrrole coating, but that the initial modulus was reduced; this was attributed to the reduction in surface friction between fibres, causing more inter-fibre slippage during extension. In related work, the same research team confirmed that the ICP coating reduced the coefficient of friction of the wool fibres and reduced the differential friction effect caused by fibre scales by coating and 'smoothing' the fibre surface (Wang *et al.*, 2005). Tensile properties of the individual fibres were largely unchanged. In an interesting off-shoot they found that the *thermal* conductivity of polypyrrole-coated wool was increased compared with that of uncoated wool (Wang *et al.*, 2006).

Hirase *et al.* (2004) examined reaction parameters for deposition of polyaniline by the Milliken technique onto wool and other fibre types. In particular they compared the performance of three oxidants (potassium iodate, ammonium peroxydisulphate and potassium dichromate) and varying polymerisation durations. Most of the composite fabrics produced had too low a conductivity to be able to draw any clear conclusion about the effect of fibre type, but one result stood out: the only conductive composite that could be formed with potassium iodate was on wool. Fourier transform infrared (FTIR) spectroscopy showed the formation of cysteic acid (presumably from cystine in wool) when potassium iodate was used, but not with the other two oxidants. The researchers postulated that aniline interacted with this species, resulting in deeper penetration into the fibre and a heavier ICP loading. Microscopic examination of the fibre cross-section supported this finding.

Modification of the wool fibre surface by atmospheric plasma treatment prior to deposition of polypyrrole was described by Garg *et al.* (2007). The aim was to establish if the adhesion of the ICP to the fibre could be improved by this technique. The uniformity of the ICP coating was improved on plasma-treated wool, resulting in higher conductivity, and when subjected to Martindale abrasion the plasma-treated specimens showed reduced loss of the ICP coating. The more hydrophilic surface generated by plasma discharge was more conducive to polymer coating than the hydrophobic unmodified wool fibre surface.

Another *in situ* polymerisation technique frequently used in the literature is vapour phase deposition. The basic technique (well exemplified for deposition onto cotton thread by Tan and Ge, 1996) is to impregnate the substrate with a solution of the oxidant and then expose this to the ICP monomer in vapour form. A continuous vapour phase deposition for coating wool yarns with polypyrrole was described by Najar *et al.* (2007). Two 40 tex yarns with different twist levels were treated; the yarn with lower twist had better conductivity due to better penetration of the reagents leading to greater ICP deposition. Yarn tensile properties were not diminished by polypyrrole coating. The technique was extended in work described in a subsequent paper (Kaynak *et al.*, 2008), to include cotton and nylon yarns, but the difference in yarn structure (ring-spun wool, rotor-spun cotton, continuous filament nylon) was probably too great to allow any clear conclusion about the influence of fibre type to be reached.

A two-stage process can also be used where the ICP monomer is dissolved in a suitable solvent before impregnation into the substrate, before or after impregnation with the oxidant. Reagent concentrations are usually much higher than those used in the Milliken technique. Richardson *et al.* (2004) reported the deposition of polypyrrole by two-stage impregnation (using iron(III) chloride as the oxidant) onto wool fibres. Conductivity varied widely between fibres, being of the order of 0.1 mS/cm. Johnston *et al.* (2006) used scanning electron microscopy (SEM) to examine the coating morphology of polypyrrole deposited by two-stage impregnation on cellulose and wool fibres. They found (on both substrates) that the coating appeared to be smooth at low magnifications, but higher magnification revealed that it was made up of 'fused spheres', approximately 200 nm in diameter.

A number of researchers have used a high reagent concentration impregnation system for the preparation of textile–polyaniline composites. In some situations this has been as two separate impregnations; more commonly it has used successive additions of precursor and oxidant to a single vessel (or 'bath') in which the specimen is placed (this latter case is still closer in principle to the two-stage impregnation technique than to the one-step Milliken method). An early example of the technique used wool and glass-fibre fabrics (Trivedi and Dhawan, 1992) immersed in a solution of aniline and an acid (either an aromatic sulphonic acid or sulphuric acid), which was then chilled, and ammonium peroxydisulphate added as an oxidant. After 2 hours the specimens were removed and rinsed in a solution of the same acid used earlier. The composites prepared in this way had resistivity in the range 400–1000 Ω/cm.

Varesano *et al.* (2005) also used a single bath with high reagent concentrations to deposit an ICP onto wool; in this case polypyrrole. Of particular interest is that the coated fibres were subsequently spun into yarns, then knitted into fabric. SEM assessment showed the coating appeared to be basically smooth with a large number of granular deposits present, which were largely removed by the mechanical processes that the fibre was subjected to as it was processed into a fabric. This suggests that the granular deposits were not well bound to the fibres, so were probably polypyrrole that formed in the solution and then accumulated on the fibre surface. The frictional forces of fibres rubbing against each other and against surfaces of the processing machinery then dislodged the particles, with the result that the processed fibre surface was relatively clean. The coating was obviously quite thin, as the familiar overlapping scale structure of the wool fibre surface was not obscured by the coating.

In the same paper (Varesano *et al.*, 2005) an extensive study of durability properties was reported. They assessed the effect on electrical properties (of wool–polypyrrole fabrics) of laundering with either perchloroethane (dry cleaning) or a typical domestic washing system; exposure to light and temperature (45 °C); and abrasion. They used the relative resistance (the resistance after a designated test, relative to its original resistance) as a simple way of tracking the change in electrical properties. There was only a small change in resistance with dry cleaning (relative resistance after five cycles was 1.2), but a very large change with domestic laundering (relative resistance increased exponentially, and was about 45 after five cycles). De-doping was identified as the likely cause of this drastic deterioration in electrical properties. Exposure to light caused a linear increase in relative resistance, to about 5.5 after 160 hours. Heat exposure increased the relative resistance asymptotically until approximately 80 hours, after which there was no further increase. At this point the relative resistance was about 3.3. Oxidation of the polypyrrole coating was identified as the likely cause of both these increases. A third mechanism for the loss of conductivity is physical erosion of the coating; after 40 000 abrasion cycles (using a Martindale-type tester) the relative resistance had increased to about 8.5.

Pyrrole variants have been one focus of recent research by a team at Deakin University. In two papers, Foitzik *et al.* (2005, 2006a) described the preparation of soluble poly(3-alkylpyrrole) and its application to wool substrates. Initially the work focused on preparing and comparing a range of poly(3-alkylpyrrole) variants, with differing carbon chain lengths (Foitzik *et al.*, 2005). As the carbon chain length increased (across the range 0 to 22), the polymer's solubility in chloroform increased, but its conductivity dropped. The optimum carbon chain length was between 10 and 14, giving a balance between solubility and conductivity. The conductivity of all variants up to a carbon chain length of 10 was comparable to that of unmodified polypyrrole.

The poly(3-alkylpyrrole) with a carbon chain length of 10 was employed in their subsequent work (Foitzik *et al.*, 2006a), which evaluated various techniques for depositing poly(3-decanylpyrrole) onto fabrics. Three solution-based techniques (spray, brush and dip coating) were compared with *in situ* polymerisation of 3-decanylpyrrole in solution polymerisation and by a mist/vapour technique. The initial resistance of the *in situ* polymerised specimens was lower, but all showed increased resistance after abrasion, as the ICP coating was worn away. The main advantage of the deposition of the ICP from solution is that the substrate is not exposed to the potentially damaging oxidising agent that is required for *in situ* polymerisation. However, the electrical properties of solution deposited poly(3-decanylpyrrole) were inferior to *in situ* polymerised.

Other pyrrole variants synthesised and polymerised onto wool by the Deakin group team include 3-iso-butyl- and 3-iso-pentyl-pyrrole (Foitzik *et al.*, 2006b) and 3-menthyl carboxylate pyrroles (Foitzik *et al.*, 2007a). Of particular interest in the latter was the creation of a monomer that polymerised onto wool to produce a dark green ICP (as opposed to the typical polypyrrole grey-to-black coloration). Crosslinked pyrrole monomers were also synthesised: α,ω-bis(3-pyrrolyl)alkanes (Foitzik *et al.*, 2007b). These are effectively two pyrrole molecules linked by a carbon chain, the length of which was varied. Monomers of this type may result in crosslinked ICPs with improved mechanical properties. The optimum chain length (taking into account conductivity, yield and stability) was eight, and ICP deposition was achieved, although the polymerisation was more difficult.

Deposition of ICPs other than polypyrrole and polyaniline is also possible, although relatively uncommonly reported for wool. Thiophene variants have been polymerised *in situ* on a range of fibre types including wool (Collie, 2007). Variants included reddish-brown 2,2,′-bithiophene and blue 3,4-ethylendioxythiophene (EDOT). An interesting aspect of this work was that the scale structure of wool is not obscured by the ICP coating. While there were some scale fragments and damage apparent, fabric tests showed no reduction in strength or flexibility. Durability assessment in the same study revealed much better durability of conductivity for ICPs based on thiophenes than for polypyrrole when polymerised by the same technique.

A group of Indian researchers experimented with a simple electrochemical

ICP deposition approach: Bhadani *et al.* (1993) simply wrapped wool, cotton and silk yarns around the anode of an electrochemical cell, with pyrrole and thiophene used as ICP precursors, and tetrabutylammonium hexachloroantimonate in 1,2-dichloroethane as the electrolyte. The ICP formed at the anode, but largely accumulated on the yarn. Resistivity with pyrrole was in the range 5–25 kΩ cm, while for thiophene it was 7–65 MΩ cm. The research was repeated with aniline as the precursor and *p*-toluenesulphonic acid as the electrolyte (Bhadani *et al.*, 1996). The effect of electrolysis time was reported; the amount of ICP deposited increased fairly steadily with time, while the resistivity decreased. Resistivity after five hours was 1.7, 1.6 and 5.2 kΩ cm for cotton, silk and wool respectively, while the increase in weight was 83%, 224% and 155% respectively. Similar experiments with pyrrole in place of aniline demonstrated similar trends, but with higher overall conductivity (Bhadani *et al.*, 1997a,b).

13.4.2 Electronic textiles incorporating wool

Electronic textiles (or wearable electronics) are the fields of intelligent textiles where most commercial outcomes have been achieved. Initially these have been in the form of textile-based control panels (such as those from Softswitch and Eleksen). Very few of these have been in wool products, primarily because the initial products have been restricted to high-performance outerwear garments, such as skiing or snowboarding jackets. The first product of this type was the Burton Analog Clone jacket, with a Softswitch keypad integrated into the sleeve, controlling a Sony MiniDisc portable audio player (Buechner *et al.*, 2002). In keeping with current trends the device now most commonly controlled by these systems is the Apple iPod, including in the iPod-enabled suit released in 2007 by Marks and Spencer in the United Kingdom (Eleksen, 2007). This system, utilising an Eleksen keypad, and manufactured for Marks and Spencer by Bagir Ltd, integrated a five-button keypad inside the lapel of the suit jacket. The jacket reportedly contained wool and elastane (Finch, 2007), and the keypad allowed control of the primary iPod functions without needing to remove it from its internal storage pocket.

The other electronic textile application where wool has been utilised is in electrically heated textiles. In a development supported by Australian Wool Innovation, New Zealand-based Canesis Network Ltd (now part of AgResearch Ltd) developed technologies for electrically heated wool socks (AWI, 2004a; Jones *et al.*, 2007) and bedding (AWI, 2004b). In both cases the resistive heating element was created from a blend of metallic fibre with Merino wool to give an appropriate resistance level for heat emission with an applied current. Higher conductivity non-wool yarns were used to provide the connection to a power supply, which in the sock situation was to be based on lightweight, high-capacity rechargeable batteries typically used in mobile phones. The heating 'element' was confined to a small region near the end of the foot to maximise warming effect and battery lifetime.

New Zealand company Weratech have also developed a textile-based electrical heating system. The Weratech system is also knitted, and the fabrics are reportedly soft and flexible, and suitable for integration into a range of garments, including gloves, footwear and coats (Tait, 2007). According to their website (www.weratech.com/Technology.html), the Weratech yarns use conductive fibres blended with fine Merino wool, or other thermally stable fibres, such as fire-retardant polyester or aramids. The fabrics rapidly heat up to as high as 180 °C (although much lower temperatures would be appropriate for apparel applications).

It must be noted that in neither of the application areas described above does the inherent properties of wool have any particular role in making the system an intelligent textile. Rather it is simply the substrate that the non-fibre specific technology has been integrated into. They do, however, illustrate that intelligent textiles are starting to permeate all areas of textiles, including the very traditional wool market sectors of socks, knitwear and worsted suits. In the case of the heated textile technologies, the concept of a heated textile enhances wool's image of being a warm, comfortable fibre to wear, while in the unlikely event of a malfunction in the system, any overheating or short-circuiting will not cause melting or burning of the textile thanks to wool's thermal stability and inherent flame-retardant behaviour.

13.4.3 Other applications

Specific applications for wool in the first two categories of intelligent textiles defined in Section 13.2 appear to be less common than in those areas discussed above. In a 2005 article, Australian Wool Innovation was reported to be examining applications for wool as a carrier for 'smart chemicals', with the potential for wool-based 'smart fibre dressings, for detecting infectious wounds, to alert medical staff when a wound becomes infected' (Anon, 2005). Such a system might make use of the sensing capability of ICPs (discussed in Section 13.4) to detect the biological agents produced in the wound environment.

The only report of wool substrates being used in actuation systems appears to be in work describing the binding of polyethylene glycols (PEGs) to a range of fibrous substrates (Vigo, 1997). Wool is mentioned in passing in this work, but no experimental results are provided. PEGs have a temperature buffering effect thanks to phase change behaviour, and a shape-change behaviour due to swelling when water is absorbed and desorbed. These behaviours are imparted to the textile substrate to which the PEG is bound.

13.5 Opportunities for the use of wool in intelligent textiles

It should be apparent from the above that wool, while not yet widely exploited in intelligent textile systems, is not fundamentally unsuited to them. There are many intelligent textile technologies where fibre type is irrelevant, and the textile is merely the substrate on which the intelligent system is constructed. So far wool has earned its place in intelligent garments because of its traditional properties, its fundamental comfort, drape and appearance, rather than through any 'smart' characteristics imparted by the work of researchers.

There are some potential drawbacks to wool's use in intelligent textile systems. For some applications, textiles need to be easily coated, and laminated to layers of other materials. Wool textiles tend to have relatively irregular surfaces, while smooth surfaces for the application of other technologies are easier to achieve with synthetic continuous filament-based substrates. Also a requirement in a multi-component smart textile system is lightness of weight. Wool is limited in the fineness of its fibres and in the minimum number of fibres required to form a yarn, and thus in the lightness of fabrics that can be produced from it. These are just minor drawbacks, so perhaps the only real disadvantage for wool in high-performance applications is the fact that functional components cannot be easily incorporated into the fibre during production as can be done with extruded synthetics. The control over the synthetic fibre's physical shape and polymer composition is undoubtedly an advantage. With wool we must start with the chemical and physical properties that nature has provided, at least until biotechnology changes this state of affairs.

Holcombe and Wallace (2002) point out that opportunities for wool in intelligent textiles exist where a new technology takes advantage of its unique characteristics. Its major strength is its rich chemical functionality, primarily relating to its constituent proteins, which provide amine, carboxylic, amide, hydroxyl and thiol/disulphide functionality. Synthetic fibres are either mono-functional or aliphatic, limiting the extent to which they can be chemically modified. The development of intelligent applications may involve applying chemical modifications to the fibres and the range of opportunities provided by wool is unparalleled in the textile fibre arena. It should also be noted that the intracellular spaces within the fibres present opportunities for hosting a range of molecules giving smart functionality.

13.6 References

ANON (1999), Smart thinking by the French. *Wool Record*, **158**, 18.

ANON (2005), Opportunities from the technical properties of wool – TTNA Members briefing. *Australasian Textiles and Fashion*, July–August, 31–32.

AWI (2004a), Hi-tech socks no mean feet. *Beyond the Bale*, Issue 9, 2004.

AWI (2004b), Safer and warmer in bed – we have the technology. *Beyond the Bale*, Issue 9, 2004.

BHADANI, S N, SEN GUPTA, S K and GUPTA, M K (1993), Electrically conducting natural fibers. *Indian Journal of Fiber and Textile Research*, **18**, 46–47.

BHADANI, S N, SEN GUPTA, S K, SAHU, G S and KUMARI, M (1996), Electrochemical preparation of conducting fibers from natural fibers. *Journal of Polymer Materials*, **13**, 61–67.

BHADANI, S N, KUMARI, M, SEN GUPTA, S K and SAHU, G C (1997a), Preparation of conducting fibers via the electrochemical polymerization of pyrrole. *Journal of Applied Polymer Science*, **64**, 1073–1077.

BHADANI, S N, KUMARI, M, SAHU, G S and SEN GUPTA, S K (1997), Electrochemical formation of conducting polyester fibers. *Journal of Polymer Materials*, December, 291–295.

BUECHNER, M M, GROSSMAN, L and HAMILTON, A (2002), Coolest inventions 2002. *Time*, 18 November, 37–59.

COLLIE, S R (2007), Development of Durable Textile-Conductive Polymer Composites, PhD thesis, Massey University.

DE ROSSI, D, DELLA SANTA, A and MAZZOLDI A (1999), Dressware: wearable hardware. *Materials Science and Engineering*, **C7**, 31–35. DOI: 10.1016/S0928-4931(98)00069-1

ELEKSEN PLC, (2007), Marks & Spencer to launch iPod suits for Fathers' Day. Media release, issued 7 June 2007. Accessed at: http://www.eleksen.com/?page=news/index.asp&newsID=73. Retrieved 7 March 2008.

FINCH, J (2007), White stripes not included: M&S's iPod suit. *Guardian*, 8 June 34.

FOITZIK, R C, KAYNAK, A, BECKMANN, J and PFEFFER, F M (2005), Soluble poly-3-alkylpyrrole polymers on films and fabrics. *Synthetic Metals*, **155**, 185–190. DOI: 10.1016/j.synthmet.2005.08.001.

FOITZIK, R C, KAYNAK, A and PFEFFER, F M (2006a), Application of soluble poly(3-alkylpyrrole) polymers on textiles. *Synthetic Metals*, **156**, 637–642. DOI: 10.1016/j.synthmet.2006.03.003.

FOITZIK, R C, KAYNAK, A, PFEFFER, F M and BECKMAN, J (2006b), Synthesis, polymerization and wool coating studies of 3-iso-butylpyrrole and 3-iso-pentylpyrrole. *Synthetic Metals*, **156**, 1333–1340. DOI: 10.1016/j.synthmet.2006.10.003.

FOITZIK, R C, BOWEN, E K, TAYLOR, A M, PFEFFER, F M and KAYNAK, A (2007a), Synthesis and polymerization studies of 3-(+) and (–)-menthyl carboxylate pyrroles. *Synthetic Metals*, **157**, 924–929. DOI: 10.1016/j.synthmet.2007.09.003.

FOITZIK, R C, KAYNAK, A and PFEFFER, F M (2007b), Conductive poly(α,ω-bis(3-pyrrolyl)alkanes)-coated wool fabrics. *Synthetic Metals*, **157**, 534–539. DOI: 10.1016/j.synthmet.2007.05.014.

GARG, S, HURREN, C and KAYNAK, A (2007), Improvement of adhesion of conductive polypyrrole coating on wool and polyester fabrics using atmospheric plasma treatment. *Synthetic Metals*, **157**, 41–47. DOI: 10.1016/j.synthmet.2006.12.004.

GREGORY, R V, KIMBRELL, W C and KUHN, H H (1989), Conductive textiles. *Synthetic Metals*, **28**, 823–835.

GUAN, B O, TAM, H Y, TAO, X M and DONG, X Y (2000), Simultaneous strain and temperature measurement using a superstructure finer Bragg grating. *IEEE Photonics Technology Letters*, **12**(6), 1–3. DOI: 10.1109/68.849081.

HIRASE, R, SHIKATA, T and SHIRAI, M (2004), Selective formation of polyaniline by chemical polymerization, using potassium iodate. *Synthetic Metals*, **146**, 73–77. DOI: 10.1016/j.synthmet.2004.06.009.

HOLCOMBE, B and WALLACE, G (2002), The brave new world of wearable intelligence. *Wool Technology and Sheep Breeding*, **50**(3), 312–318.

JOHNSON, N A G, WOOD, E J, INGHAM, P E, MCNEIL, S J and MCFARLANE, I D (2003), Wool as a technical fibre. *Journal of the Textile Institute*, **94**, Part 3, 26–41.

JOHNSTON, J H, KELLY, F M, MORAES, J, BORRMAN, T and FLYNN, D (2006), Conducting polymer composites with cellulose and protein fibres. *Current Applied Physics*, **6**, 587–590. DOI: 10.1016/j.cap.2005.11.067.

JONES, D, TASKER, A and COLLIE, S (2007). Patent: US2007278201 – Heated Garment System.

KAYNAK, A, WANG, L, HURREN, C and WANG, X (2002), Characterization of Conductive Polypyrrole Coated Wool Yarns. *Fibers and Polymers*, 3(1) 24-30.

KAYNAK, A, NADAR, S S and FOITZIK, R C (2008), Conducting nylon, cotton and wool yarns by continuous vapor polymerization of pyrrole. *Synthetic Metals*, **158** 1–5. DOI: 10.1016/j.synthmet.2007.10.016.

KUHN, H H and CHILD, A D (1997), Electrically conducting textiles. In: *Handbook of Conducting Polymers*, Skotheim, TA, Elsenbaumer, RL and Reynolds, JR (eds), Marcel Dekker, Inc, 993–1013.

KUHN, H H and KIMBRELL, W C (1989), Patent US4,803,096 – Electrically Conductive Textile Materials and Method for Making Same.

MARKS, P (2001), Shirt rolls up its own sleeves. *New Scientist*, 25 July.

MATTILLA H R (ed.) (2006), *Intelligent Textiles and Clothing*, CRC.

MUNRO, B J, STEELE, J R, CAMPBELL, T E and WALLACE, G G (2004), Wearable textile biofeedback systems. In: *Wearable eHealth Systems for Personalised Health Management*, Lymberis, A. and De Rossi, D. (eds), IOS Press, 271–277.

NAJAR, S S, KAYNAK, A and FOITZIK, R C (2007), Conductive wool yarns by continuous vapour phase polymerization of pyrrole. *Synthetic Metals*, **157** 1–4. DOI: 10.1016/j.synthmet.2006.11.003.

OTERO, T F and SANSIÑENA, J M (1998), Soft and wet conducting polymers for artificial muscles. *Advanced Materials*, **10**(6), 491–494.

RICHARDSON, M J, JOHNSTON, J H, BORRMANN, T and WISEMAN, N (2004), Conducting polymer–wood composites retain properties of both materials. In: *Proceedings of International Conference on the Science and Technology of Synthetic Metals 2004*, Wollongong, June/July.

TAIT, N (2007), Avantex/Techtextil offers extensive view of future for workwear. *Technical Textiles International*, September, 31–35.

TAN, S N and GE, H (1996), Investigation into vapour-phase formation of polypyrrole. *Polymer*, **37**(6), 965–968. DOI: 10.1016/0032-3861(96)87278-8

TAO, X M (2000), Integration of fibre-optic sensors in smart textile composites – design and fabrication. *Journal of the Textile Institute*, **91**(3), 448–459.

TAO, X M, TANG, L Q, DU, W C and CHOY, C L (2000), Internal strain measurement by fibre Bragg grating sensors in textile composites. *Journal of Composite Science and Technology*, **60**(5), 657–669. DOI: 10.1016/S0266-3538(99)00163-3.

TRIVEDI, D C and DHAWAN, S K (1992), Grafting of electrically conducting polyaniline on insulating surfaces. *Journal of Materials Chemistry*, **2**(10), 1091–1096.

VARESANO, A, DALL'ACQUA, L and TONIN, C (2005), A study in the electrical conductivity decay of polypyrrole coated wool textiles. *Polymer Degradation and Stability*, **89**, 125–132. DOI: 10.1016/j.polymdegradstab.2005.01.008.

VIGO, T L (1997), Intelligent fibrous substrates with thermal and dimensional memories. *Polymers for Advanced Technologies*, **8**, 281–288.

VIGO, T L (1999), Intelligent fibrous materials. *Journal of the Textile Institute*, **90**, Part 3, 1–13.

WANG, L, LIN, T, WANG, X and KAYNAK, A (2005), Frictional and tensile properties of conducting polymer coated wool and alpaca fibers. *Fibers and Polymers*, **6**(3), 259–262.

WANG, J, KAYNAK, A, WANG, L and LIU, X (2006), Thermal conductivity studies on wool fabrics with conductive coatings. *Journal of the Textile Institute*, **97**(3), 265–269. DOI: 10.1533/joti.2005.0298.

WILKINSON, T (2004), Electro interest rekindled. *World Sports Activewear*, March/April, 30–32.

14

Application of wool keratins ranging from industrial materials to medical devices

R. KELLY, RJK Developments Limited, New Zealand

Abstract: Wool keratin protein is well suited to the development of advanced applications in areas well outside conventional textile and fibre products. Current commercial and legislative trends encouraging the development of biopolymer alternatives to oil-derived synthetics have created additional interest in the development of keratin biopolymer products. Extending historic activities in the development of extruded keratin fibres, recent applications range from medical devices for the treatment of wounds or hard tissue injuries to nutritional applications capitalising on the unique protein composition of wool keratin. Future developments promise to capitalise on the most recent advances in medical science and biopolymer technology.

Key words: wool, keratin, biomaterials, biopolymers.

14.1 Introduction

When considering the application of wool beyond textiles and fibres it is important to first consider the materials from which wool is made up. Once the macroscopic structure as a fibre is not fundamental in defining the application, then it comes down to the base materials from which the fibre is formed to define the relevant applications.

Both macroscopic and microscopic characteristics of the wool fibre have been extensively reviewed (Gillespie, 1983; Marshall *et al.*, 1991). By far the most important material in this consideration is keratin protein, defined by the US National Library of Medicine as 'A class of fibrous proteins or scleroproteins important both as structural proteins and as keys to the study of protein conformation. The family represents the principal constituent of epidermis, hair, nails, horny tissues, and the organic matrix of tooth enamel.' It is clear from this definition that keratin proteins, in various forms, are widely present in biological systems, in hair, skin, nails and other tissue. This widespread occurrence in nature, as a tough, physical material from which to build structures, is of significance when considering advanced applications of wool. The tenacity of wool, the resilience of our nails, the barrier characteristics of our skin, tell us much about the features that we can look for in advanced applications of wool-derived keratins. While keratin may account for up to 95% of the dry matter of a wool fibre, the other components are also of note for their interesting characteristics. Considered as an ordered collection of elongated cells, rich in multiple types of keratin proteins, the wool fibre structure is brought together by the cell membrane complex, the material that defines the cell boundaries and a key element in keeping the cells together in a strong and ordered fibrous alignment. Lipids found within this region of the fibre resemble strongly those in the skin, and as such they find ready application in topical skin products to supplement skin lipids and improve a range of important cosmetic skin parameters (de Pera *et al.*, 2000).

14.2 Overview of the structure and characteristics of keratin protein

As a protein, keratin is characterised by a higher cystine content than other proteins, typically from 4 to 28 residues per 100 in keratin when compared with 0 to 2 residues per 100 for other proteins. This sulphur-rich amino acid is associated with crosslinking, and so the building of tough and strong materials. This crosslinking is of clear importance in the physical role that keratin plays in nature as fibres and horny tissue to protecting DNA on a cellular level (Hearle, 2000; Gilbert *et al.*, 2001). In addition, cystine has some known biological function in providing oxidative protection to cells and is associated with treatment of a range of disease states from chronic bronchitis to HIV and cardiovascular disease (Parcell, 2002). An ordered and fibrous character is also of significance when considering defining characteristics of wool keratins due to the tendency of this characteristic to drive keratins to organise and form structured physical materials (Hermann *et al.*, 2000).

Given the widespread occurrence of keratin in nature it is important to consider the features of wool keratins as compared with other sources. Biological characteristics and practical considerations are both of importance to evaluate future applications of wool. Horns and hooves from slaughterhouse waste have traditionally been used as a source of keratin for a variety of consumer and industrial uses. The high degree of additional animal tissue, the physical form and consumer perceptions of the form have restricted use of these materials to harsh processing methods and industrial applications of low value. The focus in this chapter is the consideration of advanced applications of wool beyond textiles and fibres.

Avian keratin sources, in particular feather keratins, also have this practical limitation of source and consumer perception. In addition, fundamental differences in protein structure and cystine content limit their functionality and use for the development of robust materials. Feather keratins have an inherently lower cystine content than wool and other hard alpha keratin sources. In addition, they have a protein structure that comprises proteins of much lower molecular weight and a lower degree of fibrous organisation. The absence of an ordered fibrous structure is a substantial disadvantage when reconstituting keratins into advanced materials and applications. However, some applications of feather proteins as bioplastics have been the focus of academic research (Schrooyen and Dijkstra, 1998).

14.3 Extraction of keratin from wool for new product development

Advanced applications of wool beyond the fibre utilise the keratin protein from which the fibre is constructed. The first step, then, in development of these materials is isolation of that protein in a versatile form that can be further manipulated into the desired application. Wool, by its nature, is a tough insoluble material evolved by nature to be robust and resistant to degradation. Its conversion to a state where it can be widely manipulated is not straightforward and this has been a key reason that keratin protein has not been commercially exploited as a material in the way that other biopolymers have been, such as collagens, alginates, celluloses and starches. The simplest way to isolate a soluble from of keratin from the source fibre is to apply conventional methods of hydrolysis, using acid, alkali or enzymatic digestion of the peptide bonds to create a mixture of peptides and amino acids that have retained none of the secondary or tertiary structure of the core protein. In most cases these processes also destroy many of the core amino acids. Although this harsh processing destroys many of the unique characteristics of the protein, it has provided a route for several developments, such as the historic industrial uses explored below.

In order to progress beyond these limited applications of keratin-derived peptides, hydrolysates and amino acids, a series of processes for the isolation of keratin proteins from wool that maintain core features of the protein and its structure have been developed (Smith *et al.*, 1997; Kelly *et al.*, 2002). Most of the processes do not hydrolyse the peptide linkages and this allows the keratin proteins that are produced to retain a form and function similar to the native keratins. By maintaining the structural and chemical complexity of the native proteins, by having substantial fibrous proteins that naturally form robust and ordered materials, and with chemically active amino acid components such as cystine present in abundance, then advanced applications in a range of areas can be developed. The development of the technology to achieve these forms of keratin is key to the creation of advanced applications of wool.

14.4 Industrial trends relating to sustainable polymers

Advanced applications of wool keratins can be considered in a similar manner to developments undertaken elsewhere into biopolymer- or biologic-derived materials. The factors influencing the developments come primarily from the commercial arena, which is in turn driven by opportunities created by innovation and importantly environmental and legislative factors. Global interest in sustainably produced materials, and full life-cycle analysis of products has driven interest and investment in biological materials. Biopolymers, as alternatives to synthetic plastics, have been the promise of white biotechnology for several decades, with Biopol™ from Monsanto an early example of a promising technology that has not yet had the appropriate alignment of technological, environmental and legislative factors to drive enough commercial interest to lead to success. NatureWorks™ from Cargill Dow, used from packaging to fibre applications, is an example of a biopolymer material for which there is rapidly increasing consumer interest. The development was made possible by advanced biotechnology, but its continued development is driven by a commercial need to find biological alternatives to oil-derived synthetic polymers because of legislative and economic pressure on oil. While price signals are important, there is increasing pressure on landfill which is being progressively occupied by synthetics derived from oil. These same drivers create strong opportunities for other natural biopolymers with an established biological function as a material. Advanced applications from wool keratin are emerging to capture this opportunity.

14.5 Advanced applications of wool keratins

If wool is considered as a source of a durable biopolymer material for which application development to parallel synthetic polymers is possible, at least in principle, then the possibilities for advanced applications of wool are boundless. Although the practical reality has thus far fallen short of the enormous potential of the material, a wide range of historic and current advanced uses of wool keratin have been developed.

Early applications included reconstituted fibres, developed in the 1940s (Wormell and Happey, 1949), in which fibrous keratin proteins were isolated from a wool source and spun into smooth, fine fibres for textile applications. This utilised emerging expertise in wet spinning gained in the rayon industry; however, the reconstituted fibres fell short of meeting the physical requirements of the target textile applications, and presumably failed to develop for this reason. While ambitious in its intentions, this work involved first creating a soluble form of the wool keratin prior to reforming it into a solid of the desired shape. Repeating that first step, of making the tough insoluble wool fibre into a soluble keratin protein (or in most cases a peptide mixture) as a result of hydrolysis, has proved an essential first, and in many cases only, step in the development of other industrial and consumer uses of wool keratin. Keratin peptides, made soluble through the harsh processes of hydrolysis, have formed a core part of many familiar products including fire fighting foams, as protein foam concentrates, as moisture retardants in building plaster, and most significantly as hair and skin care products. While there is some variation in preferences in different regions of the world, the use of keratins for enhancing the cosmetic properties of the hair and skin is extensive. Japan, South America and parts of Europe have particularly favoured the use of keratins in these applications. Given that the substrate to be enhanced, be it hair or skin, is either almost entirely keratin or at least keratin-rich, the natural affinity and active characteristics of the keratin protein, to strengthen hair, to smooth and moisturise skin, far outperform the plant-derived proteins often used to achieve the same effects. Familiar brands on supermarket and pharmacy shelves contain keratin for this purpose, with at least a portion of this derived from wool. Some market analysts project that keratin protein use in cosmetic applications will be as high as US$69 million by 2011 (Freedonia, 2000).

Beyond cosmetic enhancement of skin and hair is biological enhancement of skin and other tissues. The use of keratin materials for medical applications, to enhance the repair of tissue both within and on the surface of the body, has been the subject of both research and commercial endeavour. Applications examined have included hard tissue repair, in which the robust character of the keratins is used to integrate with damaged bone to provide a better healing outcome (Peplow *et al.*, 2003; Dias *et al.*, 2006; Tachibana *et al.*, 2006). Although keratin is not intrinsic to bone, its use for this purpose is well suited. If bone is considered as a fibrous protein–mineral composite (largely collagen and the calcium mineral hydroxyapatite), then use of a fibrous protein–mineral composite, in which the fibrous protein is derived from an alternative source, for bone augmentation, is understandable. Further, bone tissue is remodelled in the body on a relatively short time frame. Consequently a temporary construct familiar in structure and composition to the body that fills a temporary role to facilitate the remodelling process is highly desirable. The ability to prepare controlled porosity constructs using keratin protein and calcium minerals has been documented (Peplow *et al.*, 2003). The importance of composition and structure for biomimetic stimuluation of bone growth response has also been detailed (Ripamonti, 2004) and a powerful combination of structure and biology is possible as a result of using versatile keratin biopolymers in bone augmentation.

In addition, the repair and enhancement of soft tissue has provided a substantial focus of research and development. Keratin has been considered as the biological material to replace areas of soft tissue, such as intervertebrate discs in the spine, or to fill spaces for tissue enhancement, such as breast or other soft tissue augmentation (Smith *et al.*, 2001). Further to tissue repair is the use of keratin as a construct in the area of regenerative medicine. There is rapidly expanding interest in cell therapies, including stem cells, for the development of tissue constructs, and this is driving a need for biologically acceptable matrices to provide a framework on which to grow regenerated tissue. Keratin has been explored for this purpose and found to possess suitable characteristics (Furth *et al.*, 2007). In addition to using the material as a flexible and versatile biocompatible biopolymer to create the appropriate shaped constructs, the favourable biological environment would appear to assist in the growth of tissue, and potential in nerve regeneration has been identified (Sierpinski *et al.*, 2008).

Yet another field of medical development where advanced applications of wool keratin have been developed is wound care. Wound dressings derived from wool keratin have been created and applied successfully to a range of wound types (Van Dyke *et al.*, 2000), from acute to chronic, with positive healing outcomes reported (Kelly *et al.*, 2007). Keratin protein is a fundamental part of skin structure, and the role of different proteins from the skin keratin family in the healing wound has been the subject of extensive research (Kim *et al.*, 2006). This is very relevant to the consideration of wool-derived keratins owing to the high degree of homology between keratin proteins across species and protein types, with keratin intermediate filament proteins in particular having a substantial similarity in structure, and so potentially function, when compared with skin keratins.

By considering wool as a convenient assembly of proteins, with an amino acid balance rich in the sulphur-containing amino acid cystine, developers of advanced applications of wool and keratin proteins have turned their attentions to ingestible uses of the material. A variety of ingredients have emerged on the market finding use in oral supplements and liquid beverages in markets around the world (www.swansonvitamins.com). The benefits of these products to date are claimed in the area of joint health, targeting the high-growth ageing population and the popularity of joint health products such as glucosamine and chondroitin. The high sulphur content of the proteins, and the ability to protect the joint from oxidative stress, are important factors in the performance of the products. The ingestible use of wool keratins is not new, with recorded use dating back almost 100 years. Keratin was used as an enteric coating of pharmaceutical tablets, to protect a pharmaceutical from degradation in the stomach, and is recorded in this application in the British Pharmaceutical Codex of 1911 (http://www.henriettesherbal.com/eclectic/bpc1911/keratinum.html). However, the entry of keratin is missing from subsequent pharmacopea and the use of wool keratin for this purpose has not become established.

14.6 Future trends

The future advanced uses of wool keratins lie in two distinct areas. The first is medical technology. Developments will be driven by science and improved healing outcomes. Recent academic research into the previously unknown biological role of keratins in the body, identified through the work of Coulombe, Omary and others (Kim *et al.*, 2006; Omary and Ku, 2006), has determined that keratin in the body plays more than a physical role, but is essential in the transportation of signalling peptides crucial for the growth and differentiation of cells. Coupled with the developing technology for the creation of medical devices from keratin, substantial advanced uses of wool will develop in the medical area. This is already commencing in the areas of wound care and orthopaedics and the natural evolution into other areas of medical care will follow.

The second area of advanced use is in industrial technology, and will be driven by legislation and commercial necessity. With the price of oil on a sustained sharp increase over recent years, and a flow-on effect to all oil-derived materials, coupled with harsher environmental legislation being established globally relating to both carbon neutrality and source and waste management, the focus on biopolymers will naturally increase sharply. Keratin is one of nature's most successful materials; it was evolved to be versatile, tough and protective, and as a natural protein, there are wide opportunities in a range of industrial areas. Robust materials, such as plastics, to the less obvious industrial speciality chemicals that are also derived from a synthetic backbone, are all a likely development for advanced uses of wool. None of these future applications requires the fibre to maintain its fibrous form, but all utilise the complex fundamental structure that the wool fibre is based on.

14.7 Sources of further information and advice

Further information on the structural components of wool can be gained from Marshall *et al.* (1991) and Gillespie (1983). The biological role of keratin in the skin and other tissues is the subject a variety of publications including Kim *et al.* (2006) and Omary and Ku (2006). The use of keratin in applications beyond the conventional fibre is the subject of numerous patent applications (see, for example, Kelly *et al.*, 2002, and Smith *et al.*, 1997). Biopolymer trends and importance of both technology and legislation is well documented by the Greentech organisation (www.greentech.org).

14.8 References

DIAS, G. J., MAHONEY, P. M., SMITH, R. A., ALI, A., KELLY, R. J. (2006) Biological response of a resorbable keratin-based biopolymer as a bone regeneration scaffold *Australasian*

Society of Biomaterials and Australasian Polymer Society Conference Proceedings, Rotorua, New Zealand.

Freedonia (2000) *Cosmetic and Toiletry Chemicals to 2006*, Freedonia Market Research report 1540.

FURTH, M. E., ATALA, A., VAN DYKE, M. E. (2007) Smart biomaterials design for tissue engineering and regenerative medicine. *Biomaterials*. Dec; **28**(34): 5068–73.

GILBERT, S., LORANGER, A., DAIGLE, N., MARCEAU, N. (2001) Simple epithelium keratins 8 and 18 provide resistance to Fas-mediated apoptosis. The protection occurs through a receptor-targeting modulation. *J Cell Biol*. 20 Aug; **154**(4): 763–73.

GILLESPIE, J. M. (1983) The structural proteins of hair: isolation, characterisation and regulation of biosynthesis. In *Biochemistry and Physiology of the Skin*, vol 1, ed L A Goldsmith, Oxford University, Press, 475.

HEARLE, J. W. (2000) A critical review of the structural mechanics of wool and hair fibres. *Int J Biol Macromol*. 12 April; **27**(2): 123–38.

HERRMANN, H., STRELKOV, S. V., FEJA, B., ROGERS, K. R., BRETTEL, M., LUSTIG, A., HÄNER, M., PARRY, D. A., STEINERT, P. M., BURKHARD, P., AEBI, U. (2000) The intermediate filament protein consensus motif of helix 2B: its atomic structure and contribution to assembly. *J Mol Biol*. 19 May; **298**(5): 817–32.

KELLY, R. J., WORTH, G. H., RODDICK-LANZILOTTA, A. D., RANKIN, D. A., ELLIS, G. D., MESMAN, P. J. R., SUMMERS, C. G., SINGLETON, D. J. (2002) Production of soluble keratin derivativs. US Patent 7,148,327.

KELLY, R., SIGURJONSSON, F., SMITH, R. A., ALI, A., THAN, M., MCPHERSON, R., HAMMOND, C. (2007) Acceptability and efficacy of keratin-based dressings in the chronic wound, *European Wound Management Association Conference Proceedings*, Glasgow, UK.

KIM, S., WONG, P., COULOMBE P. A. (2006) A keratin cytoskeletal protein regulates protein synthesis and epithelial cell growth. *Nature*. 18 May; **441**(7091): 362–5.

MARSHALL, R. C., ORWIN, D. F. G., GILLESPIE, J. M. (1991) Structure and biochemistry of mammalian hard keratin. *Electron Microscopy Rev*., **4**, 47.

OMARY, M. B., KU, N. O. (2006) Skin care by keratins. *Nature*, 18 May; **441**(7091): 296–7.

PARCELL, N. D. (2002) Sulfur in human nutrition and applications in medicine. *Altern. Med. Rev*., **7**(1): 22–44.

PEPLOW, P. V., DIAS, S. N. G. P. J., RODDICK-LANZILOTTA, A. D., KELLY, R. J. (2003) Orthopaedic materials derived from keratin'. US Patent 7,297,342.

DE PERA, M., CODERCH, L., FONOLLOSA, J., DE LA MAZA, A., PARRA, J. L. (2000) Effect of internal wool lipid liposomes on skin repair. *Skin Pharmacol Appl Skin Physiol*. May–Aug; **13**(3–4): 188–95

RIPAMONTI, U. (2004) Soluble, insoluble and geometric signals sculpt the architecture of mineralized tissues. *J Cell. Mol. Med.* **8**, (2): 169–80

SCHROOYEN, P. M., DIJKSTRA, P. J. (1998) Biodegradable films from selectively modified keratin dispersions. *Polymer Preprints*, 2, 160.

SIERPINSKI, P., GARRETT, J., MA, J., APEL, P., KLORIG, D., SMITH, T., KOMAN, L. A., ATALA, A., VAN DYKE, M. (2008) The use of keratin biomaterials derived from human hair for the promotion of rapid regeneration of peripheral nerves. *Biomaterials*. Jan; **29**(1): 118–28.

SMITH, R. A., BLANCHARD, C. R., TIMMONS, S. F. (1997) Keratin-based hydrogel for biomedical applications and method of production. US Patent 5,932,552

SMITH, R. A., ZUCHERMAN, J. F., HSU, K. Y., YERBY, S. (2001) Implantable prosthetic or tissue expanding device. US Patent 6,783,546.

TACHIBANA, A., NISHIKAWA, Y., NISHINO, M., KANEKO, S., TANABE, T., YAMAUCHI, K. (2006) Modified keratin sponge: binding of bone morphogenetic protein-2 and osteoblast differentiation. *J Biosci Bioeng*. Nov; **102**(5): 425–9.

VAN DYKE, M. E., TIMMONS, S. F., BLANCHARD, C. R., SILLER-JACKSON, A. J., SMITH, R. A. (2000) Absorbent keratin wound dressing. US Patent 6,270,793

WORMELL, R. L., HAPPEY, F. (1949) Regenerated keratin fibres. *Nature* Jan 1; **163**(4131): 18.

Index

absorbable organic halide (AOX), 148, 150–2, 177
absorbable organo halogens, 196–8
acid dyes, 183–6
air-jet spinning
  air vortex, 100
  MVS system, 100
  MVS yarns, 103
Airflow, 30, 41–5
Airweb doffing system, 135–6
Akwatec, 279
amino groups
  nylon, 186
  silk, 186
  wool, 186
anionic dyes *see* acid dyes
antimicrobial applications, 256–7
  scanning electron micrographs, 257
  sulfur-silver colloids, 257
Australian Wool Testing Authority (AWTA) Ltd., 23

Barco, 118
Benninger Versomat, 108
Bioclip, 18
Bunte salt polymer, 152–4
  curing reactions, 154
  preparation, 153

CAD *see* computer-aided design
chlorine-Hercosett process, 197, 199
chrome dyeing, 191–8
  Diamond Black PV oxidation, 193
    reflectance spectra, 196
  dyebaths residual chromium content, 194–6
  hydrogen peroxide mechanism of action, 195
  maleic acid/chromium (III) complex, 193
  residual chromium levels, 195
CIE Ganz 82 whiteness index, 241
circular knitting machines, 123
cleaning and carding
  fiber breakage, 299
  VM wool, 299
  wool fiber, 299
  worsted cards, 299
Climalite, 279
coloration effects
  Carbon Black pigments, 259
  wool dyeing, 259
ColorClear, 240
computer-aided design, 125–6
  M1 CAD, 125
  SDS ONE CAD, 125
controller area network (CAN) bus technology, 116
Coolmax, 279
Coolmax fibers, 296
CoolWool, xxiv
core yarns, 302–3
  double-roving feeding, 302
  friction spun core yarns, 303
  monofilament, 302
  spinning zone, 302
Crochet, 126
cross-lapper, 135

cross-machine direction (CMD), 132–5

Dark Lock Sorter, 98
differential frictional effect (DFE), 149
Dilo hyperpunch system, 138–9
DNA sequence, 5
double-roving feeding, 302
drawing
 drafting systems, 301
 wool blend, 301
 yarn twist, 301
Dri-FIT, 279
drying, 275
 polypropylene, 275
 QuickDry Merino, 276

eco–wool garments, xxiii
ecolabels, 66–70
 classification, 65–6
 EU processing practice, 77–82
 low-residue wool production, 78–9
 preparing processing consignments, 79–80
 residue targets for wool, 77–8
 wool scouring requirements, 80–2
elastane core yarns, 303–4
 elastic yarns, 303
 stretch fabrics, 303
 yarn tension, 304
Elbit Vision System (EVS), 119
electron spin resonance (ESR), 233
electronic textiles, 317–18
 electric heating system, 318
 heat emission, 317
 high-performance outwear garments, 317
 non-wool yarns, 317
 shape-change behavior, 318
 temperature buffering, 318
electrostatic interaction, 188
enterprise resource planning (ERP), 118
environmental claims, 65–77
 biodynamic, 73
 Eco-wool, 66
 ecolabels and greenwash types, 65–6
 environmental accreditation, 71
 environmental control authority, 67
 environmental discharges, 69
 EU eco-label, 66–71
 ISO 14021, 65
 organic textile processing, 73–7
  chemical oxygen demand (COD), 76
  EEC regulation 2092, 75
  natural dyes, 74
  risk-based sampling, 76
  shrink resist treatments, 75
  textile processing standards, 74
 organic, biodynamic, ecological, sustainable and ethical wools, 66–77
  biodynamic wool farming, 73
  eco-wool and EU eco-label, 66–70
  organic textile processing, 73–7
  organic wool, 71–2
  organic wool farming, 72–3
  textile eco-labels, 70–1
environmental consciousness, 62–3
 retailer and consumer, 62–3
 supermarket chains, 63
 sustainability initiatives, 63
 sustainable textiles, 63
European legislation, 63–5
 REACh, 63–5
 Substances of very high concern (SVHC), 64
 World Trade Organization (WTO), 65

fabric pre-treatment, 209
 chlorination, 184
 digital printing, 208–9
fabric production, xxii
Fabricscan, 119
felting shrinkage, 285–6
 consumer demand, 285
 ratchet effect, 285
 washing machines, 285
 wool fabrics, 285
 wool knitted garments, 285
FFDA *see* Fiber Fineness Distribution Analyzer
Fiber Fineness Distribution Analyzer (FFDA), 43–5
fiber morphology, 227
 absorption spectrum, 227
 cuticle cells, 227
 Merino wool fiber, 11, 227
 phototendering, 228

sulfur-rich proteins, 228
fiber security, 293–4
  blend fabrics, 293
  fiber migration, 293
  synthetic blend, 294
flame retardants, 294–6
  natural flame retardants, 294
  Nomex, 295
  Zirpro treatment, 294
flame-retardant finishing, 165–8
  brominated flame retardants, 167
  char–formation, 167
  Zirpro treatments, 167
flat-knitting machines, 120
Free Flight system, 114
FWA-treated wool, 236

garment knitting machines, 121
genetic engineering
  molecular structure modification on wool quality, 10–17
    keratin expression alteration by transgenesis, 13–17
    wool protein structure and physical properties, 10–13
  wool production improvement, 3–19
    human hair follicle structure, 7
    targets, 5–6
genetic manipulation *see* genetic engineering
gore technique, 122
greenwash, 65–6
Grosse, 116–18
Guided Gripper system, 114

H2-Out, 279
hydroentanglement, 130, 140–2
  citrus-style microfibers, 142
HydroMove, 279
hydrophobic agents, 279

inherently conductive polymers, 311–17
  carbon chain length, 316
  composite fabrics, 313
  conductive fibers, 311
  crosslinked pyrrole monomers, 316
  polymerization, 314
  polypyrrole–coated wool, 313
  scanning electron microscopy, 314
  textile substrate, 312
  textile–polyaniline composites, 314
  thiopene, 316
  wearable electronic, 312
insect-repellent treatments, 161–5
insect-resist finishing, 161–5
  chemical modification, 162
  chemical structure, 163
  cystine linkages, 162
  digestion-affecting poisons, 162
  microbial pathogens, 162
  Microencapsulation technology, 162
insulin-like growth factor, 8–9
intelligent garments, xxiv
intelligent textiles, 308–19
  active intelligent textiles, 309
  high-performance applications, 308
  passive intelligent textiles, 309
  wool textiles, 310–18
    biomimetics, 311
    electronic textiles, 317–18
    inherently conductive polymers, 311–17
    intelligent textile development, 311
    performance attributes, 310
    polyethylene glycols (PEG), 318
    warm–blooded organism, 310
  wool utilization, 319
intermingling filaments
  ring spinning machine, 304
  surface hairiness, 304
International Wool Textile Organisation, 31–5
  information by greasy wool measurements, 34
  IWTO Blue Book, 35
  IWTO Regulations, 35

Jacquard shedding mechanism, 116–17

knit fabric, 268
KnitExpert, 124
knitting, 106–28
  advances, benefits and applications, 119–27
    future trends, 127–8
    knitted wool fabric structures and constructions, 119–20
    warp knitting, 126–7

weft knitting, 120–6
machinery, 280
circular knitting machines, 123
flat-knitting machines, 120
garment knitting machines, 121
Køpke crease angle method, 202

Lanalbin APB, 241
lightweight fabrics, 291–2
LincSpun yarn technology, 292
SiroSpun, 291
synthetic filaments, 292
Treotek, 292
weaving, 291
wool rovings, 291
LincSpun yarn technology, 292
liquid moisture transport, 273
fiber surface chemistry, 275
surface energies, 274
sweat movement, 273
LITE, 293
low-temperature plasma (LTP), 155, 174
Lycra Power fabrics, 291

machine direction (MD), 132–5
machine-washable wool, 148–58
dimensional stability, 148
shrink-resistant finishing, 148
Maliwatt, 139
melanin, 218
Merino wool fiber, 227
metabolic heat loss, 267
exothermic system, 267
heat production, 267
skin temperature, 267
sweat evaporation, 268
Milanese, 126
Model software, 124
moisture transport
Coolmax fibers, 296
hydrophilicities, 296
moisture vapor transport
buffering effect, 273
wear conditions, 273

nanotechnology, 248–9, xxiv
carbon nanotubes, 254
clay nanoparticles, 254
composite fibers, 253
cyclodextrins, 260
fiber production, 253
lubricants, 261
microencapsulation, 254
nanoHorizons SmartSilver, 260
nanoparticles and nano-emulsions, 249–52
disc-shaped particle, 250
percentage of surface atoms, 251
shape factors, 251
nanostructured films, 252–3
film densities, 252
self-assembled monolayers, 252–3
textile fiber surface, 252
principles, 249
risks, 255–6
ecological risks, 255–6
explosion risks, 255
human health risks, 255
toxicity, 255
stain-resist finishes, 260
textile and wool applications, 248–9
antimicrobial applications, 256–7
coloration effects, 259
self-cleaning wool textiles, 259
stain blocking, 257–9
water repellence, 258–9
Ultra–Fresh Silpure, 260
value-added performance, 261
wool products, 248
natural fibres, 61, xxi
consumer trends, 62
eco-fibers, 62
wool, 61
needle bed
cylinder, 123
Nomex, 295
nonwoven wool
advances in manufacture, 130–45
bonding systems
chemical bonding, 142–3
finishing, 144–5
future trends, 145
hydroentanglement, 140–2
kiss roller application, 143
needle punch, 137–9
point-bonding pattern, 145
stitch bonding, 139–40
thermal bonding, 143–4

processes and recent innovations, 131–7
air-lay systems, 136
basic non-woven card, 132
cross-lapper, 135
cross-lapping, 134
non-woven card with condenser rollers, 133
random card, 134
web drafting, 134–6
web formation, 131–4
wet-laid systems, 136–7
NSC, 135

odor management, 276
antimicrobial agents, 277
fungal growth, 276
OFDA100, 47
open-end rotor spinning, 301
automation, 301
ring spinning system, 301
Organic Guarantee System, 71–2
organic textiles *see also* organic wool
processing, 73–7
organic wool, 71–2
farming, 72–3
EEC 2092/91, 72
national certification, 73

photobleaching
fluorescent whitening agents (FWA), 217
sunlight, 217
photostability, 217–44
amino acids, 233
chromophores in photomodified peptide sequences, 235
CIE Ganz 82 whiteness index, 241
ColorClear, 240
colorless fluorescent dyes, 236
effect of processing, 222–5
scouring and mechanical processing, 222
wet processing and finishing, 222–3
wool colour vs cotton and synthetic fibres, 223–5
electron spin resonance (ESR), 233
future trends, 241–3
high-pressure mercury, 231
Lanalbin APB, 241
methods of improving whiteness, 239–41
natural chromophores, 226
photobleaching, 217, 229–31
simulated sunlight exposure, 230
wool colour, 218–21
colour measurement, 221
environmental and storage effects, 220–1
evolution and genetics of natural white colour, 218–20
wool colour chemistry, 226–39
fibre morphology and physical properties, 227–9
FWA-treated wool photoyellowing, 236–9
mechanism of photoyellowing, 231–5
natural chromophores, 226–7
photoreduction mechanism, 237
photoyellowing and photobleaching, 229–31
stilbene formation, 238
yellowness index, 232
yellow chromophores, 234
polyester, 278
Prima Vision, 124
protease immobilization, 175
soluble polymer, 175

QSC *see* quick style change
QualiMaster, 119
quick style change, 118–19
Elbit Vision System (EVS), 119
enterprise resource planning (ERP), 118
Fabricscan, 119
Loomdata, 118
pneumatic shaft lock system, 118
rates of filling insertions (RFI), 115
WeaveMaster, 118
QuickDry Merino, 276

Rasche, 126–7
reactive dyes, 198–9

scouring, 80–2, 88–9

raw wool, 88
wool consignments, 88
scrim, 140
self-twist spinning, 103–4
WinPro, 103
WinTwist, 103
yarn strength, 103
selvage, 121
Serfil, 94
shed geometry, 114
shrink resistance, xxiv
shrink-resist treatment, 222
Siroclear, 97–8
Sirofil, 94
Sirolan Laserscan, 33, 45–6
SiroSpun, 92, 291
skin sensory behavior, 268
coarse fibers, 269
diameter distributions, 269
heat flow, 268
knit fabric, 268
smart textiles *see* intelligent textiles
Solospun, 92, 113
spacer fabrics, 127
Spinnbau, 135
spinning technology, 86–104
advances in topmaking
greasy wool conversion, 87–8
scouring, 88–9
future trends, 100–103
air-jet spinning, 100
Murata Vortex Spinning system, 101
MVS twist insertion principle, 102
self-twist spinning, 103–4
worsted spinning, 91–100
abrasion resistance, 91
Compact spinning system, 93
self-twist spinning, 94
Selfil, 94
Sirofil, 94
SiroSpun, 92
Solospun rollers, 92–3
twisting, 99–100
winding and clearing, 95–9
yarn hairiness, 93
Yarnspec, 90–1
algorithms, 90
antistatic agents, 90
spinning performance, 90–1
sports apparel
wool performance, 265–83
apparel fibers, 266
apparel-based devices, 281
fabric structure, 280
functionality, 282–3
future trends, 280–2
human system, 266–9
knitting machines, 280
product differentiation, 266
solid-state devices, 281
Sportwool, 279–80
synthetic fibers, 265
textile properties, 269–77
textiles, 277–9
Sportwool, 279, xxiv
density gradient structure, 279
double-knit structure, 279
evaporative cooling, 280
surface properties, 279
sweat absorption behavior, 280
spunlace *see also* hydroentanglement
products
filters, 141
home furnishings, 141
protective apparel, 141
sanitary products, 141
synthetic leather, 141
wipes, 141
SRS, 7
stain blocking, 257–9
Advanced Teflon, 259–60
micro-sized nodule, 258
nano-sized hairs, 258
Nanoguard-GOR, 259
super-hydrophobic surfaces, 259
stain-release finishing, 158–61
stain-release properties, 159
stain-repellent finishing, 158–61
fluorochemical repellent finishes, 159
hydrophobic interaction, 159
oil-based stain resistance, 159
surface energy, 159
Stäubli, 116–17
stretch and fit, 290–1
fatigue, 290
Lycra Power fabrics, 291
wrinkle recovery, 291

Sumo, 116
sustainable polymers, 326
synthetic fibers, xxi

TEAM system, 30–1
Tension Sensor Feeler (TSF) device, 114
textile properties, 269–77
  drying, 275–6
  fiber moisture sorption, 271–2
  heat of sorption, 272
  liquid moisture transport, 273–5
  moisture vapor transport, 273
  odor management, 276–7
  thermal resistance, 269–70
  vapor resistance, 271
textile supply chains, xxiii
thermal bonding
  sheath-core and side-by-side, 144
  through-air bonding oven, 144
thermal insulation garments, 292–3
  LITE, 293
  synthetic fibers, 293
  thermal conductivities, 292
thermal resistance, 269
  domestic heating, 270
  fabric thickness, 270
  heat transfer, 269
  knitwear, 269
Thermosplicer, 97
thread tensioning device
  dynatens, 110
  isotens, 110
  megatens, 110
through-air bonding oven, 144
topmaking, 87
  ERA, 89
  greasy wool, 87
  Millennium, 89
  TEAM formulae, 87
  Topmaker, 88
  TopSpec, 88
transgenesis, 3–5
  keratin expression alteration, 13–17
  methods, 4–5
    DNA microinjection, 4
    liposome-mediated transfer, 4
    nuclear transfer, 4–5
    sperm-mediated DNA, 4
transmission control protocol/Internet protocol (TCP/IP), 118
Treotek, 292
Tricot, 126–7
Turbo doffing system, 135–6
Twinsplicer, 95–7
  mechanical splicer operation, 96
two-layered self-assembled monolayer fiber coating, 253

Unished, 117
Unival 100, 117

vapor resistance
  steady-state conditions, 271
  textile materials, 271
vegetable matter (VM), 131
velour needle punch, 138
VM wool, 299

warp knitting, 126–7
  Crochet, 126
  Milanese, 126
  Rasche, 126–7
  spacer fabrics, 127
  Tricot, 126–7
warp preparation, 107–13
  creel and yarn feeding, 109–10
    automation, 110
    thread tensioning device selection, 110
    warp-stop motions ionisation, 109–10
    waxing devices, 110
  drawing-in, 110–1
  dropper–pinning, 112
  sectional warping, 108–9
    automatic leasing device, 108–9
  Benninger Versomat, 108
  electrical beaming, 109
  transportation and loading, 112–13
    electro-hydraulic loading device, 112–13
  tying–in, 111–2
warp-stop motions, 109–10
  ionization, 109
Warplink, 111
waxing device, 110

wearable electronics *see* electronic textiles
WeaveMaster, 118
weaving, 106–28
advances in yarn preparation, 107
advances, benefits and applications, 113–19
air-jet machines, 115–16
Jacquard shedding mechanism, 116–17
quick style changes, 118–19
rapier machines, 114–15
woven wool fabric structure and construction, 113–14
future trends, 127–8
warp preparation, 107–13
creel and yarn feeding, 109–10
drawing-in, tying-in and dropper pinning, 110–2
sectional warping, 108–9
transportation and loading, 112–13
worsted, 107
web drafter, 135
weft knitting, 120–6
circular knitting machines, 123–5
computer-aided design, 125–6
computerized complete garment knitting machines, 121–3
flat-knitting machines, 120–1
Weratech system, 318
wet-laid systems, 136–7
common products, 137
process schematic, 137
WholeGarment, 122, 125
winding and clearing, 95–9
electronic tagging, 99
fabric inspection, 97
glass transition temperature, 97
knotting, 95
Loepfe technology, 98
mechanical splicer operation, 96
Siroclear, 98
synthetic fibers, 99
Thermosplicer, 95
wool *see also* nonwoven wool
applications of metrology
commercial trading, 23
quality control, 26
research, 26
commercial regulations applications, 35
developments in metrology, 55–6
dark and medullated fibres, 55–6
fiber length, 55
new staple measurements, 6
enhancing products with nanotechnology
advantages, 253–4
application, 256–60
commercial applications, 260
future trends, 261
nanoparticles and nano-emulsions, 249–52
nanostructured films, 252–3
principles, 249
risks, 255–6
follicle structure, 6
genetic engineering for improved production, 3–19
future trends, 19
targets, 5–6
improving photostability and whiteness, 217–44
colour, 218–21
colour chemistry, 226–39
effect of processing on colour, 222–5
future trends, 241–3
methodology, 239–41
increase in yield, 6–10
biochemical pathways, 9
growth factors overexpression, 8–9
reaction–diffusion mechanism, 7–8
SRS, 7
measurement advances
colour, 52–3
coloured and medullated fibres, 53–4
fibre diameter, 41–7
fibres length in top, 54
moisture, 37–8
sampling, 35–7
staple measurements, 47–52
style, 54–5
yield, 38–41
metrology principles, 23
microstructure, xxv

molecular structure effect on quality, 10–17
keratin expression alteration by transgenesis, 13–17
protein structure and physical properties, 10–13
natural renewable fibre, 61–83
environmental claims, 65–77
environmentally aware retailer and consumer, 62–3
EU eco-label processing, 77–82
European legislation, 63–5
future trends, 82–3
objective fibre quality measurement, 22–59
development, 30–1
future trends, 57–9
standardization, 31–5
statistical methods, 26–30
accuracy and precision, 27
analytical method precision, 28
bias, 27–9
detection limit, 29
range, 29
selectivity, 30
sensitivity, 29
techniques, 4–5
technological advances, xxi–xxv
CoolWool, xxiv
eco-wool and organic wool, xxiii
finishing and coloration, xxiii–iv
nanotechnology, xxiv
spinning, weaving and knitting, xxiii
testing and modeling, xxiii
terminology, 23
transgenesis effect on production, 17
wool blends, 284–304
benefits and applications, 285–96
bulk, cover and warmth, 292–3
felting shrinkage, 285–6
fibre security, 293–4
flame retardance, 294–6
lightweight fabrics, 291–2
moisture transport, 296
prickle reduction, 288–9
softness to handle, 289–90
stretch and fit, 290–1
wrinkle recovery and easy care, 286–8
cotton spinning system, 298–301
cleaning and carding, 299
drawing, roving and ring spinning, 301
open-end rotor spinning, 301
wool tops, 300–301
thermal insulation, 284
water-repellent surface, 284
wool–filament composite yarns, 301–4
core yarns, 302–3
elastane core yarns, 303–4
intermingling filaments with staple fibres, 304
worsted system, 297–8
yarn production, 296–7
wool coloration *see* wool dyeing
wool dyeing, 183–209
acid dyes, 183–6
acid leveling dyes, 185
acid milling dyes, 184–5
pre-metallized dyes, 185
anti-setting agents, 200
electrophilic reagents, 206–7
oxidants, 205–6
dye diffusion, 189–91
reactive/non-reactive dye models, 190
environmental impact, 191–8
absorbable organo halogens, 196–8
chrome dyeing, 191–96
ink-jet printing, 207–9
pre-treatments and ink formulations, 209
printing machinery, 208–9
non-covalent interactions, 187
electrostatic interactions, 187–8
solvophobic or hydrophobic, 187, 189
Van der Waals (VW), 187–8
physical chemistry, 186–9
reactive dyes, 198–9
amphoteric levelling agents, 199
effect of time and temperature, 203
Supramino Orange R, 198
thioether dye formation, 204

theoretical basis, 184–91
acid dyes, 184–6
dye diffusion pathway, 189–91
physical chemistry, 186–9
wool protection, 200–7
crosslink formation, 202
cysteine/cysteine representation, 200
electrophilic reagents, 206
elimination reactions, 201
wool fabrics, 207–9, xxii
ink-jet printing, 207–8
ink formulation, 209
pre-treatments and ink formulations, 209
printing machinery, 208–9
wool fiber, 166, 323
flammability properties, 166
quality
conformance, 22
definitions, 23, 24–5
design, 22
objective measurement, 22–3
reactive dye systems, 199
wool finishing, 147–77
chlorfenapyr, 165
dimensional stability and shrink-resist finishing, 148–57
Bunte salt curing reaction, 154
Bunte salt polymer preparation, 153
isocyanate and bisulphite adduct reaction, 155
plasma treatment, 157
SEM of wool fibre, 149
wool cuticle schematic diagram, 150
enzymes applications, 167–76
EC classification system, 170
kinetics, 172
peptidases classification, 171
protease immobilisation, 175
wool fibres treated with proteases, 174
flame-retardant, 165–7
commercial flame retardant, 168–9
flammability properties of wool fibre, 166
future trends, 176–7
insect-repellent treatments, 161–5
insect resist agents, 164
insecticide chemical structures, 163
wool-damaging insect species, 161
stain-repellent and -release finishing, 158–61
fluorine-based hydrophobic and hydrophilic hybrid copolymer, 160
fluorochemical polymer molecule, 159
wool industry, xxi
wool keratins, 323, xxiv
advanced applications, 326–8
biological enhancement, 327
remodeling process, 327
wound dressings, 328
expression alteration by transgenesis, 13–17
transgenic sheep wool fiber, 16
extraction for product development, 325
future trends, 329
industrial technology, 329
medical technology, 329
industrial trends in relation to sustainable polymers, 326
protein conformation, 323
structure and characteristics, 324–5
wool protein structure, 10–13
cuticle cell protein layers, 12
disulphide bonds between matrix molecules, 12
Merino wool fiber, 11, 227
overlapping flattened cuticle cells, 12
wool knitting
warp (*see* warp knitting)
weft (*see* weft knitting)
wool measurement *see* wool metrology
wool metrology
applications
quality control, 26
research, 26
color, 52–3
CIE color space, 52–3
colored and medullated fibers, 53–4
fiber contamination, 53

illumination method, 53
commercial regulations, 35
IWTO blue book, 35
IWTO regulations, 35
core sampling, 36–7
fiber diameter, 41–6
Airflow instrument, 30
carding machines, 42
Fiber Fineness Distribution Analyzer, 43–5
Sonic Airflow, 43
fiber length
Schlumberger Analyzer, 54
future trends, 57–8
industry trends, 57
new measurement technologies, 57–8
grab sampling, 35–6
grab machines, 36
latest developments, 55–7
dark and medullated fibers, 55
fiber length, 55
new staple measurements, 56
measurement systems, 30–1
grab sampling systems, 30
measurement instruments, 30
moisture, 37–8
electrical capacitance, 37
on-line measurements, 38
scoured wool, 37
standardization of measurement systems, 31–5
draft test methods, 32–3
IWTO test methods, 32–3, 41, 45, 58
mean fiber curvature (MFC), 34, 45
standard deviation, 33
vegetable matter base (VMB), 25, 34, 51
staple measurements, 47
ATLAS, 49–50
mechanical tuft sampling (MTS) machine, 48–9
staple length, 52
staple strength, 47
statistical methods, 26–30
accuracy and precision, 27
bias, 27–9
detection limit, 29
objective determination, 26
range, 29
selectivity, 30
sensitivity, 29
style, 54–5
yield, 38–41
wool processing, 222–5
cotton and synthetic fibers, 223–5
finishing, 222–3
mechanical processing, 222
Merino wools, 222
photoyellowing, 222
pyrazoline, 223
reflectance spectrophotometers, 223
scouring, 222
shrink-resist treatment, 222
wet processing, 222–3
whiteness values, 225
Wool Research Organization of New Zealand *see* WRONZ
wool testing, xxiii
wool tops
mean fiber length, 300
polyester blends, 300
staple fibers, 300
wool trade, xxi
worsted weaving, 107
wrinkle recovery and easy care, 286–8
fabric resistance, 286
glass transition temperature, 287
natural cotton fabrics, 287
WRONZ, 241

yarn preparation, 107
Yarnspec, 90–1

Zipro treatment, 294